D0065403

WATER ANALYSIS

VOLUME III   ORGANIC SPECIES

CONTRIBUTORS

KARL J. BOMBAUGH

WILLIAM J. COOPER

W. GIGER

K. GROB, JR.

J. RONALD HASS

FRED KATSUMI KAWAHARA

JERRY A. LEENHEER

DANIEL L. NORWOOD

R. P. SCHWARZENBACH

JAMES C. YOUNG

WATER ANALYSIS

VOLUME III   ORGANIC SPECIES

Edited by

ROGER A. MINEAR

Department of Civil Engineering
The University of Tennessee
Knoxville, Tennessee

LAWRENCE H. KEITH

Radian Corporation
Austin, Texas

 1984

ACADEMIC PRESS, INC.
(Harcourt Brace Jovanovich, Publishers)

Orlando   San Diego   New York   London
Toronto   Montreal   Sydney   Tokyo

ACADEMIC PRESS, INC.
Orlando, Florida 32887

United Kingdom Edition published by
ACADEMIC PRESS, INC. (LONDON) LTD.
24/28 Oval Road, London NW1 7DX

Library of Congress Cataloging in Publication Data

Main entry under title:

Water analysis.

  Includes bibliographies and index.
  Contents: v. 1. Inorganic species, part 1 -- / -- v. 3.
Organic species.
  1. Water--Analysis. I. Minear, R. A. II. Keith,
Lawrence H., Date   .
QD142.W36 1982   628.1'61   82-1755
ISBN 0-12-498303-0 (v. 3)

PRINTED IN THE UNITED STATES OF AMERICA

84 85 86 87   9 8 7 6 5 4 3 2 1

CONTENTS

4   GAS CHROMATOGRAPHY

R. P. SCHWARZENBACH, W. GIGER, AND
K. GROB, JR.

5   ORGANIC MASS SPECTROMETRY

J. RONALD HASS AND DANIEL L. NORWOOD

6   THE USE OF HPLC FOR WATER ANALYSIS

KARL J. BOMBAUGH

7   INFRARED SPECTROPHOTOMETRY OF POLLUTANTS IN WATER SYSTEMS

FRED KATSUMI KAWAHARA

Contents

LIST OF CONTRIBUTORS

Numbers in parentheses indicate the pages on which the authors' contributions begin.

KARL J. BOMBAUGH (317), Radian Corporation, Austin, Texas 78766

WILLIAM J. COOPER (41), Drinking Water Research Center, Florida International University, Miami, Florida 33199

W. GIGER (167), Swiss Federal Institute for Water Resources, and Water Pollution Control (EAWAG), Dübendorf CH-8600, Switzerland

K. GROB, JR. (167), Kantonales Laboratorium, Zürich CH-8030, Switzerland

J. RONALD HASS (253), National Institute of Environmental Health Sciences, Research Triangle Park, North Carolina 27709

FRED KATSUMI KAWAHARA (381), U.S. Environmental Protection Agency, A. W. Breidenbach Environmental Research Center, Cincinnati, Ohio 45268

JERRY A. LEENHEER (83), Water Resources Division, U.S. Geological Survey, Denver, Colorado 80002

DANIEL L. NORWOOD (253), Department of Environmental Sciences and Engineering, School of Public Health, University of North Carolina, Chapel Hill, North Carolina 27514

R. P. SCHWARZENBACH (167), Swiss Federal Institute for Water Resources, and Water Pollution Control (EAWAG), Dübendorf CH-8600, Switzerland

JAMES C. YOUNG (1, 41), Department of Civil Engineering, University of Arkansas, Fayetteville, Arkansas 72701

PREFACE

This is the final volume of this treatise. Whereas the first two volumes focused on inorganic constituents in water, this volume is centered around the organic constituents.

The first two chapters deal with waste strength and waste pollution parameters of a nonspecific variety. These include biochemical oxygen demand (BOD), chemical oxygen demand (COD), total organic carbon (TOC), spectroscopic measurements, electrochemical methods, and a number of other techniques that provide chemical class determinations.

The third chapter provides an in-depth review of the current methods available for isolating organic constituents from water and for concentrating and partitioning them. This is a critical step in the analytical procedure, and there are a multitude of techniques to choose from, each with advantages and disadvantages for different types of organic pollutants. A wrong decision at this point negates all of the following work because, no matter how accurate and precise the analysis is, if the sample is not representative, the data are in error.

Chapter 4 deals with gas chromatographic (GC) separations and analyses, one of the most popular and rapidly advancing techniques in the field of complex mixture analyses. Both capillary and packed-column techniques are discussed in depth in this well-illustrated chapter. In addition, injector techniques, derivatizations, detector types, qualitative and quantitative analyses, and a representative list of applications are presented.

Mass spectrometry (MS) is the method most often used to identify unknown organic compounds in environmental samples. Chapter 5 presents a comprehensive discussion on the principles of organic mass spectrometry, mass analysis, ion detection, chromatography/mass spectrometry, tandem mass spectrometry, qualitative and quantitative analysis, and selected applications.

Chapter 6 describes the principles and applications of using high performance liquid chromatography (HPLC) for water analyses. This is the most valuable technique available for separation and analysis of nonvolatile organic constituents from water. The necessary equipment, a description of the chromatographic process, and practical use and optimization of the method are presented in detail.

The final chapter covers the use of infrared (IR) spectrophotometry for analyzing for organic pollutants in water. Both theoretical considerations and practical applications of this technique are discussed in depth.

The chapters in this volume represent a collection of the state-of-the-art methodology in current use, which accounts for the major portion of the methods used for analysis of organic constituents in water. Together with the chapters on analysis of the inorganic constituents in water from the first two volumes, this treatise represents a comprehensive work on water analysis. Thus, we have attempted to provide, in a single source, a collection of detailed theoretical and applied treatments on each of the important methodologies necessary for understanding and determining the trace level components found in both natural and polluted waters.

WATER ANALYSIS

VOLUME III ORGANIC SPECIES

1 WASTE STRENGTH AND WATER POLLUTION PARAMETERS

James C. Young

Department of Civil Engineering
University of Arkansas
Fayetteville, Arkansas

WATER ANALYSIS, VOL. III
Copyright © 1984 by Academic Press, Inc.
All rights of reproduction in any form reserved.
ISBN 0-12-498303-0

1

I. INTRODUCTION

Analysis of the organic content of wastewaters is essential to properly determine their treatability, to design water pollution control plants and monitor their performance, and to evaluate the impact of wastewater discharges on lakes and streams. The number and variety of compounds, their highly variable concentrations, the low concentrations of specific compounds, and the difficulty of identifying and measuring each compound have led to the general use of nonspecific or indirect tests to indicate the gross amount of organic material in water or wastewaters.

Biochemical oxygen demand (BOD) is perhaps the most widely used of these nonspecific tests. Basically, the BOD test measures the amount of oxygen used by microorganisms during aerobic decomposition of organic pollutants. The amount of oxygen used is an indirect measure of the amount of biodegradable organic material present in a given sample. Despite their importance, BOD tests and their results are poorly understood and widely misused.

The BOD test procedure is time consuming and may produce relatively imprecise results, and the interpretation of test data often is subjective. Because of these potential problems, nonbiological test procedures—for example, chemical oxygen demand (COD) and total organic carbon (TOC)—have been proposed as alternatives to the BOD test. These tests also provide nonspecific measures of organic content, but in terms of the chemical equivalent of the amount of oxygen required to oxidize the organic matter. Although these tests have the advantage of being more rapid and precise than the BOD test, they have the major disadvantage of not distinguishing between biodegradable and nonbiodegradable organic matter. BOD is the only parameter that provides an indication of the amount of biodegradable organic materials in wastewaters and in streams receiving wastewater discharges.

II. OXYGEN DEMAND AS A POLLUTION CONTROL PARAMETER

Oxygen demand was recognized in the late nineteenth century as being an important effect of the decomposition of organic material in natural waters. Samples of water from streams showed a definite depletion of oxygen if those streams received significant amounts of wastewater. As waste loads increased, the oxygen demand frequently caused the streams to become septic with resulting production of odors and occasional fish kills. Early attempts to quantify this effect involved placing samples of stream water in capped bottles and measuring the dissolved oxygen concentration (DO) over a period of time. Complete uptake of dissolved oxygen in 5 to 7 days indicated a significantly polluted stream while longer persistence of a DO residual signified acceptable stream conditions. Continued research on oxygen demand led to the formalization early in this century of the BOD test (Theriault, 1927; Phelps, 1944; Buswell *et al.*, 1954; O'Brien, 1962). Although research has continued, the currently recognized standard procedure was adopted in 1936, and few substantial changes have been made since that time.

The BOD reaction consists of two major phases as illustrated in Fig. 1. The first is a carbonaceous phase in which organic material is oxidized as shown by Eq. (1):

$$\text{Organic material} + O_2 \xrightarrow{\text{microorganisms}} CO_2 + H_2O + \text{microbiological solids} \qquad (1)$$

The oxygen required for complete conversion of organic material to carbon dioxide and water is defined as the theoretical oxygen demand (THOD).

The accumulated first-stage oxygen consumption after a 20- to 30-day (theoretically infinite) period of incubation represents the ultimate BOD (BOD_u). Because some organic material remains as nonbiodegradable cell residue, it is not possible biologically to oxidize all the organic material

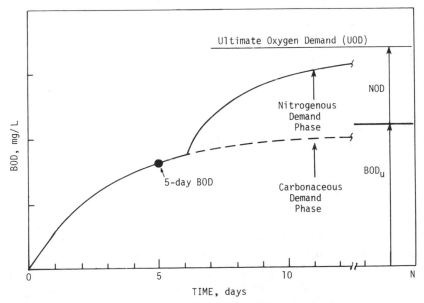

Fig. 1. Schematic representation of a typical BOD curve showing carbonaceous and nitrogenous BOD reactions.

originally present in a sample. Consequently, BOD_u is somewhat less than the THOD.

In the second phase, ammonia and nitrite are converted to nitrate as follows:

$$\text{a. } NH_3 + \tfrac{3}{2} O_2 \xrightarrow{\textit{Nitrosomonas}} HNO_2 + H_2O$$

$$\text{b. } HNO_2 + \tfrac{1}{2} O_2 \xrightarrow{\textit{Nitrobacter}} HNO_3$$

$$\text{Net: } NH_3 + 2\,O_2 \longrightarrow HNO_3 + H_2O \qquad (2)$$

These reactions collectively are defined as nitrification. Nitrifying organisms, especially *Nitrosomonas,* exhibit a lower maximum growth rate than the carbon-decomposing organisms, often causing nitrification to be delayed until much of the organic material has been oxidized. Both *Nitrosomonas* and *Nitrobacter* use carbon dioxide as their source of carbon for growth, and the nitrogenous compounds serve as electron acceptors. *Nitrosomonas* grow more slowly than *Nitrobacter* so that the conversion of ammonia to nitrite usually controls the rate of nitrification.

Theoretically, if the reactions in Eq. (2) are carried to completion, 3.43 g of molecular oxygen are used per gram of ammonia nitrogen oxidized to nitrite, and 1.14 g of oxygen are used for each gram of nitrite nitrogen converted to nitrate. However, some of the reduced nitrogen is assimilated as cell material by the nitrifying organisms, which reduces the amount of oxygen theoretically needed for nitrification. Although no means exists to determine the exact amounts of oxygen used in each of the above reactions, the following equation [Eq. (3)] predicts the nitrogenous oxygen demand (NOD) quite accurately (Wezernak and Gannon, 1968; Montgomery and Borne, 1966).

$$NOD = 3.22\,(NH_3\text{-}N \longrightarrow NO_2^-\text{-}N) + 1.1\,(NO_2^-\text{-}N \longrightarrow NO_3^-\text{-}N) \qquad (3)$$

Therefore, if ammonia nitrogen exists in appreciable quantities, as it does in domestic and many industrial wastes, a significant part of the potential oxygen demand can be attributed to nitrification.

The BOD reaction is not as simple as Eqs. (1) and (2) imply but is a complex interaction between many species of microorganisms and a number of organic and nitrogen compounds. The initial stages of organic decomposition involve hydrolysis of high molecular weight sugars, proteins, fats, and starches, and synthesis of the products of these reactions. Other species of microorganisms grow in response to by-products released by these initial metabolic reactions. Eventually, predator organisms such as protozoa, amoebae, and stalked ciliates begin to graze on bacterial cells and other particulate organic material. Throughout these reactions, soluble organic by-products are released and become available to support further bacterial growth, and living microorganisms use oxygen for cell maintenance or endogenous respiration (Young and Clark, 1965; Bhatla and Gaudy, 1965; Gates and Ghosh, 1971).

A. Ultimate Oxygen Demand

When the total NOD is added to BOD_u, the result is known as the ultimate oxygen demand (UOD), or

$$UOD = BOD_u + NOD \qquad (4)$$

The NOD can be calculated from Eq. (3) or measured by BOD test procedures. Although UOD has little utility in wastewater treatment, it can be used as a measure of the maximum oxygen-demanding load in wastewater streams discharged to receiving waters.

B. The 5-Day BOD

The oxygen uptake at 5 days of incubation and at a test temperature of 20°C is recognized in the United States and many other countries as the standard BOD. BOD_5 has become an important parameter for designing and evaluating the performance of water pollution control plants and for monitoring the quality of treated effluents. Response of receiving streams to oxygen-demanding waste loads commonly is determined from stream models using oxygen uptake rate and oxygen demand measurements as input parameters. The dependability of these models is closely related to the reliability of BOD_5 measurements (Dobbins, 1964). U.S. Public Law 92-500 (Water Pollution Control Act Amendments of 1972) and subsequent rules and regulations identified BOD_5 as one of the parameters to be used for establishing effluent and stream standards for point-source discharges. This means that the 5-day 20°C BOD takes on a legal meaning that originally was not considered in the test objective. Factors affecting the reliability and interpretation of BOD_5 measurements are discussed in the succeeding sections.

C. The Plateau

A plateau often is observed in the carbonaceous BOD reaction within 24 h of incubation (Fig. 2). The plateau generally represents the effect of a change between major phases of the carbonaceous metabolic reaction (Bhatla and Gaudy, 1965; Gates and Ghosh, 1971; Busch, 1958). The first phase involves the uptake of soluble organic substrate initially present in the sample; part is oxidized and part is synthesized into bacterial cell

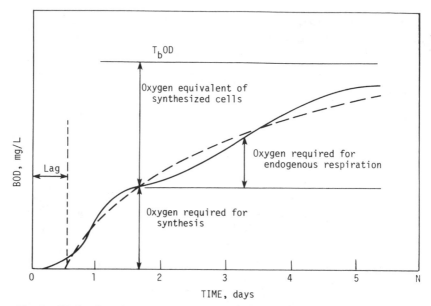

Fig. 2. Illustration of the occurrence of a plateau in the BOD reaction. The dashed line represents the first-order approximation of the reaction.

solids. The second phase is associated primarily with endogenous respiration of bacterial cells, growth of predator organisms that graze on bacterial cells produced during the first growth phase, lysis of microbial cells, and regrowth of bacteria on the soluble products released. A plateau also may be an environmentally induced artifact produced by chemical influences such as a pH change or by the release of organic by-products that temporarily inhibit bacterial growth or delay the onset of predator growth.

A plateau may not always be observed. Individual microbial reactions may be superimposed in such a manner that different metabolic phases are not distinguishable. Particulate organic matter is generally decomposed more slowly than soluble organics so that the BOD reaction for fresh raw domestic wastewaters may exhibit no plateau. Carbonaceous oxygen demand in samples of effluent from secondary or tertiary biological treatment plants is associated primarily with endogenous respiration. A plateau, therefore, is not expected to occur with samples of these wastewaters.

D. Lag

A lag occurs in the BOD reaction when an insufficient number of acclimated microorganisms is present (see Fig. 2). This lag can be reduced by adding seed microorganisms to the test sample. In some instances,

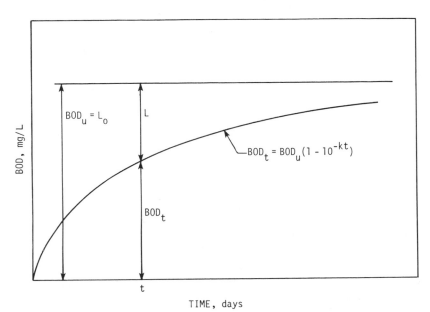

Fig. 3. Graphical illustration of the first-order BOD curve and equation.

organisms must acclimate to new environmental conditions, such as a change in temperature, pH, or waste concentration. Such acclimation usually occurs within a few hours after seeding a sample.

Industrial wastewaters containing compounds resistant to biological decomposition or inhibitory to growth often require that acclimated cultures be developed as seed for BOD analyses. Several days of continuous exposure of the seed culture to the waste usually are required to develop this type of acclimation (Young and Baumann, 1976; Hartman and Singrun, 1968; Young and Affleck, 1974). Continuous exposure is essential with many compounds; otherwise the microorganisms lose their capacity to decompose that compound. This is characteristic of cultures adapted to phenol and phenol derivatives (Hartmann and Singrun, 1968).

III. MATHEMATICAL MODELING OF THE BOD REACTION

A. The First-Order Equation

The BOD reaction may be modeled mathematically by assuming that the rate of removal of oxygen-demanding material is first order with respect to the concentration of unoxidized material L remaining at any time t (Fig. 3), or

$$\frac{dL}{dt} = -kL \tag{5}$$

where $L = BOD_u - BOD_t$. Integration of Eq. (5) with boundary conditions of

$$L = BOD_u \text{ at } t = 0 \qquad \text{and} \qquad L = BOD_t \text{ at } t = t$$

and converting from natural logarithms to base 10 gives

$$BOD_L = BOD_u (1 - 10^{-k_{10}t}) \tag{6}$$

where k_{10} is a first-order rate coefficient (base 10).

This first-order equation represents an empirical description of the BOD reaction. If the three major oxidation reactions—synthesis, respiration and nitrification—could be distinguished, each no doubt would have its individual and different reaction rate. The first-order equation fits BOD test data reasonably well but only because of the averaging effect of simultaneous biological reactions.

Deriving the reaction-rate coefficient and ultimate BOD in the first-order equation from BOD data presents a number of problems. Although regression methods exist (Reed and Theriault, 1931), other graphical and mathematical procedures have been developed. These include the Thomas slope method (Thomas, 1940), Moore's method of moments (Moore *et al.*, 1950), and the log-difference method (Fair, 1936). Each method generates different values of k_{10} and BOD_u. The Thomas slope method overestimates BOD_u and underestimates k_{10}, while the method of moments gives results in close agreement with the regression method. The method of moments has an advantage in that the use of a computer is not required (Marske and Polkowski, 1972). Other more simplified methods have been proposed but are not used as widely (Hammer, 1975; Bagchi and Chaudhuri, 1970; Gaudy *et al.*, 1967).

Historically, k_{10} values for domestic sewage (as measured by the standard dilution BOD method) have been considered to average ~0.1/day (at 20°C test temperature). This means that the BOD_u would be 1.5 times the 5-day 20°C BOD. In practice, however, k_{10} values are so variable that estimation of BOD_u from BOD_5 measurements without direct evaluation of k_{10} seldom is justified (Young and Baumann, 1972; Lee and Oswald, 1954; Ali and Bewtra, 1972). There is no reason to assume that k_{10} values should be the same for all wastewater samples (especially domestic sewage). When cell growth or synthesis is the predominant reaction, the k_{10} values are greater than 0.20/day and vary greatly depending on the type of organic waste materials present, freshness of the sample, and history of sample storage and handling. If cell respiration predominates, the k_{10} rate averages closer to 0.15–0.20/day. Nitrification occurring a day or two after the beginning of a BOD measurement can cause the overall k_{10} value to be lower than 0.10/day, while the reaction rate for either the carbonaceous and nitrogenous BOD reaction is greater than 0.10/day. This occurs because the nitrification reaction spreads the total oxygen uptake over a longer period of time, thereby giving a false indication of the individual rates for the major oxygen-demanding reactions. In general, this latter problem is more severe with stream waters and treatment plant effluents (Ruchhoft *et al.*, 1948).

The National Research Council concluded in 1946 that "In a certain way the variability of k invalidates the usual assumption that the 5-day BOD is directly proportional to the strength of the sewage." In spite of this warning, the 5-day BOD is still assumed to represent a valid measure of the BOD_u or the total amount of biodegradable organic matter in wastewaters. This may be a reasonable assumption for a long-term average but a dangerous assumption when using only a single BOD measurement. For example, consider two wastewaters A and B; each has the same 5-day BOD but wastewater A has a reaction rate of 0.1/day whereas wastewater B has a rate of 0.05/day. The BOD_u for wastewater B is then 1.56 times the BOD_u for wastewater A. Treatment plant B then receives a total waste load 56% greater than treatment plant A if both are designed to receive the same BOD_5 loading rate. The simple assumption that BOD_u for both wastewaters is 1.5 times the BOD_5 then means that treatment plant B is severely underdesigned compared to treatment plant A.

The 20°C test temperature is accepted as standard, but occasionally it is necessary or convenient to measure oxygen uptake at another temperature. The resulting first-order reaction rates are related to 20°C values by

$$k_T = k_{20}\theta^{T-20} \tag{7}$$

where k_{20} is the first-order reaction rate at 20°C and k_T is the first-order reaction rate at the actual test temperature T. The coefficient θ averages $\sim$1.05 for temperatures between 15 and 35°C for normal domestic wastes (Table I). Other equations have been suggested for making this temperature adjustment. These equations and their associated coefficients are summarized in Table I.

Although some investigations have indicated small increases in BOD_u with temperature increases, most studies have shown no significant effect. Note that Eq. (7) assumes that BOD_u does not vary with temperature changes.

TABLE I

Equations Used to Adjust First-Order BOD Reaction Coefficients for Temperature Differences

Equation	Waste-water	Coefficients			Temperature range (°C)	References
		θ	a	b		
$k_T = k_{20}\theta^{T-20}$	Sewage	1.049	—	—	20–30	Theriault, 1927
	Sewage	1.053	—	—	<20	Theriault, 1927
	Sewage	1.042	—	—	20–35	Young and Clark, 1965
	Synthetic	1.050	—	—	20–35	Young and Clark, 1965
	Sewage	1.131	—	—	5–10	Gotaas, 1948
	Sewage	1.032	—	—	25–30	Gotaas, 1948
	Sewage	1.047	—	—	—	Rich, 1963
$k_T = aT^b$	Sewage	—	0.0115	0.932	5–30	Gotaas, 1948
$BOD_{u-T} = BOD_{u-20} \times [1 + a^{T-20}]$	Sewage	—	0.02	—	5–30	Theriault, 1927
	Sewage	—	<0.01	—	5–30	Gotaas, 1948
	Sewage	—	0	—	20–35	Young and Clark, 1965
	Synthetic	—	0	—	20–35	Young and Clark, 1965

B. The Second-Order Equation

A second-order equation has been proposed to better model the BOD reaction. The equation for this model is

$$BOD_t = \frac{t}{a + bt} \qquad (8)$$

where BOD_t is the BOD at time t, t is the time of incubation, $1/a$ is the initial oxygen uptake rate (mg/L/h), and $1/b$ is the BOD_u. One advantage of this equation is that it can be solved conveniently by regression methods (Young and Clark, 1965; Marske and Polkowski, 1972; Woodward, 1953).

Although the first-order or second-order equations may be fit to data from tests lasting <10 days, the second-order equation provides a better fit of data collected from long-term oxidation tests. The first-order equation has a covariant property of giving satisfactory fit to most data because of a compensating relationship between k_{10} and BOD_u. That is, an increase in one coefficient compensates for a decrease in the other coefficient, producing a number of coefficient pairs that describe a given set of data with equal precision (Young and Clark, 1965; Dougal and Baumann, 1967; Berthouex and Hunter, 1971).

C. Other Models

More rigorous analyses of the kinetics of bacterial growth have not produced general equations that better describe the BOD reaction (Re-Velle et al., 1965; Gates and Ghosh, 1971). However, examination of the classical biological growth equation developed by Monod (1949) for batch-culture systems can be used to illustrate the removal of any nutrient or substrate in the BOD reaction. This equation can be expressed as

$$\frac{dS}{dt} = \frac{k'SM}{K_s + S} \tag{9}$$

where dS/dt is the rate of waste utilization (mg/L/unit time), M is the concentration of microorganisms (mg/L), k' is the maximum rate of waste utilization (mg/L/unit time/unit microorganism concentration), K_s is the Michaelis–Menten or half-velocity coefficient or the S concentration at which $(dS/dt)/M = 0.5k'$, and S is the concentration of a limiting nutrient or growth factor (mg/L).

This equation has been used successfully for describing substrate removal when using pure cultures and single substrates. When used to describe mixed-culture growth in the presence of numerous substrates, the equation form may fit measured data but may represent the slowest single biological reaction or a weighted average for all biological reactions (Irvine and Schaezler, 1971; Isaacs, 1969; Chen, 1970). Examination of Eq. (9) reveals that when the substrate concentration S is much greater than K_s,

$$\frac{dS}{dt} = k'M \tag{10}$$

That is, substrate use rate is a function only of the mass of microorganisms and the reaction approaches zero order (Fig. 4). If oxygen uptake is directly proportional to substrate removal, the BOD reaction also approaches zero order. When S is small compared to K_s,

$$\frac{dS}{dt} = \frac{k'}{K_s} SM \tag{11}$$

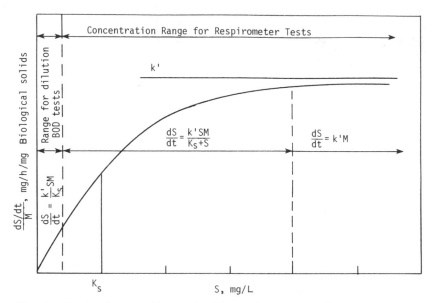

Fig. 4. Graphical presentation of the "Monod" substrate utilization equation. The reaction approaches first order at low substrate concentrations and zero order at high substrate concentrations.

If oxygen uptake is proportional to substrate removal, the oxygen uptake rate becomes a function of the concentration of the limiting nutrient and the mass of microorganisms present and the reaction approaches second order. This nutrient can be nitrogen or phosphorus when the $BOD_5 : N : P$ ratio is $>100 : 5 : 1$ (Lewis and Busch, 1965; Schaezler *et al.,* 1969). If the mass of microorganisms produced by growth is small compared to that mass provided as initial seed organisms, Eq. (11) reduces to a first-order equation and is identical to Eq. (5). If the microorganism concentration changes significantly during the reaction and S remains much less than K_s, the rate of oxidation is proportional to the product of the concentrations of oxygen-demanding organic material and oxygen-demanding microorganisms, and a second-order equation may best describe the reaction.

When all soluble organic materials are removed and only biological solids remain, the oxygen uptake reaction essentially consists of endogenous respiration with some limited cell lysis and resynthesis. In this case, the rate of decay is expressed as

$$dM/dt = -bM \qquad (12)$$

where M is the concentration of active microorganisms and b is a decay or respiration rate. If oxygen uptake is proportional to the mass decay rate, the BOD reaction during decay approaches first order.

Thus the first- and second-order equations [Eqs. (6) and (8)] provide approximations of more complex biochemical reactions under specific test conditions and stages of microorganism growth and decay.

D. Nitrification Reactions

Equations (9)–(12) also apply to nitrification reactions where S is the concentration of ammonia or nitrate nitrogen and M is the mass concentration of *Nitrosomonas* or *Nitrobacter*, respectively. K_s in Eq. (9) for both the conversion of ammonia to nitrite and nitrite to nitrate is <1 mg/L at 20°C [U.S. Environmental Protection Agency (U.S.E.P.A.), 1975], so that in the dilution test the reaction is expected to approach first order. In respirometers, where the ammonia and nitrite concentrations may be several times greater than 1 mg/L, the reaction approaches zero order as described by Eq. (10) (Young and Cameron, 1975; U.S.E.P.A., 1975).

IV. BOD TEST METHODS

A. The Standard Dilution Test

The third edition of "Standard Methods" [American Public Health Association (APHA), 1917] presented the first standard BOD measuring procedure. In 1936, the eighth edition contained a procedure that has been essentially unchanged since that time. Briefly, the test involves dilution of the sample, if necessary, so that the oxygen consumed during the test period is not greater than the amount of DO contained in the diluted sample. For purposes of standardization, a number of control procedures have been established:

1. The diluted sample is stored during testing in a filled and sealed bottle and placed in a darkened, temperature-controlled environment.
2. The water used for sample dilution must be free of biodegradable organics and toxic minerals, but should contain sufficient inorganic nutrients and trace minerals to support biological growth and a buffer to control pH.
3. The sample may require addition of seed microorganisms if the sample contains few organisms initially; samples containing industrial wastes may require the addition of organisms acclimated to that waste.

4. Dissolved-oxygen concentrations must be measured by a suitable method; the initial DO should be as close to saturation as possible but not above saturation; the final DO must be above the lower limit at which bacterial growth is adversely affected; and the difference between initial and final DO must be sufficiently high that DO measurement errors are minimized.

Detailed procedures for BOD tests are given in standard methodology manuals (APHA, 1980; U.S.E.P.A., 1979). Figure 5 shows laboratory materials needed to set up dilution BOD tests. Because of the historical importance of BOD to design and operational evaluation of treatment plants and natural waters, the dilution test has become the only officially

Fig. 5. Typical laboratory setup and apparatus for conducting dilution BOD tests. (A) Storage reservoir for dilution water; (B) sample container and magnetic stirrer for mixing the sample; (C) graduated cylinder for making dilutions; (D) 300-mL BOD bottles; (E) probe apparatus for measuring DO; (F) filtration equipment for measuring DO. Other equipment includes pipets, thermometer, beakers, etc.

recognized biochemical procedure for monitoring treatment-plant effluents and receiving water quality for discharge permit compliance and
many other purposes (Federal Register, 1976).

B. Reaeration Technique

A modification of the dilution test involves placing a sample of undi-
luted wastewater in a large jug or bottle and aerating it so that the DO is
near saturation. Aliquots are transferred to small sealed bottles in which
the depletion of oxygen is measured periodically with a DO probe or by
withdrawing small aliquots of the sample for chemical analysis for DO.
When the DO in the small bottles approaches zero, their contents are
added back to the large sample and reaerated to near-saturation DO, and
the transfer and oxygen depletion measurement process is repeated. Continued reaeration and DO depletion measurement provide points for plotting an oxygen uptake curve (Orford *et al.*, 1953) (Fig. 6).

The advantages of the reaeration technique are that large samples can
be used and little or no dilution is required. The reaeration technique is
most suited for measurement of oxygen uptake for low-strength wastewaters such as treated effluent or stream samples.

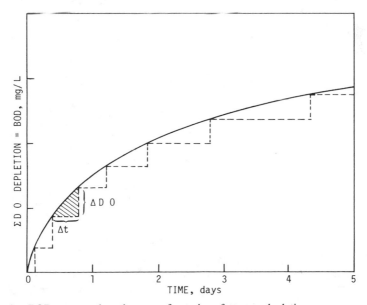

Fig. 6.   BOD measured as the sum of a series of oxygen depletion measurements.

C. Short-Term T_bOD Test

A short-term total biochemical oxygen demand (T_bOD) test has been developed using the concept that all soluble organic material is removed from solution at the plateau in the BOD reaction (see Fig. 2) (Busch, 1961). Thus the BOD exerted up to the time of the plateau, as measured by the dilution or other appropriate means, plus the oxygen demand equivalent of the organic solids including microbial cells remaining at the plateau, equals the T_bOD.

D. Respirometers for BOD Measurement

Manual measurement of oxygen uptake at frequent intervals throughout a test is quite time consuming and requires numerous test bottles and DO measurements. Thus there has been interest in developing a more direct and continuous method for measuring oxygen uptake in biochemical reactions. Respirometers of various types have been used for this purpose since at least 1908, and extensive studies of the use of respirometers have been published (Caldwell and Langelier, 1948; Gellman and Heukelekian, 1951; Heukelekian and Gellman, 1951; Lee and Oswald, 1954; Jenkins, 1960; Montgomery, 1967; Barnhart and Kehrberger, 1969; Simpson and Anderson, 1969; Young and Baumann, 1972, 1976).

Respirometers have been used for determining the specific influence of various environmental factors on biological reactions and for determining waste treatability or toxicity. Studies have been directed at using respirometric measurements of BOD to replace standard dilution methods, but to date there is no standard respirometric procedure for BOD determination. Work did progress to the point that a tentative procedure was published in the 10th and 11th editions of "Standard Methods" (APHA, 1955, 1959). Attempts to establish an acceptable standard seemed to have failed because of little justification for change, conflict with established procedures, inability to develop a consistent correlation between respirometric and dilution test data, and the variety of types of commercially available instruments. This should not, however, deter the use of respirometers because they provide useful biological data that cannot be obtained from the dilution method.

Commercially available respirometers fall into three general classifications: manometric, electrolytic, and DO depletion devices. The basic principle of operation of the first two types is the determination of oxygen weight changes in a closed system by measuring or responding to oxygen pressure changes at constant temperature and volume or volume changes at constant pressure. The third type uses a DO probe to make direct measurements of depletion of dissolved oxygen from solution. Operational procedures for respirometers vary widely, but in general the output is a continuous curve of oxygen uptake. Inherent advantages of respirometric measurements, as compared to the dilution method, are that

1. The sample may not require dilution and therefore its oxygen uptake characteristics are measured in a more natural state.
2. Larger sample volumes can be used so that more representative samples are obtained and sampling errors are minimized.
3. The samples are mixed continuously to provide uniform contact of microorganisms, substrate and oxygen.
4. No chemical titrations are required.
5. A continuous record of oxygen uptake is possible.

Respirometers also are convenient for measuring the effect of various factors, such as dilution, substrate type, temperature, and the presence of toxins, and for determination of bacterial growth and substrate removal coefficients. A summary of the major features of commercially available respirometers is given in Table II.

1. Warburg Respirometer

The Warburg respirometer is a constant-volume design consisting of a U-tube manometer attached to a reaction flask (Fig. 7). One end of the manometer is open to the atmosphere. The liquid level in the two legs of the manometer changes in response to oxygen removed from the enclosed atmosphere in the flask. Carbon dioxide is adsorbed in a potassium hydroxide solution placed in a small container within the flask.

The amount of oxygen used (ΔO_2) is related to the differential pressure ΔP, across the two legs of the manometer so that

$$\Delta O_2 = k_r(\Delta P) \tag{13}$$

where k_r is a respirometer calibration coefficient and is constant for a given gas, instrument, and flask size. The value of k_r is found by calibration procedures specific to each instrument type.

The differential pressure also is affected by barometric pressure and temperature changes, and therefore tests are conducted at constant temperature; one or more flasks containing only distilled water provide an indication of the influence of barometric pressure changes.

2. Barcroft Respirometer

The Barcroft respirometer is similar in principle to a Warburg respirometer except that a second flask is attached to the end of the manometer

TABLE II

Features of Commercial Respirometers

Manufacturer and/or instrument type class	Type of instrument	Sample size	Oxygen uptake measuring device or method	Detection or electrolyte description	Pressure compensation	Oxygen uptake limits
Warburg respirometer	Warburg constant volume, laboratory	Up to 50 mL	Pressure manometer	None	No	Weight of oxygen in reaction flask
Barcroft respirometer	Constant volume, laboratory	Up to 25 mL	Pressure manometer	None	Yes	Weight of oxygen in reaction flask
A.R.F. Products, Inc. (Raton, New Mexico)	Electrolytic, laboratory	Up to 2000 mL	Electrolytic production	Acidic or basic solution	No	128 mg/h
Oceanography International Corp. (College Station, Texas)	Electrolytic, laboratory	Up to 2000 mL	Electrolytic production	Acidic or basic solution	No	60 mg/h

Manufacturer	Type	Volume	Measurement principle	Sensor		Output
J.M. Voigt Gmbl "Sapromat" (Heidenheim, West Germany)	Electrolytic, laboratory	Up to 250 mL	Electrolytic production	Acidic copper sulfate solution	Yes	30 mg/h
Robertshaw, Controls Co. (Anaheim, California)	DO depletion continuous flow on-line	3 L residence chamber	DO difference across residence chamber	Silver–lead galvanic electrode	Not needed	30 mg/L·h
Arthur Brothers (Fond-du-lac, Wisconsin)	DO depletion, fill and draw, on-line or laboratory	1–4 L	Pressure change	Pressure transducer	Not needed	Weight of oxygen in reactor
Hach Chemical Co. (Ames, Iowa)	Manometric	Up to 157 mL	Mercury manometer	—	Yes	Volume of oxygen in bottle

WARBURG RESPIROMETER   BARCROFT RESPIROMETER

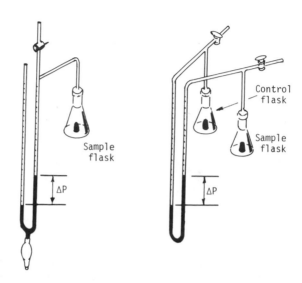

Fig. 7. Schematic diagram showing the basic features of the manometer and reaction flask for the Warburg respirometer (left) and Barcroft respirometer (right). (Adapted from Tool, 1967; reprinted with permission of Scranton Gillette Communications, Inc.)

opposite the reaction flask to provide pressure compensation during barometric pressure changes (Fig. 7). The Barcroft type is more difficult to calibrate than the Warburg, but nomograms can be prepared (Dixon, 1951) and usually are provided by the instrument manufacturer.

3. Other Manometric Respirometers

Various modifications of manometric respirometers include large-volume respirometers and respirometers with air or pure oxygen injection to increase the oxygen storage capacity within the reaction flask (Arthur, 1974). Specially designed flasks are available for tests requiring the addition or removal of samples during a test. Each of these features improves or modifies the design and operating characteristics of the Warburg or Barcroft respirometers but does not change their basic function (Jenkins, 1960).

4. Oxygen-Depletion Respirometers

Considerable effort has been expended to promote the use of DO depletion, as measured by DO probes, to monitor the progress of biochemical reactions. Basically, the equipment for making these measurements consists of a sample container, a mixing device, and one or more DO probes.

Equipment can vary from a standard 300-mL BOD bottle fitted with a DO probe attachment to large-volume units having automatic controls for sampling and recording (see Fig. 5). In batch-fed units, the sample is inserted into the container and mixed to provide a positive DO concentration. The oxygen source is turned off and the DO depletion is measured over a fixed period of time or until the DO becomes less than a specified minimum value. When this limit is reached, the sample is again mixed to provide reaeration and the DO depletion is measured over another time period. Accumulation of the oxygen depletion with time, as described in Section IV,B for the reaeration method, provides a measure of total oxygen uptake or BOD curve (see Fig. 6). One source of error with such measurements is that there is no accounting for the BOD exerted during the reaeration period.

Some DO depletion units (Robertshaw Controls Co., Anaheim, California) operate with continuous flow through a detention chamber. The sample is aerated at the inlet end, and the DO is measured in the flowing streams at both the inlet and outlet ends of the reaction chamber. The difference in DO between the two detectors represents oxygen uptake during the reaction period.

One application of oxygen-depletion respirometers is for monitoring activated-sludge basins to detect the presence of toxins or loss of mixed-liquor solids that can cause plant failure (Fig. 8) (Arthur, 1974; Young, 1981). In batch-fed units, the sample is replaced with a fresh sample after each DO depletion measurement. Changes in oxygen uptake rate expressed in mg/L/hr or mg/h-g of volatile suspended solids (VSS) indicate the status of the oxygen-demanding reaction in the sample source.

5. Electrolytic Respirometers

A recent addition to the family of respirometers is the electrolytic respirometer (Fig. 9) (Young et al., 1965; Liebmann and Offhaus, 1966; Montgomery et al., 1971; Young and Baumann, 1976). These units operate basically the same as do manometric respirometers except that when a pressure drop (usually <1 mm of water pressure) caused by oxygen consumption is detected, a power supply unit is activated and oxygen is

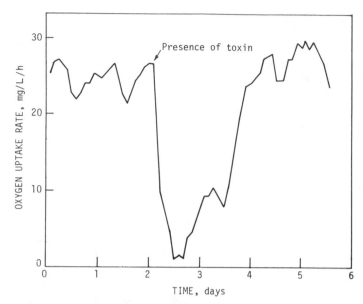

Fig. 8. Oxygen uptake rates provided from respirometers showing effect of the presence of a toxin. (Adapted from data presented by Robertshaw Controls Co., Anaheim, California.)

generated by electrolysis in a multichambered cell (Fig. 9). This oxygen is produced at a constant and precise rate and increases the oxygen pressure within the reaction vessel. When the gas pressure within the reaction vessel equals the original pressure, the power unit is deactivated. The BOD for a given sample size V is directly proportional to the total time of electrolysis, or

$$BOD = 4.97 \; It/V \quad mg/L \tag{14}$$

where I is electrical current (DC) in amperes, t is the time of oxygen production in minutes, and V is the volume of sample in liters.

One major advantage of electrolytic respirometers is that they can be automated to provide digital or binary-coded output data for direct recording or processing. Several investigators have identified operating characteristics and limitations of electrolytic respirometers (Young and Baumann, 1976; Montgomery *et al.*, 1971; Fuhs, 1968; Bridié, 1969).

E. Rapid or Alternative BOD Tests

The possibility of reducing the BOD incubation period or replacing or supplementing the dilution test with improved procedures has some inter-

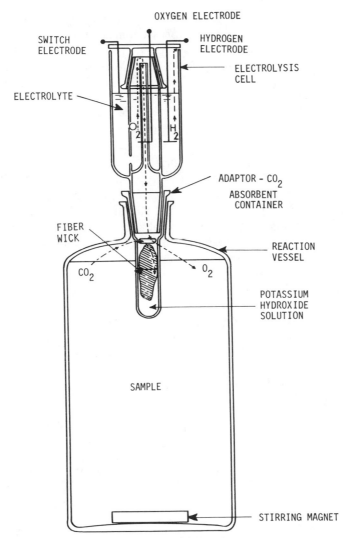

OXYGEN ELECTRODE

SWITCH ELECTRODE

HYDROGEN ELECTRODE

ELECTROLYSIS CELL

ELECTROLYTE

ADAPTOR - CO_2 ABSORBENT CONTAINER

FIBER WICK

REACTION VESSEL

CO_2 O_2

POTASSIUM HYDROXIDE SOLUTION

SAMPLE

STIRRING MAGNET

Fig. 9. Schematic diagram showing the basic operation of an electrolytic respirometer. (From Young and Baumann, 1972; reprinted with permission from Pergamon Press Ltd.)

esting and practical implications: more tests can be run in a given amount of time and data can be obtained more rapidly and possibly can be of increased value to the user. Attempts at developing short-term or rapid BOD test methods often have involved correlations of 1-, 2-, or 3-day dilution BOD with the 5-day BOD and using this correlation to convert short-term measurements to a 5-day BOD value (LeBlanc, 1974). Generally, the 20°C incubation period can be reduced to 2 days with only a small sacrifice in precision of both the short-term measured BOD and the projected BOD_5 (Orford and Matusky, 1959; Ballinger and Lishka, 1962). Increasing the incubation temperature increases the rate of biochemical oxidation, and tests have shown that at 30 to 35°C, the 2.5- to 3-day measured BOD agrees reasonably well with 5-day 20°C values (LeBlanc, 1974; Young and Clark, 1965).

In some cases the oxygen uptake rate measured by DO depletion or respirometer can be correlated with reasonable precision to the dilution BOD_5 of wastewaters (Arthur and Hursta, 1968) (Fig. 10). Generally, such correlations are best for low-strength wastewaters. This use of very short-term data presents considerable risk, however, because lag or immediate oxygen demand can introduce extreme variation in the short-term measurements. It seems that the best use of such short-term oxygen uptake measurements is as a rapid indicator of changes in oxygen uptake rate and not as a substitute for the 5-day BOD test.

The question of replacing or supplementing the standard dilution test with short-term respirometer tests has been considered repeatedly. New equipment developments, better understanding of oxygen demand reactions, which show a need for more than one point on the BOD curve, and continued use of respirometers for studying biological reactions will help resolve this question. If respirometers are considered as a substitute for the dilution BOD test, some guidelines must be established as to which test criteria are most important. If precision is the controlling criterion,

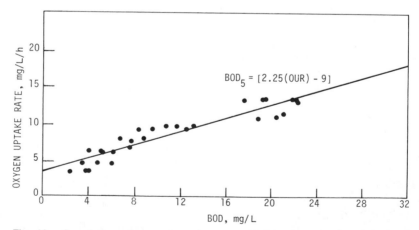

Fig. 10. Correlation of oxygen uptake as provided by a continuous-flow dissolved oxygen depletion respirometer to the BOD_5 of a secondary effluent. (Adapted from data presented by Robertshaw Controls Co., Anaheim, California.)

both laboratory and field tests have shown that respirometers give more precise measurements at 2 days of incubation than is claimed for the dilution test at 5 days (Young and Baumann, 1972, 1976). If the single-point BOD at two-thirds of ultimate BOD is the controlling criterion, respirometers can provide this measurement in ~2.5 days. This shorter time period is expected to create fewer problems in data interpretation than with the 5-day dilution test, because the continuous readout provided by respirometers more easily permits identification of lag, toxicity, nonbiodegradability, and other uncommon biological growth characteristics. Table III summarizes some correlations between dilution and respirometric BOD measurements.

V. FACTORS AFFECTING BOD MEASUREMENT

Many factors can cause variation in BOD test data and affect the reliability of BOD measurements. Sampling procedure, sample storage, varying characteristics of the sample, lack of stirring, dilution, seeding, method of dissolved oxygen measurement, and nitrification can contribute significant variation to BOD test measurements. These sources of variation are somewhat common to all methods of measuring BOD and only vary in the measured magnitude of the effect.

A. Sampling Procedure

Reliability of test results depends significantly on the procedure for collecting samples. In some cases, it is difficult to collect a sample that is representative of the entire wastewater stream. This problem is greatest with raw domestic wastewaters containing appreciable amounts of suspended organic solids and wastewater streams having a highly variable flow or waste concentration. Automatic samplers can provide sample aliquots collected at close time intervals, but commercial samplers are susceptible to plugging with wastewater solids, causing some sample aliquots to be too small or missed entirely. The most reliable although labor-intensive method is to collect samples manually at uniform time intervals for subsequent compositing in proportion to flow rate.

B. Sample Storage and Preservation

Most samples must be stored for some time prior to analysis. Although storage for longer than 24 h is not recommended, this is not always possible and some means of sample preservation must be used. Storage proce-

TABLE III

Typical Relationship between Dilution BOD$_5$ and Respirometric BOD Measurements

Comparison (Y : X)	Type of sample	Correlation (Y = A + Bx)		Correlation coefficient (R, %)	N	References
		A	B			
Dilution BOD$_5$: 18-h Warburg BOD	Glucose	—	1.59	0.91	28	Dillingham and Jose, 1960
Dilution BOD$_5$: 18-h Warburg BOD	River water + glucose	—	0.84	0.83	85	Dillingham and Jose, 1960
Dilution BOD$_5$: 24-h Warburg BOD	Domestic sewage	13.6	1.27	0.97	10	Caldwell and Langelier, 1948
Dilution BOD$_5$: 24-h Warburg BOD	Industrial wastes	−48.0	1.57	0.98	10	Caldwell and Langelier, 1948
Dilution BOD$_5$: Warburg BOD$_5$	Pure compounds	30.7	0.82	0.97	6	Gellman and Heukelekian, 1951
Dilution BOD$_5$: Warburg BOD$_5$	Domestic sewage	−4.52	0.91	0.92	11	Lee and Oswald, 1954
Dilution BOD$_5$: electrolytic BOD$_5$	Domestic sewage					
	plant A	−6.7	0.85	0.98	74	Young and Baumann, 1976
	plant M	−17.4	0.94	0.92	61	Young and Baumann, 1976
	plant D	−12.8	0.86	0.98	20	Young and Baumann, 1972

dures vary with the type of analysis to be conducted, and guidelines for various analytical purposes have been given in procedures manuals (U.S.E.P.A., 1979; APHA, 1980). A common procedure for storing samples for BOD analysis is to hold them without alteration at 3–5°C. If holding times exceed 6 h, the data should be so identified. Samples requiring storage for more than 24 h often are frozen or preserved by adding acid (usually sulfuric) to bring the pH below about 4 so that biological growth is inhibited. (A lower pH may cause excessive hydrolysis of proteins and carbohydrates.) Heavy metals such as copper, mercury, or silver also have been used for preservation purposes (U.S.E.P.A., 1979).

No preservation technique is ideal, and BOD data measured from preserved samples should be identified as such to assist interpretation of the data. For example, samples preserved with acid and subsequently neutralized and reseeded may not show significant nitrification in the 5-day test period, whereas BOD analyses using unaltered samples of the same wastewaters could show significant nitrification. As another example, samples having high concentrations of organic material may turn septic during storage thereby causing anaerobic decomposition producing organic acids and changes in pH and microorganism activity.

C. Characteristics of the Sample

Sample characteristics such as the ratio of soluble to particulate matter, initial pH or temperature, and freshness can introduce variation in BOD test measurements. The dilution BOD test is susceptible to reduced oxygen uptake rates caused by particulate solids settling to the bottom of the bottle or floating to the top. Oxygen reaches these solids primarily by diffusion through the quiescent water.

Chemical reducing agents such as ferrous iron, sulfite, sulfide, and certain organic reducing agents cause an initial high rate of oxygen uptake if present in a wastewater sample. This immediate dissolved oxygen demand (IDOD) is not biological and may represent a significant part of the total oxygen demand of raw or stored wastewater samples that have been septic and contain sulfides.

D. Mixing

The standard dilution BOD test procedure does not include mixing of the test solutions. Mixing has been shown to affect both the rate and magnitude of oxygen uptake in BOD determinations, but the effect has not been consistent in all measurements reported (Young and Baumann, 1972; Isaacs and Gaudy, 1967; Gannon, 1966; Jennelle and Gaudy, 1970;

Ali and Bewtra, 1972). As a result of this mixing effect, the usual dilution BOD technique may have little practical value in predicting the kinetics of oxygen uptake in streams receiving wastewater discharges. Lack of mixing also limits contact between organic material and microorganisms, and microbial activity is more likely to occur on the surface of the test bottle. Separation of solids by settling or flotation can further reduce opportunity for contact. One of the major advantages of respirometers is that mixing is provided during the test period.

E. Seeding

Wastewaters that are relatively fresh, industrial wastes requiring acclimated organisms, and samples that have been preserved often require seeding prior to analysis. The important factor in seeding is to provide sufficient microorganisms to eliminate the lag period without changing the dominant oxygen uptake reaction. A common seed source is settled domestic wastewater that has been stored at 20°C for 24–36 h. Updated procedures for the dilution test state that seed should be obtained from a treatment process receiving the waste for which BOD measurements are being made (APHA, 1980). Freezing and storing seed at −18°C helps reduce variations in seed quality (Tyler and Hargrave, 1965).

Seed organisms customarily have been added to the stock solution used to dilute wastewater samples, but with care the seed solution can be added by pipetting small samples of well-mixed seed stock to each BOD bottle. A blank sample containing only seed organisms should be run to determine the BOD of the seed added. The quantity of seed added is important. *Standard Methods* (APHA, 1980) recommends adding only sufficient seed stock to each liter of diluted sample to give at least 0.6 mg/L DO depletion in the dilution-test seed control.

Continuous growth of seed cultures may be necessary if frequent analysis of wastewaters requiring acclimated or adapted cultures is anticipated. Such cultures may be grown and maintained by feeding small amounts of the wastes to be analyzed to continuous- or batch-feed units. Batch-culture units are easily maintained by adding small aliquots of new waste each day to small (2–5 L) aerated containers. Solids retention times are controlled by withdrawing mixed liquor while the contents are being thoroughly mixed. New waste and sufficient water are then added either periodically by hand or continually by pump to reestablish the original volume. Small amounts of soil or material that have been in frequent contact with the waste being tested may be added to accelerate development of an acclimated culture.

Developing and using an acclimated or adapted seed culture may require trial of several methods before a suitable one is achieved. The objective is to develop a culture that gives consistent reaction characteristics.

The contribution of the seed culture to the measured oxygen uptake must be removed to provide an indication of sample BOD. This usually has been done by simply subtracting the BOD measured for a seed sample only from the BOD of seed plus sample after both have been corrected for dilution. This procedure has some associated risks because the oxygen uptake reaction for microorganisms in the seed culture alone may not be the same as in the seeded sample. This is especially true if the chemical characteristics of the sample are not similar to that for the seed stock. For example, the BOD reaction of highly soluble organic waste or waste containing toxic materials may be appreciably different from that of the seed culture. Nitrification may not occur at the same time with the seeded sample as with the seed culture alone. Seed corrections for samples that have been preserved, reconstituted, and then reseeded with a seed culture from a similar wastewater source are most likely to be valid.

Seed corrections should be made with considerable judgment only after considering the need for correction and the potential effect on data interpretation. In all cases, analysts should identify data for samples that have been seeded.

F. Dilution

Except for samples having an oxygen uptake less than ~ 7 mg/L in 5 days, the dilution test requires that the original sample be diluted. This changes the sample environment by reducing the concentration of organic material in contact with microorganisms, diluting the concentration of nutrients such as nitrogen and phosphorus available to microorganisms, and changing the ionic strength or salt balance surrounding the microorganisms. If a substance in the original sample is toxic, dilution may reduce its concentration to below the toxic level.

A basic assumption of the dilution test is that dilution reduces the magnitude of oxygen uptake in proportion to the amount of dilution so that the rate of oxygen uptake is not affected. Although the effect of dilution has been studied extensively by a number of investigators, there seems to be no consensus as to its effect. Some investigators found no appreciable effect of dilution (Heukelekian and Gellman, 1951), whereas others concluded that high concentrations gave higher reaction rates (Jennelle and Gaudy, 1970; Isaacs and Gaudy, 1967; Isaacs, 1969).

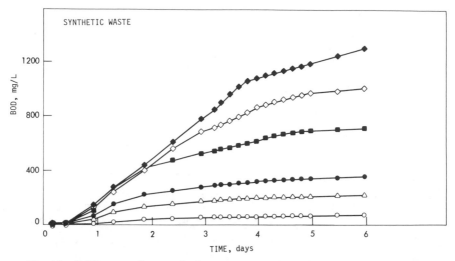

Fig. 11. BOD curves for a synthetic wastewater at theoretical oxygen demands ranging from 100 to 2000 mg/L. ○, 100 mg/L; △, 300 mg/L; ●, 500 mg/L; ■, 1000 mg/L; ◇, 1500 mg/L; ◆, 2000 mg/L. (Young and Baumann, 1972, 1976; reprinted with permission from Pergamon Press Ltd.)

This effect is illustrated in Fig. 11, which shows respirometer BOD data for various concentrations of synthetic waste (Young and Baumann, 1972). The reaction rate increases as the sample THOD increases up to ~500 mg/L and then remains constant as the substrate concentration becomes high enough to make the reaction approach zero order (Fig. 12). The values shown are for the given test conditions and could be different for other substrates and seed cultures. Reaction trends, however, are expected to be consistent from test to test. Thus, while the 5-day BOD may not be affected greatly by dilution, the rate of oxidation is affected significantly.

Calculations requiring measurements of oxygen uptake rate, such as oxygen sag determinations for streams and oxygen transfer rates for sizing aerators, may then be in error unless the oxygen uptake rate is determined at substrate concentrations close to those expected in the stream or treatment plant environment. This is not always possible with the dilution BOD test because of the limited DO capacity of the bottles. In fact, reaction rates for dilution tests may fall into a narrow range because the substrate and microorganism concentrations in the test bottles can vary only over a similar narrow range. Thus, use of respirometers may be required to obtain realistic measurements of oxygen uptake rates.

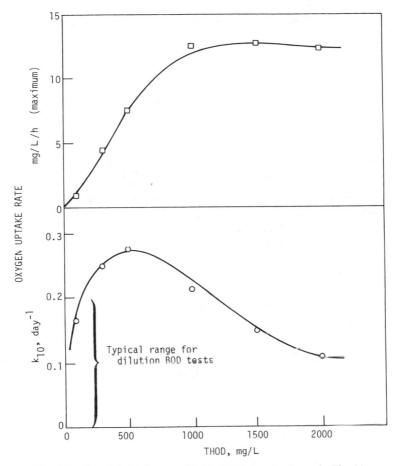

Fig. 12. Oxygen uptake rates for the BOD curves shown in Fig. 11.

G. Dilution Technique

The dilution technique must be controlled closely to eliminate extraneous variation in test data. Samples may be diluted prior to transfer to BOD test bottles or by adding a measured volume of sample to each test bottle and then carefully filling the bottle with the prepared dilution water. Dilution prior to transfer is most likely to produce the least error because larger samples can be transferred, and each test replicate is more likely to have the same initial DO and microorganism concentration. Test variation usually is greater as sample volumes become smaller, because of the greater errors associated with measuring and transferring small volumes and the improbability of obtaining representative samples with each transfer.

H. Dilution Water Quality

Ideally, dilution BOD blanks run with dilution water only should not show an oxygen uptake of more than 0.2 mg/L, but if dilution water has had buffer and other chemicals added more than 1 day prior to its use, or becomes contaminated with organic material, the oxygen uptake of the blank may be >0.2 mg/L. Waters from deionizing resins generally are not suitable for BOD-dilution water, because organic materials tend to leach into the water and the resin bed is ideal for culturing microorganisms. Even waters from single-stage distilling units often contain appreciable amounts of organic material or ammonia nitrogen.

If the DO depletion of dilution blanks exceeds 0.2 mg/L, the analyst should identify and correct the source of the problem rather than make corrections by subtracting this depletion from that of the wastewater dilutions.

Dilution water should never be permitted to come in contact with copper or brass tubing, and even the metal connections on fritted glass aeration stones can permit enough copper to be dissolved to be toxic. Tap water is not recommended for use as dilution water even in respirometer tests because of possible contamination with metals dissolved from the piping system.

I. pH

BOD tests should be conducted at pH near 7.0, with the range 6.5–8.0 generally providing acceptable results. The buffer for the standard dilution BOD test (0.32 mM as P) is designed to maintain the pH at 7.2. However, when wastes are more acidic than pH 6.5 or more alkaline than pH 8.3, adjustment to pH 7.2 is advisable before beginning BOD tests.

Samples tested in respirometers generally require a greater buffer concentration than that used with the dilution test because of the greater concentration of organic material. A phosphate buffer concentration of 0.03–0.04 M (as P) has been found adequate to maintain pH between 6.7 and 7.3 for BOD_u concentrations up to ~1000 mg/L (Young and Baumann, 1972, 1976). This buffer stock solution can be prepared by neutralizing 207 g of $NaH_2PO_4 \cdot H_2O$ (1.5 M) to pH 7.2 with potassium hydroxide and diluting to 1 L. Buffer concentrations greater than 0.04 M may be required for higher strength wastes. It may not be possible to use phosphate buffers with samples containing calcium or magnesium salts in excess of 100–200 mg/L (as the ion), and more suitable buffers may need to be developed or the sample diluted.

J. Mineral Nutrients

Adequate nutrients must be present in all BOD test solutions or oxygen uptake rates may be limited. The dilution water for the standard dilution test is designed to provide adequate iron, nitrogen, calcium, magnesium, nitrogen, and phosphorus to satisfy the needs of almost any sample. There may be, however, some nutrient limitations, specifically nitrogen, when measuring the BOD of relatively weak wastes that require little dilution. In this case, and in respirometer tests, the BOD : N : P ratio should be no greater than 100 : 5 : 1 for satisfactory test purposes.

In some cases, stream waters that are to receive wastewater discharges are used for dilution water after they have been filtered and stored with aeration for several days prior to use. Usually no nutrients are added to this water because the objective is to determine the BOD reaction as it might occur in the receiving stream.

K. Method of Dissolved Oxygen Measurement

The standard procedure for the dilution BOD test permits DO to be measured by either a chemical (Winkler) method or by dissolved-oxygen probes (see Fig. 5). Analytical testing has established that the probe provides DO and therefore BOD measurements essentially equal to the chemical methods. The various commercial probes, however, require that the sample be agitated during DO measurement. Thus if a probe is used to determine DO at intervals throughout the test in an attempt to establish a reaction curve or rate, the mixing provided with each measurement may cause significant deviation from similar chemical DO measurements made using separate unmixed BOD test bottles for each determination.

L. Dissolved Oxygen Concentration

At low DO, the rate of oxidation of organic material may become oxygen limited; that is, the limiting nutrient for substrate utilization is oxygen and not the organic material being oxidized. In respirometers, the oxygen limited uptake curves resemble a straight-line zero-order kinetic reaction in which only the rate of oxygen transfer into solution is being measured; the curve does not represent potential maximum oxygen uptake by the microorganisms in the sample.

The exact minimum DO below which oxygen becomes the limiting nutrient is a function of the characteristics of the suspended solids in the sample, the rate of mixing the sample, and the geometric shape of the test apparatus used by each investigator. Minimum recommended DO for the dilution BOD test is 1.0 mg/L (APHA, 1980).

Concentrations from 0.5 to 35 mg/L normally do not affect respiration rates so long as the cells remain dispersed (Kalinske, 1971). Samples with low mixing rates require higher minimum DO concentrations because flocculation of cells occurs more readily than with rapid mixing. The DOs in respirometer test samples change in response to the oxygen uptake rate and the minimum DO achieved is related to the mixing rate, which controls the maximum rate that oxygen can be transferred across the air–liquid interface. Using air (21% oxygen) in respirometers provides adequate oxygen transfer for samples containing from 500 to 1000 mg/L BOD_u depending on the rate of uptake required (Young and Baumann, 1972) (Bryant et al., 1967). Use of pure oxygen increases the capacity of the oxygen transfer rate by a factor of about five.

M. Toxicity

The presence of toxic substances in wastewaters results in a decrease in the rate of oxidation or produces a lag period until tolerant microorganisms develop. The absence of toxic materials must be established before BOD results can be considered valid. This can be done by comparing BOD values for several dilutions. If toxic materials are present, the 5-day BOD increases as dilution increases until the effect of the toxin is diluted out. Often toxic effects do not show up in the dilution test because dilution reduces the concentration of toxins to below the threshold toxic level.

A common toxin occurring in wastewater analysis is the chlorine used for disinfecting treated effluents prior to discharge to a receiving water. The effect of chlorination is generally a reduction in carbonaceous oxygen uptake rate and delay of nitrification (Susag, 1968; Young and Cameron, 1975; Zaloum and Murphy, 1974). Chlorine often can be dissipated by letting the sample stand for 1–2 h before analysis or can be neutralized by adding stoichometric amounts of sodium thiosulfate. Excess thiosulfate should not be added because it exerts an oxygen demand. The chlorine-free sample must then be seeded.

Chlorine or other agents in quantities added for disinfection do not oxidize significant amounts of organic material, and adding a disinfected effluent to a receiving water causes it to be reseeded so that the full oxygen demand load is exerted in the stream (Young and Cameron, 1975; McCallion, 1977). Therefore, less test variation occurs and a more realistic measure of the true BOD load to the stream is obtained if the BOD is measured immediately prior to disinfection. This is true when using chlorine, ozone, ultraviolet radiation, or other suitable disinfectants.

N. Nitrification

Nitrification represents one of the most confusing factors in making BOD measurements and is possibly the source of greatest variation in BOD test data. Traditionally, the beginning of nitrification has been considered delayed until after 5 days of incubation at 20°C; this reason is often cited as justification for not using a longer incubation period. However, this delay does not always occur, and carbonaceous and nitrogenous oxygen demand can procede simultaneously if environmental conditions are suitable (Courchaine, 1968; Patrie *et al.*, 1966; Siddiqi *et al.*, 1967; Young, 1973; Young and Baumann, 1976).

This problem was recognized in early research work with the BOD reaction, but not until 1946 was there a method of controlling nitrification. At that time, pasteurization of the sample was proposed as a means of killing nitrifying organisms. The sample was then cooled and reseeded with a culture less likely to contain nitrifying organisms (Sawyer and Bradney, 1946). Acidification to pH 4, reneutralization and reseeding also was shown to be about as effective as pasteurization for nitrification control (Hurwitz *et al.*, 1947). Chlorine, methylene blue, ammonia nitrogen, and other chemicals have been shown to inhibit the nitrification reaction, but none seem to be entirely effective. These chemicals introduce considerable variation into the test or actually inhibit the carbonaceous BOD reaction (Young, 1978).

Two chemicals, allylthiourea (ATU) and 2-chloro-6-(trichloromethyl) pyridine, have been found to inhibit the nitrification reaction without interfering with the carbonaceous reaction (Montgomery and Borne, 1966; Young, 1973). With the use of suitable chemical inhibitors, nitrification can be controlled effectively in BOD tests so that only carbonaceous BOD is measured. The potential NOD can be determined by comparing the BOD of samples run with and without inhibitor, or the ammonia and nitrite-nitrogen concentrations in the sample can be measured and the NOD calculated using Eq. (3). Much more objective interpretation of BOD data can be made if the contributions of both carbonaceous and nitrogenous oxygen demands are known.

VI. ACCURACY AND PRECISION OF BOD TESTS

The factors mentioned in Section V that affect BOD can contribute to considerable variation in test results and lead to inconsistent and imprecise measurements. One of the problems with BOD tests is that there is no standard against which to compare the results of a particular measurement. Thus there is no standard for establishing accuracy. One method for checking the dilution procedure in general and for identifying problems with seeding or measurement technique involves measuring the BOD_5 of a solution containing equal amounts of glucose and glutamic acid. Various interlaboratory tests using this mixture produced the results summarized in Table IV. Unpublished results of tests conducted by the U.S.E.P.A. (1978) have shown that the recovery and precision for samples containing a 1 : 1 mixture of glucose and glutamic acid over the total concentration range 5–340 mg/L can be expressed as

$$\bar{X} = 0.665[S] - 0.149 \tag{15}$$

and

$$S_d = 0.120[S] + 1.04 \tag{16}$$

TABLE IV
Precision of BOD_5 Measurements

	References			
	Method Research Study 3, U.S.E.P.A., 1971 (dilution test)		Ballinger and Lishka, 1962	Young and Baumann, 1976
Parameter	Low level	High level	(dilution test)	(respirometer)
Theoretical or expected value (mg/L)	2.2 (BOD_5)	194 (BOD_5)	308 (THOD)	350 (THOD)
Mean of measured BOD_5 values (mg/L)	2.12	175	214	296
Recovery (BOD_5 as % of THOD)	96	90	69	85
Coefficient of variability[a] (%)	33.2	15.0	19.5	5.8
Number of analyses/number of laboratories	74/56	73/56	34/34	4/4

[a] Standard deviation expressed as a percentage of the mean value.

where $\bar{X}$ is the measured 5-day BOD (mg/L), [S] is the added level of glucose + glutamic acid (mg/L) and S_d is the standard deviation (mg/L). Analysts, who use the glucose–glutamic acid check and find that their measured BOD_5, after seed correction, falls within the range $\bar{X} + S_d$, can feel with some confidence that their technique is correct and that there is no toxic substance in the dilution water.

Measurement precision (as indicated by coefficient of variability) for a given analyst usually should be considerably better than shown in Table IV or by Eqs. (15) and (16) because of the availability of a single seed source and use of a more consistent setup procedure.

Much lower coefficients of variability can be obtained with respirometers because of the ability to use larger and more representative samples. Coefficients of variability less than 1% have been reported for 3- to 7-day respirometric BOD measurements conducted by an individual analyst and less than 6% among analysts from a number of laboratories measuring the BOD of a standardized synthetic waste and using their own seed source and setup procedure (Table IV) (Young and Baumann, 1976).

REFERENCES

Ali, H. I., and Bewtra, J. K. (1972). *J. Water Pollut. Contr. Fed.* **44**, 1798–1807.
American Public Health Association (APHA) (1917). "Standard Methods for the Examination of Water and Wastewater," 3rd ed. APHA, New York.
American Public Health Association (APHA) (1936). "Standard Methods for the Examination of Water and Wastewater," 8th ed. APHA, New York.
American Public Health Association (APHA) (1955). "Standard Methods for the Examination of Water and Wastewater," 10th ed. APHA, New York.
American Public Health Association (APHA) (1959). "Standard Methods for the Examination of Water and Wastewater," 11th ed. APHA, New York.
American Public Health Association (APHA) (1976). "Standard Methods for the Examination of Water and Wastewater," 14th ed. APHA, New York.
American Public Health Association (APHA) (1980). "Standard Methods for the Examination of Water and Wastewater," 15th ed. APHA, New York.
Arthur, R. M. (1974). *Water Sewage Works* **121**, 100–102.
Arthur, R. M., and Hursta, W. N. (1968). *Proc. Ind. Waste Conf.* **23**, 242–249.
Bagchi, D., and Chaudhuri, N. (1970). *J. Water Pollut. Control Fed.* **42**, R136–R139.
Ballinger, D. G., and Lishka, R. J. (1962). *J. Water Pollut. Control Fed.* **34**, 470–474.
Barnhart, E. L., and Kehrberger, G. L. (1969). *Proc. Ind. Waste Conf.* **24**, 1385–1391.
Berthouex, P. M., and Hunter, W. G. (1971). *J. Sanit. Eng. Div., Am. Soc. Civ. Eng.* **97**, 333–334, 393–406.
Bhatla, M. N., and Gaudy, A. F. (1965). *J. Sanit. Eng Div., Am. Soc. Civ. Eng.* **91**, 63–87.
Bridié, A.L.A.M. (1969). *Water Res.* **3**, 157–165.
Bryant, J. O., Akers, W. W., and Busch, A. W. (1967). *Proc. Ind. Waste Conf.* **22**, 686–692.
Busch, A. W. (1958). *Sewage Ind. Wastes* **30**, 1336–1349.
Busch, A. W. (1961). *Water Sewage Works* *108*, 255–259.
Buswell, A. M., Mueller, H. F., and Van Meter, I. (1954). *Sewage Ind. Wastes* **26**, 276–285.

Caldwell, D. H., and Langelier, W. F. (1948). *Sewage Works J.* **20**, 202–218.

Chen, C. W. (1970). *J. Sanit. Eng. Div., Am. Soc. Civ. Eng.* **96**, 1085–1097.

Courchaine, R. J. (1968). *J. Water Pollut. Control Fed.* **40**, 835–942.

Dillingham, E. O., and Jose, A. G. (1960). *Tappi* **43**, 626–630.

Dixon, M. (1951). "Manometric Methods as Applied to the Measurement of Cell Respiration and Other Processes," 3rd ed. Cambridge Univ. Press, London and New York.

Dobbins, W. E. (1964). *J. Sanit. Eng. Div., Am. Soc. Civ. Eng.* **90**, 53–57.

Dougal, M. D., and Baumann, E. R. (1967). *Proc. Am. Water Resour. Conf. 3rd, 1967* pp. 207–212.

Fair, G. M. (1936). *Sewage Works J.* **8**, 430–438.

Fuhs, G. W. (1968). *Wasser Abwasser Forch.* **5**, 161–168.

Gannon, J. J. (1966). *J. Sanit. Eng. Div., Am. Soc. Civ. Eng.* **92**, 135–142.

Gates, W. E., and Ghosh, S. (1971). *J. Sanit. Eng. Div., Am. Soc. Civ. Eng.* **97**, 287–309.

Gaudy, A. F. Jr., Komolrit, K., Follett, R. H., Kincannon, D. F., and Modesitt, D. E. (1967). Methods for Evaluating the First Order Constants K_1 and L For BOD Exertion, Report M-1, Center For Water Research in Engineering, Oklahoma State Univ., Stillwater.

Gellman, I., and Heukelekian, H. (1951). *Sewage Ind. Wastes* **23**, 1267–1281.

Gotaas, H. B. (1948). *Sewage Works J.* **20**, 441–448.

Hammer, M. J. (1975). "Water and Waste-Water Technology." Wiley, New York.

Hartmann, L., and Singrun, M. E. (1968). *Water Sewage Works* **115**, 289–294.

Heukelekian, H., and Gellman, I. (1951). *Sewage Ind. Wastes* **23**, 1546–1563.

Hurwitz, H., Barnett, G. R., Beaudoin, R. E., and Kramer, H. P. (1947). *Sewage Works J.* **19**, 995–999.

Irvine, R. L., and Schaezler, D. J. (1971). *J. Sanit. Eng. Div., Am. Soc. Civ. Eng.* **97**, 409–424.

Isaacs, W. P. (1969). *Proc. Ind. Waste Conf.* **24**, 52–59.

Isaacs, W. P., and Gaudy, A. F., Jr. (1967). *Proc. Ind. Waste Conf.* **22**, 165–182.

Jennelle, E. M., and Gaudy, A. F., Jr. (1970). *Biotechnol. Bioeng.* **12**, 519–539.

Jenkins, D. (1960). *In Waste Treatment,* pp. 99–121. Pergamon, Oxford.

Kalinske, A. F. (1971). *J. Water Pollut. Control Fed.* **43**, 73–80.

LeBlanc, P. J. (1974). *J. Water Pollut. Control Fed.* **46**, 2207–2208.

Lee, E. W., and Oswald, W. J. (1954). *J. Water Pollut. Control Fed.* **26**, 1097–1108.

Lewis, J. W., and Busch, A. W. (1965). *Water Sewage Works* **112**, 106–109, 139–143, 185–188, 209–211.

Liebmann, H., and Offhaus, K. (1966). *Abwassertechnik* **17**, 4–6.

McCallion, C. D. (1977). M.S. Thesis, Iowa State University, Ames (unpublished).

Marske, D. M., and Polkowski, B. (1972). *J. Water Pollut. Control Fed.* **44**, 1987–2000.

Monod, J. (1949). *Annu. Rev. Microbiol.* **3**, 371–379.

Montgomery, H. A. C. (1967). *Water Res.* **1**, 631–662.

Montgomery, H. A. C., and Borne, B. J. (1966). *J. Proc. Inst. Sewage Purif.* Part 4, 3–14.

Montgomery, H. A. C., Oaten, A. B., and Gardiner, D. K. (1971). *Effluent Water Treat. J.* **11**, 23–31.

Moore, W. A., Thomas, H. A., Jr., and Snow, W. B. (1950). *Sewage Ind. Wastes* **22**, 1343–1349.

O'Brien, W. J. (1962). Master of Science Thesis, New Mexico State University, Las Cruces (unpublished).

Orford, H. E., and Matusky, F. E. (1959). *Sewage Ind. Wastes* **31**, 259–267.

Orford, H. E., Rand, M. C., and Gellman, I. (1953). *Sewage Ind. Wastes* **25**, 259–264.

Patrie, B. A., Keshavan, K., and Woodward, F. E. (1966). *Proc. Ind. Waste Conf.* **21,** 869–879.

Phelps, E. B. (1944). "Stream Sanitation." Wiley, New York.

Reed, L. J., and Theriault, E. J. (1931). *J. Phys. Chem.* **35,** 950–971.

ReVelle, C. S., Lynn, W. R., and Rivera, M. A. (1965). *J. Water Pollut. Control Fed.* **37,** 1679–1692.

Rich, L. G. (1963). "Unit Processes of Sanitary Engineering." Wiley, New York.

Robertshaw Controls, Co., Anaheim, CA, Bulletin 90-01, Aug. 1976.

Ruchhoft, C. C., Placak, O. R., and Ettinger, M. B. (1948). *Sewage Works J.* **20,** 832–839.

Sawyer, C. N., and Bradney, L. (1946). *Sewage Works J.* **18,** 1113–1120.

Schaezler, D. J., Busch, A. W., and Wood, C. H. (1969). *Proc. Ind. Waste Conf.* **24,** 607–614.

Siddiqi, R. H., Speece, R. E., Engelbrecht, R. S., and Schmidt, J. W. (1967). *J. Water Pollut. Control Fed.* **39,** 579–586.

Simpson, J. R., and Anderson, G. K. (1969). *Prog. Ind. Microbiol.* **5,** 141–167.

Susag, R. H. (1968). *J. Water Pollut. Control Fed.* **40,** R434–R441.

Theriault, E. J. (1927). *U.S. Public Health Serv. Bull.* No. 173.

Thomas, H. A. (1940). *Sewage Works J.* **12,** 504–512.

Tool, H. R. (1967). *Water Sewage Works* **114,** 211–218.

Tyler, L. P., and Hargrave, E. C. (1965). *Water Sewage Works* **112,** 181–184.

U.S. Environmental Protection Agency (U.S.E.P.A.) (1976). *Fed. Regist.* **41,** (232), 52780–52786.

U.S. Environmental Protection Agency (U.S.E.P.A.) (1971). "Method Research Study 3, Demand Analysis." U.S.E.P.A., Analytical Quality Control Laboratory, Cincinnati, Ohio.

U.S. Environmental Protection Agency (U.S.E.P.A.) (1975). "Nitrogen Control Process Design Manual." U.S.E.P.A., Office of Technology Transfer, Washington, D.C.

U.S. Environmental Protection Agency (U.S.E.P.A.) (1979). "Methods for Chemical Analysis of Water and Wastes," Rep. No. EPA-600/4-79-020. U.S.E.P.A., Washington, D.C.

U.S. Environmental Protection Agency (U.S.E.P.A.) (1978). "Water Pollution Studies 002,003,604." U.S.E.P.A., Environmental Monitoring and Support Laboratory, Cincinnati, Ohio.

U.S. Public Law 92-500. Water Pollution Control Act Amendment (1972). U.S. Govt. Printing Office, Washington, D.C.

Wezernak, C. T., and Gannon, J. J. (1968). *J. Sanit. Eng. Div., Am. Soc. Civ. Eng.* **94,** 883–891.

Woodward, R. L. (1953). *Sewage Ind. Wastes* **25,** 918–919.

Young, J. C. (1973). *J. Water Pollut. Control Fed.* **33,** 637–646.

Young, J. C. (1981). *Wat. Sci. Tech.* **17,** 397–403.

Young, J. C., and Affleck, S. B. (1974). *Proc. Ind. Waste Conf.* **29.**

Young, J. C., and Baumann, E. R. (1972). "Demonstration of the Electrolysis Method for Measuring BOD" Rep. ERI 72153. Engineering Research Institute, Iowa State University, Ames.

Young, J. C., and Baumann, E. R. (1976). *Water Res.* **10,** 1031–1040.

Young, J. C., and Cameron, W. S. (1975). *Proc. Ind. Waste Conf.* **30,** 885–896.

Young, J. C., and Clark, J. W. (1965). *Water Sewage Works* **112,** 251–257, 341–345.

Young, J. C., Garner, W., and Clark, J. W. (1965). *Anal. Chem.* **37,** 784.

Zaloum, R., and Murphy, L. L. (1974). *J. Water Pollut. Control Fed.* **46,** 2770–2779.

2 CHEMICAL NONSPECIFIC ORGANICS ANALYSIS

William J. Cooper

Drinking Water Research Center
Florida International University
Miami, Florida

James C. Young

Department of Civil Engineering
University of Arkansas
Fayetteville, Arkansas

I. INTRODUCTION

Chemical and instrumental analyses of specific organic compounds or groups of compounds in aqueous solution have undergone a tremendous increase in sophistication since the early 1960s. Increased sophistication has also resulted in expensive instrumentation and tedious analytical procedures. Because of the expense in analyst time and equipment, there is an ever increasing need for nonspecific organic analyses using chemical and instrumental methods for the collective measurements of a large number of organic compounds (Cooper and Suffet, 1981).

Nonspecific chemical analyses can be used to assess water quality (Malcolm and Leenheer, 1973), monitor unit operations (Suffet, 1980; Sylvia and Donlan, 1980), water reclamation plants (Van Rensburg *et al.*, 1981), and wastewater treatment (Michail and Idelovitch, 1981) systems, to name a few. This analysis can also screen for the need for extensive individual compound analysis, therefore minimizing the use of complex analyses for individual or groups of compounds.

Examples of nonspecific, chemical methods for organics measurements are shown in Table I. They appear to be relatively simple and less expensive than the more complicated specific organic analytical procedures. As in all analytical measurements, problems can be encountered and the analyst needs to consider the question of accuracy and precision in the use of nonspecific analyses. The question of accuracy also is a function of the proper sampling, storage, and handling procedures prior to analysis. The purpose of this chapter is to review and point out limitations of the chemical and instrumental methods of nonspecific organics analysis.

II. ELEMENTAL PARAMETERS

A. Total Organic Carbon

The amount of organic carbon in an aqueous sample can be categorized into the following fractions (MacKinnon, 1981):

1. TOC, total organic carbon, defined as the total amount of organically bound carbon in a sample
2. DOC, dissolved organic carbon, defined as the organic material that passes a 0.45-μm filter and is not lost by evaporation during the analytical procedure used for the measurement
3. POC, particulate organic carbon, defined as that organic material retained by a 0.45-μm filter
4. VOC, volatile organic carbon, defined operationally, not quantitatively, as the amount of organic material subject to removal from aqueous solution by air stripping and based on the instrumentation or analytical procedures employed

There are other definitions used in describing TOC and a fractional organic component or both, and consistency in nomenclature becomes a

TABLE I

Examples of Chemical Nonspecific Measurements of
Organic Chemicals in Aqueous Solution

Measurements	Examples
I. Elemental parameters	Organic carbon
	Organic halogen
	Organic nitrogen
II. Oxygen demand	Chemical oxygen demand (COD)
	Total oxygen demand (TOD)
III. Spectrophotometric methods	Ultraviolet/visible
	Absorption
	Fluorescence
IV. Analysis of chemical classes (functional group analysis)	Amines, carbohydrates
	Phenols, oil, and grease
V. Electrochemical methods	—

problem. For example, based on the above definition, TOC is the sum of POC and DOC if no prefilter is used before analysis and DOC only if a filter is used. Some authors refer to POC as purgeable organic carbon, which is operationally correct but leads to confusion with the particulate fraction. Throughout this section these definitions are used, and where difficulties arise, an attempt is made to reconcile other nomenclature with them.

1. Measurement of Organic Carbon

The basis of the measurement of organic carbon is the initial oxidation of organic carbon to CO_2 [Eq. (1)].

$$C_aH_bN_cO_d + \left(a + \frac{b}{4} - \frac{d}{2}\right) O_2 \longrightarrow a\,CO_2 + \frac{b}{2}\,H_2O + \tfrac{1}{2}\,N_2 \qquad (1)$$

The CO_2 can then be measured directly or reduced to CH_4 and determined. Numerous methods are available for the direct determination of CO_2. Duursma (1961) determined CO_2 using volumetric, gravimetric, conductometric, titrimetric, or coulometric methods. Menzel and Vaccaro (1964) used nondispersive infrared spectrometry, although gas chromatography (West, 1964), mass spectrometry (Games and Hayes, 1976) and plasma-emission spectrometry (Mitchell *et al.*, 1977) also have been used.

The reduction of CO_2 over a catalyst in the presence of H_2 is quantitative at high temperatures. The CH_4 detected by a flame ionization can then be used as a quantitative determination of organic carbon (Takahashi *et al.*, 1973; Takahashi, 1979).

Three techniques for the oxidation of organic matter are used:

1. *Wet oxidation* processes utilize (a) chemical oxidants, such as $KMnO_4$ or $K_2Cr_2O_7$ (Kay, 1954), silver-catalyzed dichromate (Duursma, 1961), $K_2S_2O_8$ (Wilson, 1961; Menzel and Vaccaro, 1964), or silver-catalyzed persulfate (Baldwin and McAtee, 1974; Goulden and Brooksbank, 1975); (b) photooxidation with UV irradiation (Armstrong and Tibbits, 1968; Ehrhardt, 1969; Baker *et al.*, 1974; Wolfel and Sontheimer, 1974; Collins and Williams, 1977); or (c) a combination of UV–persulfate (Goulden and Brooksbank, 1975; West *et al.*, 1976, 1977; Takahashi, 1979; Van Steenderen *et al.*, 1979; Lantz *et al.*, 1981; Van Steenderen and Lin, 1981).

2. *Dry combustion* has been developed primarily for analysis of seawater samples. Once dried, the salts are oxidized in a high-temperature (700–900°C) oxygen atmosphere or in an elemental analyzer, and the CO_2 is determined. Concentration techniques most commonly utilize evaporation (Skopintsev, 1960, from MacKinnon, 1981), freeze-drying (Gordon and Sutcliff, 1973), or vacuum evaporation (MacKinnon, 1978) for the elimination of water.

3. *Direct injection, high-temperature oxidation* has been used extensively for water containing greater than 2 mg/L TOC. The direct-injection techniques, limited by sample size, generally are considered applicable to waters with TOC >2 mg/L. The initial studies using this technique were reported by Van Hall *et al.* (1965) and Van Hall and Stenger (1967) and are the basis of several commercial instruments. This method has been automated by Van Steenderen (1976a,b, 1979).

In general, the wet oxidation and dry combustion techniques have found application in waters in which the total organic carbon is less than 2 mg/L. Reported wide discrepancies in the values obtained using the two different procedures for seawater analysis appear to result from contamination or blank corrections in earlier work (MacKinnon, 1981). MacKinnon (1978) and Gershey *et al.* (1979) compared the two procedures and showed that they agree within 15% when proper care is taken with the dry combustion technique.

Although the analytical determination of nonspecific organics appears to be simple and straightforward, MacKinnon (1981) listed several considerations that complicate the measurement of organic carbon in seawater. These considerations apply to many freshwaters as well and are

1. Low concentration of organic matter
2. High concentration of inorganic carbon as carbonates
3. Contamination during sampling, storage, preparation, and analysis
4. Difficulty in accurately determining reagent or method blanks

5. Incompleteness of oxidation
6. Lack of certified standards or well-characterized reference materials for interlaboratory method comparisons
7. Length of time of analysis

2. Carbon Analyzers

Carbon analyzers (Table II) were developed commercially in the 1960s and have been used extensively for the nonspecific determination of the amount of organic matter in wastewater samples (Van Hall *et al.,* 1963; Van Hall and Stenger, 1967; Arin, 1974). A small sample of the wastewater to be tested, usually less than 50 μL (Beckman), is injected into a stream of carrier gas (oxygen or air) directly into or onto a catalyst-coated porous medium maintained at temperatures ranging from 900 to 1000°C (Fig. 1). At these temperatures, carbon is oxidized almost instantaneously to form carbon dioxide, which is measured directly by a nondispersive infrared (NDIR) analyzer.

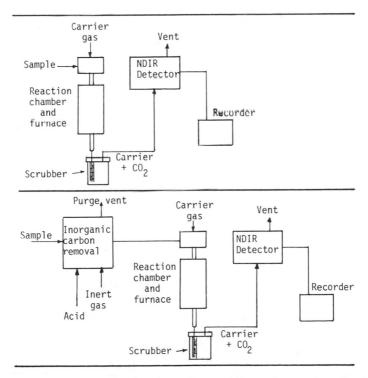

Fig. 1. Schematic flow diagrams for carbon analyzers using nondispersive infrared detection methods. (Adapted from material presented by Ionics, Inc., Watertown, Massachusetts.)

TABLE II

Features of Commercial Total Carbon (TC) and Total Organic Carbon (TOC) Analyzers

Manufacturer	Analyzer application (model)	Sample input method and size	Inorganic-carbon separation method	Carbon conversion reaction
Dohrmann Div. Envirotech Corp. (Mountain View, California)	Laboratory (Models DC-50, DC-52)	Placed by micropipet into an injection boat, 10–100 μL	Vaporization in acid at 90–100°C on CuO catalyst; not detected	Pyrolysis at 850°C followed by catalytic reduction at 350°C on nickel or aluminum
	DC-80	Automated injection	Acidify, purge	As above
Dohrmann Div. Envirotech Corp. (Mountain View, California)	On-line process monitor (Model DC-60)	Mechanical displacement injector (0.125–0.25 mL/cycle)	Purged with air in multiple-stage separator	Catalyst in ceramic tube, 900°C
Oceanography International Corp. (College Station, Texas)	Laboratory (Model ATO)	Ampule with up to 10-mL sample	Purge manually before analysis	Heated in ampule with persulfate oxidant
Oceanography International Corp. (College Station, Texas)	Laboratory (Model DIM)	Direct injection with syringe, 20–50 μL	Purge manually before analysis	Asbestos mat, no catalyst
Beckman Instruments, Inc. (Fullerton, California)	Laboratory (Model 915, 915A)	Syringe injection, 20–200 μL	Separate column packed with acid-coated quartz chips, 150°C	Cobalt oxide on inert packing, 950°C
Beckman Instruments, Inc. (Fullerton, California)	On-line process monitor	Mechanical plunger from flowing stream, 40 μL	None	Cobalt oxide on inert packing, 950°C
Astro Ecology Corp. (Houston, Texas) (Models, 1000, 1100, 1150)	Laboratory or On-line	Direct injection, syringe or by continuous metering	Sample acidified and scrubbed automatically	Fixed bed, metal catalyst, 900°C
Automated Environmental Systems, Inc. (Woodbury, New York)	On-line process monitor (Models 1600, TOC, 1610 TC)	Continuous metering by positive displacement pump, 3 mL/min	Sample acidified and scrubbed automatically with nitrogen	Fluidized aluminum oxide bed, 850°C
Ionics, Inc. (Watertown, Massachusetts)	On-line process monitor or laboratory (1200 series)	Slide plate injection valves	Sample acidified and scrubbed automatically with nitrogen (Model 1858)	Fixed bed, acid salt, palladium, 900°C
Ionics Inc. (Watertown, Massachusetts)	Laboratory (Model 445)	Syringe	Primary, fixed bed, acid salt, 170°C	Fixed bed, acid salt, palladium, 900°C
Delta Scientific Corp. (Undenhurst, New Jersey) (Model 8155)	On-line (Model 8155)	Syringe, 0.25–2.0 mL	Automatic acidification and purge	Fixed bed, copper oxide, 900°C

Detector	Carrier	Response time (min)	Readout method	Maximum concentration range (mg/L of C)	Repeatability claimed by manufacturer (% of full scale)
FID	Helium	5–10	Digital meter	1–2,000 (DC-50); 0.05–6,000 (DC-52)	±2 or less
NDIR	Oxygen	5–10	Meter and/or recorder	0.010–20 Variable, up to 0–5000	±5 ±2
NDIR	Oxygen	2 min + prep time (8 min)	Meter	0.3–3	±5
NDIR	Oxygen	1	Meter	0–5	±0.5
NDIR	Air or oxygen	2–4		Variable, 0–5 up to 0–4000	±2 at 50–4000 mg/L of C; ±5 at 5 mg/L of C
NDIR	Air or oxygen	2–4	Meter and recorder		
NDIR	Air	10	4 in. meter and 6 in. recorder	Variable ranges 0–250 0–1000 0–5000	±2
NDIR	Oxygen	5–15	4 in. meter, 6 in. strip recorder, punched tape, magnetic tape	0–50 0–4000	±2
NDIR	Nitrogen plus oxygen	5	Recorder		
NDIR	Oxygen	5	Meter or recorder		
Nephelometer (measures turbidity of barium compound)	Air	8–12		0–10 0–5000	±5

Some instruments pyrolyze the organic material to form volatile products and carbon dioxide that are converted catalytically to methane, which is measured by a flame ionization detector (FID) (Dohrmann), or the sample organic compounds are volatilized directly into an FID (Carle). At least one instrument design (Oceanography International Corp., Model ATO) has a feature that permits the sample to be heated in a sealed ampule with an oxidizing agent before it is passed through the analyzer.

Carbon analyzers may measure either total carbon (TC), which includes a measure of all inorganic and organic carbon present in a sample, or total organic carbon (TOC), which is a measure of only the organic carbon. There are two methods of differentiating between TC and TOC. One method eliminates the inorganic carbon by acidifying the sample and purging with an inert gas, such as nitrogen, or by boiling. This can be done manually prior to injection of the sample into the instrument or may be accomplished automatically within the instrument. Either of these procedures can result in loss of VOC or liquid. Dohrman instruments incorporate a purge cycle after acidification. The inorganic carbon dioxide is trapped out on lithium hydroxide, and the volatile organics are measured by an FID.

Some carbon analyzers are designed to determine the inorganic carbon by passing a sample through a catalyst bed maintained at low pH (2–4) and low temperature (90–150°C), causing the inorganic carbon to be converted to carbon dioxide. The carbon dioxide produced is measured quantitatively by the instrument. A second sample is injected onto the high-temperature catalyst to produce a measure of total carbon. The difference between the two measurements gives the organic carbon.

One analyzer (Delta Scientific) acidifies and purges the sample within the instrument to remove carbon dioxide and then catalytically oxidizes the organic material. The carbon dioxide produced is passed through a solution of barium hydroxide and precipitated as barium carbonate. Carbon is then determined quantitatively from the turbidity of the solution.

3. Low-Level Carbon Analysis

A great deal of work has been done to develop carbon analyzers capable of detecting carbon concentrations less than 5–10 mg/L for analysis of potable water supplies, stream and lake samples, and purified process streams (U.S. Environmental Protection Agency, 1971a) (Jones and Degaford, 1968; Goulden and Brooksbank, 1975; Kehoe, 1975). Some instruments (Dohrman's Model DC-52/54 and 80, and Oceanography International Corp.'s ATO) give improved measurement precision at less than 5 mg/L of carbon.

When making low-level carbon measurements, analysts must use extreme care to eliminate interferences caused by contamination of the reference standard and injection devices. Care must also be taken in preparing organic-free distilled water.

4. Carbon Analyzers: Accuracy and Precision

Tests have been made to determine accuracy and precision of carbon analyses by sending prepared samples of known composition (equal-weight mixtures of glucose and glutamic acid) to a number of participating laboratories. Results of these tests, summarized in Table III, show good recovery at the higher standard concentration but not at the low concentration. Some of the error with the lower concentration sample can be attributed to lower sensitivity of the instruments used in this range, contamination by dust, lint, cellulose, etc., and inaccuracies in volumetric measurement.

Numerous tests with pure compounds have shown that an analyst can obtain essentially complete recovery with standard deviations of less than 1%, if care is exercised in handling samples and in making calibrations (Van Hall et al., 1963). In general, it can be assumed that 100% recovery of the injected carbon can be accomplished by organic carbon analyzers unless some carbon is lost by volatilization in a preliminary purge step. No organic chemicals have been reported to be only partially oxidized by any of the catalytic combustion processes used in commercially available carbon analyzers. Most instrument manufacturers claim that repeatability of ±2% of full-scale readings can be obtained. The majority of the greater variability shown in Table III can be attributed to analyst bias and transfer errors.

TABLE III

Accuracy and Precision of Organic Carbon Analyses[a]

Parameter	Low level	High level
Theoretical or expected TOC values (mg/L)	4.9	107
Mean of measured TOC values (mg/L)	5.65	108.1
Recovery (%)	115.3	101.0
Coefficient of variability (%)	33.5	5.6
Number of analyses	27	26

[a] From U.S.E.P.A. (1971b).

5. Carbon Analyzers: Sources of Error

Inorganic salts and acids may introduce small errors into carbon analyses, because they or their oxidation products may be falsely detected as carbon dioxide by the infrared analyzers. Some instruments use filters to reduce this interference (Van Hall and Stenger, 1967).

Major errors can occur in organic carbon measurements when samples containing volatile organics are acidified and purged to remove inorganic carbon. Compounds such as benzene, toluene, and cyclohexane may be lost completely (Ford, 1968).

Heavy metals injected in samples may eventually foul the catalysts used to promote oxidation and therefore create a need to replace or repack the catalyst matrix. Low instrument sensitivity or poor reproducibility signal the need for such action. Other sources of error include

1. Different responses of various instruments in different range settings
2. Different injection volumes and techniques
3. Variations in dilution techniques

B. Organic Halogen

Organic halogen (OX) compounds have natural and anthropogenic origins. Of particular concern in drinking water is the fact that as a group they are generally suspected of causing adverse health effects (National Academy of Sciences, 1977; Cotruvo, 1978). Analytical procedures for individual or classes of OX compounds have been developed. In general, the measurement of specific OX compounds accounts for less than 25% of the organohalogens in drinking water (Dressman *et al.*, 1979). Therefore, a need exists to assess the total OX in drinking water and other waters. The OX can then be used as an indicator of water quality. In addition to water quality, OX can be used to indicate effects of various water treatment unit processes on water quality. Dressman *et al.* (1979) has proposed three abbreviations in an attempt to standardize the nomenclature: (a) OX, to indicate measurement of organohalides as a group, (b) POX, to indicate the purgeable fraction of OX, and (c) NPOX, to indicate the nonpurgeable fraction of OX.

In the development of procedures, several difficulties exist that are related to OX in aqueous solution. These difficulties should be noted by the analyst and taken into consideration in interpreting results or when selecting a method for OX analysis. The difficulties of measuring OX in water were summarized by Takahashi *et al.* (1981):

1. The range of OX in surface water is 1–50 μg/L as chlorine (0.03–1.5 μmol/L). This requires either a highly sensitive detector or concentration of the sample without losing a significant fraction of organics.
2. None of the detection methods, except neutron activation which requires highly sophisticated instrumentation, responds equally to all types of halogenated compounds. Thus, all halogenated organics must be converted to a common species before detection.
3. The inorganic halide (IX) concentration in surface waters is typically 1,000–50,000 times higher than the OX. Thus OX methods must have an IX rejection ratio of at least 50,000 : 1.

1. OX Analytical Methods

Several techniques exist for determining OX in aqueous samples. In general, the procedures involve concentration, removal of inorganic halogens, followed by detection of the halide(s).

Solvent extraction is commonly used in the determination of specific OX compounds and is discussed in detail in several papers by Keith (1982). The use of solvent extraction followed by pyrolysis of an aliquot and determination by coulometric titration was reported by Van Steenderen (1981). The analysis was completed in 3 to 4 min with a minimum detection limit in water of 1 μg/L. It is not uncommon to account for less than 25% of the OX by using liquid–liquid extraction followed by gas chromatography. Thus the procedure by Van Steenderen, although rapid, is limited to solvent-soluble OX.

An alternative detection of solvent-soluble OX, nondestructive neutron-activation analysis, has been utilized by Lunde et al. (1977) and Ahnoff et al. (1979) and Gether et al. (1979) to determine solvent-extractable organochlorine and organobromine. A highly efficient liquid–liquid extraction resulted in a lower limit of detection of 0.5 ng/L for Cl and 0.1 ng/L for Br. The sample size, 200 L, and the expense of neutron-activation analysis are limiting factors. However, given access to the type of equipment needed, nondestructive neutron activation offers a very attractive alternative for exploratory work in extending this to a more general OX measurement.

Organic compounds in aqueous solutions can be separated on the basis of molecular weight using ultrafiltration membranes. These separations are not absolute as some organic matter can be retained as a result of electrical charges on the organics. McCahill et al. (1980, 1981) used this concept to separate OX according to molecular weight. Following separation, the solutions were irradiated with UV radiation to cleave the

C—X bond. The Cl^- was determined by ion-specific electrode; ion chromatography was used to determine the Cl^- and Br^- simultaneously. The photolysis required 1.5 h followed by a typical ion-chromatographic analysis time of 0.3 to 0.5 h. No detection limits were given; however, the ion chromatograph can detect 10^{-7}–10^{-6} M halogens with a 1 mL injection.

Glaze *et al.* (1977) described a method for the determination of OX using adsorption on XAD-2 or XAD-4 to concentrate OX. The OX was eluted by ether, and microcoulometry analysis of a 20-μL aliquot provides the OX measurement. A precision of 17.3% was reported for 10^0–10^1 ng of Cl and 2.3% for 10^3–10^4 ng of Cl. In reviewing analytical procedures for OX, Dressman *et al.* (1979) stated that because of poor adsorption of some OX, the method of Glaze *et al.* (1977) has limited applicability.

Presently, the method of choice appears to be modifications of the original carbon-adsorption technique reported by Kuhn and Sontheimer (1973, 1974) and evaluated by Dressman *et al.* (1977). This method was modified by Jekel and Roberts (1980) to provide a POX and OX measurement and was extended to wastewater samples, with special reference to waters intended for reuse. Takahashi *et al.* (1981) described an analyzer using the carbon-adsorption approach to measure POX and OX.

2. OX Analyzers

Dressman *et al.* (1979), in a thorough review of analytical techniques for OX in water, recommended the carbon-adsorption procedure of Kuhn and Sontheimer (1973, 1974) with several modifications. The result has been the development of an instrumental method consisting of two parts (Takahashi *et al.*, 1981):

1. Organics are adsorbed onto granular activated carbon (GAC) packed in columns, which simplifies the adsorption step and minimizes the loss of purgeables.
2. The GAC is then combusted by an improved technique ensuring quantitative conversion of halides to HX, which are then coulometrically titrated with silver ions.

Figure 2 is a representation of the carbon-adsorption apparatus with the plumbing diagram shown in Fig. 3 (Takahashi *et al.*, 1981; Dohrman DX-20). The GAC should have (a) low halogen background, (b) low affinity for IX, and (c) high absorptivity for all organics in solution.

3. OX Analyzers: Accuracy and Precision

The accuracy and precision of the OX method are affected by the halide background of the GAC, typically 100–400 μg Cl/40 mg of GAC. A

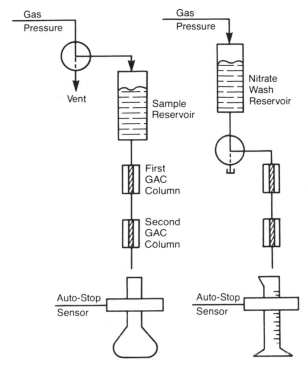

Fig. 2.   Carbon-adsorption apparatus for the determination of OX in adqueous solution. [Reprinted with permission from "Chemistry in Water Reuse," Vol. 2 (W. J. Cooper, ed.). Ann Arbor Science Publ., Inc., 1981.]

100-mL water sample typically contains 1–5 μg/L of OX. The standard deviation of several test organics was found to be 1.5 μg/L or $\pm 2\%$ of the OX, whichever was greater (Takahashi *et al.*, 1981). A summary of the results is given in Table IV.

Dressman *et al.* (1979) noted that when using the carbon-adsorption OX procedure for drinking water, dechlorination with sodium thiosulfate followed by acidification to pH 2 produced an erratic positive bias that increased on standing. They also cautioned that NO_3^- does not displace all mercuric chloride from carbon and that combustion of mercuric chloride poisons the microcoulometric titration cell. Thus, samples should not be preserved with mercuric chloride.

Although no direct comparison has been made between the Kuhn and Sontheimer (1973, 1974) method modified by Dressman *et al.* (1977) and the Dohrman DX-20 method, previous comparisons to a predecessor of the DX-20 instrumentation showed good agreement (Dressman *et al.*, 1979).

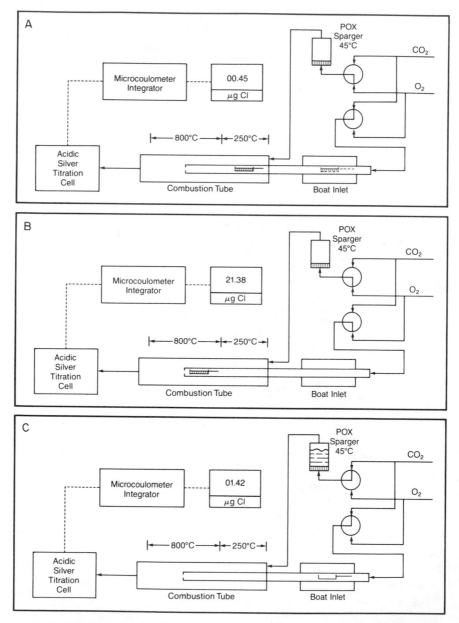

Fig. 3. OX determination steps. (A) Vaporization; (B) combustion; (C) POX determination. [Reprinted with permission from "Chemistry in Water Reuse," Vol. 2 (W. J. Cooper, ed.). Ann Arbor Science Publ., Inc., 1981.]

TABLE IV

Accuracy and Precision of OX Method[a]

Sample	Number of determinations	TOX (μg Cl/L)		SD
		Calculated	Found	
Deionized water	5	—	1.5	1.1
Trichlorophenol	7	10	11.6	0.5
Trichlorophenol	4	100	106.5	3.1
Dichlorobenzene	2	2.5	2.4	—
Dichlorobenzene	2	10	10.2	—
Dichlorobenzene	2	25	24.8	—
Dichlorobenzene	2	100	97.2	—
Deionized water[b]	38	—	1.4	1.2
Trichlorophenol[b]	20	100	101.0	4.0

[a] The lower limit of detection was reported to be 3 μg Cl/L (Takahashi et al., 1981).
[b] Obtained over a 4-month period.

III. OXYGEN DEMAND

A. Chemical Oxygen Demand

Weaknesses and limitations of biochemical tests, and perhaps more importantly the long test times required, have caused many investigators to search for and turn to chemical methods for water quality monitoring and control. The first and most widely used of these methods is the chemical oxygen demand (COD) test. The COD test uses chemical oxidation to carry out the same reaction with organic material supposedly expressed by bacteria in the BOD reaction or

$$\text{organic material} + \text{oxidant} \longrightarrow CO_2 + H_2O \qquad (2)$$

Potassium dichromate is the oxidant used in standard COD test procedures [Moore et al., 1949; American Public Health Association (APHA), 1981], although potassium permanganate is used in a similar permanganate value (PV) test (Nasr and MacDonald, 1975; Foulds and Lunsford, 1968). Dichromate generally gives a higher degree of oxidation, especially with higher molecular weight compounds.

If the organic materials in a waste are readily biodegradable and easily oxidized by dichromate, COD test results closely approximate the BOD_u value. Consequently, the change in COD (ΔCOD) produced by a treatment unit may be used in some cases to approximate the change in ultimate biological oxygen demand (BOD$_u$). This approximation offers the advantage of giving a rapid indication of the BOD-removal performance of a treatment system (Gaudy and Gaudy, 1972).

Detailed procedures for measuring COD are given in standard methodology manuals (American Society for Testing and Materials, 1980; APHA, 1981; U.S.E.P.A., 1979). Basic steps in the procedure include

1. Selecting sample size for the analysis
2. Adding mercuric sulfate to the sample to precipitate chlorides
3. Adding a mixture of oxidant, water, acid, and catalyst to the sample
4. Refluxing or heating the sample and chemicals for a specified time followed by cooling
5. Titration of the unreacted oxidant

Figure 4 shows the laboratory setup needed to conduct dichromate COD tests.

The standard COD test procedure requires the use of a 50-mL sample to which is added a strong mixture of dichromate (0.25 N), silver sulfate catalyst, and sulfuric acid to give a final acid : water mixture of 1 : 1. This mixture boils at ~180°C. British practice involves the use of a 1.2 : 1 acid : water ratio, which boils at a higher temperature. Although the higher temperature may cause a greater percentage of oxidation, the difference is slight if a 2-h heating period is used. Other acid : water mixtures are used in automated and rapid COD measuring methods to regulate the boiling point for test control purposes.

The large quantity and high concentrations of reagents do not permit precise measurements for samples containing less than ~50 mg/L COD. Test modifications for measuring the COD of lower strength samples have been described (APHA 1981; U.S.E.P.A., 1979) and generally involve the use of a more dilute dichromate solution (0.025 N) and titrant (0.10 N). This method improves the precision of low-level COD tests but requires careful laboratory control to minimize contamination of glassware with small amounts of organic material.

Costs of COD analyses have increased dramatically because of the large amounts of silver, mercury, and acid used. This has caused increased interest in using smaller sample sizes and correspondingly lesser amounts of reagents.

Microscale procedures using only 2.5- to 5-mL samples and small volumes of reagents have been developed, tested, and shown to provide accuracy and precision equal to the standard procedure (Jenkins *et al.*, 1965; Harkness *et al.*, 1972; Jirka and Carter, 1975). Use of these small sample sizes and reagent quantities permits the oxidation reaction to be carried out in sealed test tubes and heated in an oven at 150°C for 2 h.

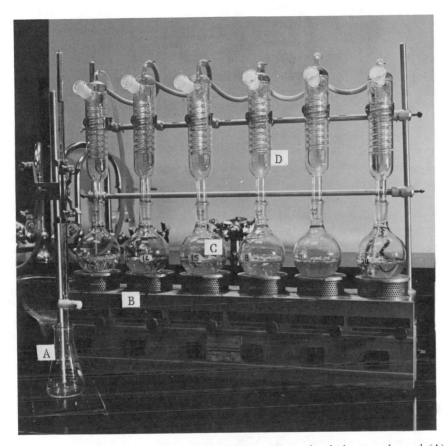

Fig. 4. Typical laboratory equipment needed to measure chemical oxygen demand. (A) Flask, buret, and buret stand for making titrations; (B) heater; (C) reaction flask; (D) reflux column.

Quantities of oxidant used can be measured by titration or colorimetric observations.

1. Rapid COD Tests

Rapid COD test procedures have been developed that involve the addition of a sample to a tube or flask containing a dichromate–sulfuric acid–phosphoric acid–silver sulfate mixture. Mercuric sulfate may be added to the sample if chloride interference is anticipated (Jeris, 1969). The reaction vessel then is heated to 165°C and held at this temperature for ~5 min. It is then cooled rapidly and the excess dichromate determined by titration.

Test results have shown that, with minor exceptions, the rapid test provides CODs averaging ~96% of the standard COD test results with a measurement precision comparable to that of the standard method (Jeris, 1969; Wells, 1970; McLean and Spicher, 1973).

Another simplified and rapid COD procedure serves as the basis of the oxygen demand index (ODI) test. The ODI is the ratio of the rapid COD of a test sample to the rapid COD of a standard glucose solution having a known (or assumed) BOD_5 concentration. This test was designed to give an estimate of the BOD_5 of a wastewater sample (Shriver and Young, 1972; Reynolds and Goellner, 1974; Westerhold, 1965).

The ODI chemical test procedure is similar to the rapid COD test, but the samples are heated to boiling in a water bath, held for 15 min and cooled; the excess dichromate is measured colorimetrically. The ODI test suffers from the same interferences and errors as the COD test, and because of the lower reaction temperature and shorter heating time, can be expected to provide less complete oxidation of resistant organic compounds than the standard or rapid COD tests.

2. Automated COD Methods

Automated measurement of COD by dichromate has been made possible through equipment developments. One semiautomatic method involves small-volume digestion of samples in test tubes that are sealed and heated at 150°C for 2 h (Jirka and Carter, 1975). The digested samples are then placed on the rotating sample tray of a Technicon Autoanalyzer (Technicon Industrial Systems/Tarrytown, New York) and automatically pumped into a small-diameter continuous tube through a colorimeter that measures the concentration of dichromate blank through the colorimeter. Sample injections are separated by passage of distilled water through the instrument. This semiautomatic method is best suited for laboratory use where numerous COD tests are conducted.

Another automatic COD instrument (Ionics Model 335) involves batch sampling and injection into small reaction flasks. Dichromate, acid, and silver and mercury salts are metered to the sample, which is allowed to reflux for 20 to 120 min. The amount of dichromate used is measured colorimetrically throughout the reaction. After one reaction has been completed, the reaction flask is flushed with clean water to prepare for another sample. This automated system has the advantage of providing a dichromate COD from a continuously flowing stream with minimum operator attention and time and is convenient for on-line monitoring.

One commercial instrument (Precision Scientific "AquaRator") measures COD by causing organic materials to be oxidized on a platinum catalyst at 900°C by the reduction of carbon dioxide to carbon monoxide, producing a mixture of carbon dioxide and carbon monoxide that is numerically equal to the number of atoms of oxygen that is required for chemical oxidation, or

$$C_aH_bN_cO_2 + m \, CO_2 \longrightarrow (m + a) \, CO + \frac{b}{2}H_2O + \frac{c}{2}N_2 \qquad (3)$$

Therefore, instrument readings of carbon monoxide are related to COD and are most accurate when the number of oxygen atoms in the organic test compounds is equal to one-half the number of hydrogen atoms. Samples of 20-μL size are injected onto the catalyst, and oxidation products are carried through the instrument in a stream of purified carbon dioxide. The instrument is standardized with sodium acetate trihydrate. Features of commercially available COD-measuring instruments are summarized in Table V.

3. Sources of Error in COD Test

a. *Incomplete Oxidation of Organic Compounds*

Several types of compounds are only partially oxidized by dichromate in the COD test. The degree of oxidation of a chemical depends on its structure. Chemicals containing hydroxyl, carboxyl, keto, or amino groups are more readily oxidized than are unsubstituted hydrocarbons or ring compounds. Pyridine and compounds having three valences of a nitrogen atom in a ring compound usually are not oxidized. Straight-chain alcohols and organic acids are oxidized so slowly that oxidation usually is not completed in the standard 2-h reflux period. Silver catalyst helps oxidize the alcohols and acids but may not permit their complete destruction and is not of much use with aromatic hydrocarbons (Ford *et al.*, 1971). Ceric sulfate improves the oxidation of some compounds while reducing the oxidation of others, but it is not as good as dichromate for use as a general-purpose oxidant (El-Dib and Ramada, 1966).

b. *Vaporization of Compounds*

Volatile compounds, particularly those not miscible with water, may be lost when adding the sulfuric acid reagent to the sample or through the reflux column during heating, and therefore they may not be detected completely by the COD test (Wolff, 1975). This is fairly typical of the aromatics such as benzene, toluene, and even some of the alcohols. This loss may be minimized by adding the acid slowly so that the sample remains cool, by placing the sample in an ice bath while acid is added, or by adding the acid reagents through the open end of the reflux condenser.

TABLE V

Features of Commercial Chemical Oxygen Demand (COD) Analyzers

Manufacturer	Analyzer application	Sample injection method	Oxidative reaction stage	Analytical measurement method	Analysis time (min)	Maximum concentration range (mg COD/L)	Reproducibility claimed by manufacturer (% full scale)
Ionic, Inc. (Watertown, Massachusetts)	On-line process monitor	10-ml sample drawn automatically from flowing stream	Dichromate oxidation, 150°C	Colorimetric	20–120	<1000 (or dilute sample)	±3
Technician Autoanalyzer	Laboratory semiautomatic	Dichromate COD solution injected into instrument after digestion	Dichromate oxidation, micro method 150°C	Colorimetric	120	<1000 (or dilute sample)	—
Precision Scientific Co. (Chicago, Illinois)	Laboratory (AquaRator)	Syringe	Fixed-bed platinum catalyst, 900°C	Infra red analysis of CO	2	10–300	±3
Hach Chemical Co. (Ames, Iowa)	Laboratory (Package test-tube method)	Pipet to test tubes precharged with reagents	Dichromate oxidation micro method 150°C	Colorimetric	120	<1000 (or dilute sample)	±3
Oceanography International Corp. (College Station, Texas)	Laboratory (Packaged precharged vial)	Pipet to vials precharged with reagents	Dichromate oxidation micro method 150°C	Colorimetric or Titrametric	120	2.5 ml-sample	±5

c. Reflux or Heating Time

The 2-h reflux time specified for the standard test is a compromise between reaction time and completion of organic oxidation. Normally, ~95% of the organic material is oxidized in the 2-h period, but this depends on the type of compounds present (Fig. 5). The rapid method may provide as little as 60% of the standard method COD while the ODI test may oxidize a lower percentage (Ford *et al.*, 1971; Foulds and Lunsford, 1968; Wells, 1970). Unless the chemical nature of test samples is known, the extent of degradation may be obtained only by using long-term reaction times. Instrumental measurement of organic carbon may help provide this information. Short reaction time may prove to be a major limitation of some automated COD-measuring instruments (Fleet *et al.*, 1972).

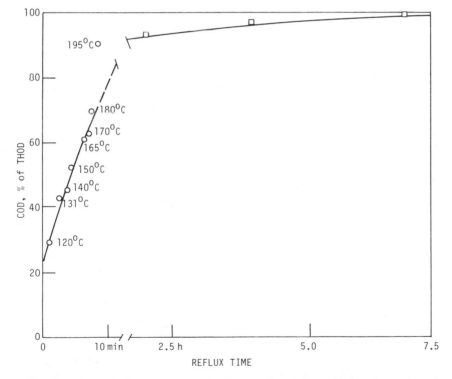

Fig. 5. Effect of reflux time and temperature on the relative oxidation of organic compounds in the chemical oxygen demand (COD) tests. ○, rapid method; □, standard method. (Foulds and Lunsford, 1968.)

d. Sample Size and Preparation

Each COD test procedure requires a sample size unique to that test, and test procedures or results should not be compared when using sample sizes not appropriate to the individual tests. The COD procedure described in *Standard Methods* (APHA, 1981) specifies a 10- to 50-mL sample, whereas most rapid tests use a 5- or 10-mL sample.

Generally, larger samples permit greater test precision because a more representative sample is obtained. Homogenizing the sample by blending prior to removing aliquots for analysis also improves test precision (McLean and Spicher, 1973).

e. Inorganic Interferences

Interference is caused by oxidation of chlorides present in a wastewater sample. This interference is controlled to a large extent by adding 1 g of mercurous sulfate for each 100 mg of chloride in a sample (APHA, 1981). Nitrite exerts a COD of 1.1 mg/mg N, but this oxygen demand can be controlled by adding 10 mg of sulfamic acid per mg of nitrite N to the reflux flask. Equal amounts of chemicals used to control interferences should be added to control flasks. Reduced inorganics, such as sulfites and sulfides, may be oxidized chemically. Gaseous sulfide can be eliminated by acid purging before adding oxidizing agent, but corrections for oxidation of other reduced inorganic material may require specific analysis for such compounds.

4. COD Test: Accuracy and Precision

Controlled tests involving sending standardized oxidizable material (equal weight mixtures of glucose and glumatic acid) to each of a number of participating laboratories have been conducted to determine accuracy and precision of the COD test (U.S.E.P.A., 1971b; APHA, 1981). Test results are summarized in Table VI. Those errors causing deviations include analytical bias and differences in oxidation and titration errors but do not include sampling and transfer errors that might be experienced when using wastes containing significant amounts of particulate solids that settle or float. By carefully preparing and manipulating samples and by closely controlling test conditions to insure that all samples are treated equally, a given analyst can expect to reduce this standard deviation to <2% of the average.

Variation in test results are greatest when the test conditions are such that oxidation of organic compounds is incomplete. This occurs because in a set of replicate samples the extent of oxidation may differ from one sample to the next. This may explain much of the greater variation in test COD results when using the rapid procedure and especially the ODI test.

TABLE VI

Accuracy and Precision of Dichromate COD Measurements

| Parameter | U.S.E.P.A., 1971b | | Ballinger and Lishka, 1962 | APHA, 1976 | |
	Low level	High level		Level A	Level B
Theoretical or expected value (mg/L)	12.3	270	308	160	200
Mean of measured values (mg/L)	12.34	257.4	281	—	—
Recovery (%)	100.3	95.3	91	—	—
Coefficient of variability (%)	33.6	6.9	8.2	10.8	6.5
Number of analyses/ number of laboratories	86/56	82/56	24/34	—/74	—/74

B.   Total Oxygen Demand

One disadvantage of carbon analyzers, especially when comparing the results to oxidative measurements such as COD or BOD, is that the oxidative potential of the organically bound hydrogen, nitrogen, and sulfur is not measured. Thus, the ratio of theoretical oxygen demand (or the COD) to carbon varies with the percentage of carbon in the organic compound and may range from ~2.7 for carbohydrates to 5.33 for methane.

Total oxygen demand (TOD) overcomes this disadvantage to a large extent by measuring the oxygen required to convert each element to its highest stable state, or

$$C + O_2 \longrightarrow CO_2 \tag{4}$$

$$4 H + O_2 \longrightarrow 2 H_2O \tag{5}$$

$$3 N^{3-} + 3 O_2 \longrightarrow 2 NO_3^- \tag{6}$$

$$S^{2-} + 2 O_2 \longrightarrow SO_4^{2-} \tag{7}$$

$$SO_3^{2-} + \tfrac{1}{2} O_2 \longrightarrow SO_4^{2-} \tag{8}$$

$$N_2 + O_2 \longrightarrow \text{no reaction} \tag{9}$$

Thus, the TOD measurement more nearly approximates the theoretical oxygen demand of a sample.

TOD Analyzers

Like organic carbon analyses, TOD measurements are carried out instrumentally by injecting the sample into a catalyst-coated medium at about 900°C; however, instead of measuring oxidation products, the analyzer monitors oxygen depletion in the carrier gas using zirconium oxide detectors or platinum–lead fuel cells (Fig. 6, Table VII) (Butzelaar and Hoogeveen, 1975; Clifford, 1968).

a. Sources of Error

Sources of error in instrumented or automated analyzers are primarily associated with sample preparation and manipulation. There is some instrument error, although this usually is relatively small because most instruments are calibrated against a standard of known composition. Incomplete oxidation of some compounds may occur, but this is uncommon.

TOD analyzers seem to be more susceptible to interference by inorganic compounds than are carbon analyzers. Nitrates cause the greatest effect because each 2 mol of nitrate potentially can yield 1.5 mol of molecular oxygen. Carbonates and bicarbonates also show a slight positive bias whereas sulfate and chlorides show a slight negative bias (Clifford, 1968). Free dissolved oxygen lowers the TOD by an amount equivalent to the mg/L of dissolved oxygen (DO) added with the sample.

b. Accuracy and Precision

Comprehensive tests to define accuracy and precision capabilities of TOD analyzers have been rather limited. Original tests by the equipment

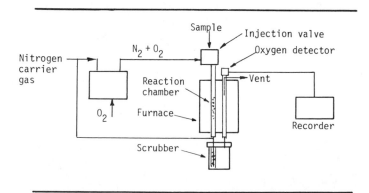

Fig. 6.   Schematic diagram of a total oxygen demand (TOD) analyzer. (Figure adapted from material presented by Ionics, Incorporated, Watertown, Massachusetts.)

TABLE VII

Features of Commercial Total Oxygen Demand (TOD) Analyzers

Manufacturer	Analyzer application	Sample injection method	Oxidation reaction stage	Detector	Carrier	Response time (min)	Maximum concentration range (mg TOD/L)
Astro Ecology Corp. (Houston, Texas)	Laboratory	Metered	Fixed bed, nonprecious metal, 900°C	Zirconium oxide cell	—	10	0–30,000
Astro Ecology Corp. (Houston, Texas)	Process monitor	Automatically metered (<100 μm particulates)	Fixed bed, nonprecious metal, 900°C	Zirconium oxide cell	—	10	0–20,000
Ionics, Inc. (Watertown, Massachusetts)	Laboratory	Automatic from container, slide valve	Fixed bed, platinum, 900°C	Platinum–lead fuel cell	Nitrogen, air	—	0–1,000
Ionics, Inc. (Watertown, Massachusetts)	On-line process analyzer	Automatic from flowing stream, rotary valve (<200 μm particulates)	Fixed bed, platinized quartz, 900°C	Zirconium oxide	Nitrogen	3–5	0–10,000
Enviro Control, Inc. (Washington, D.C.)	On-line process analyzer	Metered (<100 μm particulates)	Fluidized bed, aluminum oxide, 900°C	Polarographic sensor	Air	8	0–4,000

developers showed essentially complete recovery of theoretical TOD for a variety of pure compounds (Clifford, 1968; Goldstein *et al.*, 1968). Most test efforts have been directed at comparing TOD measurements with dichromate COD values because the TOD instruments essentially measure the same reaction. A more thorough analysis of these comparisons is presented in the following sections.

IV. SPECTROPHOTOMETRIC METHODS

The theoretical basis and instrumentation for molecular absorption spectrophotometry was reviewed by Jones *et al.* and fluorescence spectrophotometry by Wehry in Volume II, this series. Both absorption and fluorescence have been used to determine organic matter in aqueous solution. In this section, we refer to the determination of organic matter directly and discuss the use of spectrophotometry for class determinations in Section VI. For practical laboratory work, the wavelength range is limited to 200–800 nm in aqueous solutions.

A. Absorption

Ultraviolet/visible (UV/vis) absorption spectrophotometry has been used routinely to aid the organic chemist in structural elucidation and as a tool to follow the course of reactions. UV/vis has also been used to determine the amount of organic matter in natural waters. There are two methods that can be used, single-wavelength and multiple-wavelength determination. Single-wavelength determinations have been used with the scanning instruments that are unable to scan multiple wavelengths rapidly. With the advent of diode-array instrumentation, complete spectra of 200 through 800 nm can be obtained in 1 s or less.

1. Single Wavelength Determinations

Most organic compounds absorb in the 200–350 nm wavelength range in aqueous solution. A number of wavelengths have been used to correlate absorbance to organic carbon, the most common being 254 nm, the mercury line. Table VIII is a summary of the wavelengths used and their applications.

An extension in the use of single wavelength measurements has been to ratio one or more wavelengths. The use of absorbance at 254 and 546 nm correlated well with DOC in the nearshore marine environment (Mattson *et al.*, 1974). Mrkva (1975) found that measurement of absorbance at 254 nm could be used in various waters if the absorbance was corrected for

TABLE VIII

Summary of Ultraviolet/Visible Absorption as a
Nonspecific Measurement of Organic Carbon

Wavelength (nm)	Application	Reference
275	Water and wastewater	Dobbs et al., 1972
		Briggs and Melbourne, 1968
250	Monitor removal of organics by physiochemical treatment	Dobbs et al., 1972
254	Coke-oven plant wastewaters (from Lake Michigan) treatment	Bramer et al., 1966
254	Water treatment	Scheuch and Edzwald, 1981
250	Lake water organics fractionated by gell permeation	Stewart and Wetzel, 1981a
250	Lake water	Wetzel, 1974
254	River water, municipal effluents, industrial sewage	Mrkva, 1975
280	River water	Mrkva, 1969
274	Municipal effluents	Helfgolt, 1977
254	Wastewater for reuse	Michail and Idelovitch, 1981
254	Surface waters to detect accidental organic pollution	Schalekamp et al., 1978

turbidity as measured at 545 nm; use of 510 nm as a correction factor has also been suggested (Briggs et al., 1976). Michail and Idelovitch (1981) found correlation between the absorbance at 254 nm and several other nonspecific measures of organic carbon, and when the effect of turbidity was subtracted (absorbance at 545 nm) all the correlations were improved.

Several investigators have proposed the use of UV monitors for pollution monitoring (Briggs et al., 1976) and for water treatment (Kolle and Sontheimer, 1973).

2.   Multiple Wavelength Determinations

The alternative to single wavelength determinations is to use multiple wavelengths within the 200–350 nm bandwidth. Multiple wavelength measurements have been used to determine mixtures of arylsulfuric acids (Arends et al., 1964) and for evaluating pK values (Metzler et al., 1977) and equilibrium constants (Kankare, 1970). Maier and Conroy (1981) extended this methodology for use in monitoring trace organics in river water.

Diode-array spectrophotometric detectors not only allow rapid acquisition of spectra, but present the possibility of full wavelength (1 nm resolution) spectra for determining the organic content of water.

3. Potential Problems

Both Br^- and NO_3^- absorb in the UV. The presence of one or both of these could present problems in the use of UV/vis spectrophotometry for determining organic carbon in aqueous solutions. Because the compound(s) of interest absorb light, problems can arise in quantitative determination resulting from spectral changes in the natural environment (Stewart and Wetzel, 1981a,b; Zika, University of Miami, Coral Gables, Florida, personal communication, 1982). These spectral changes result presumably from the absorption of light energy and the exact nature of these changes are not characterized.

B. Fluorescence

The use of fluorescence as a nonspecific analytical procedure was first reported by Kalle (1949), who examined the organic matter of nearshore seawater. The mathematical relationship for absorbance and fluorescence was developed by Duursma and Rommets (1961). The nonlinear behavior of fluorescence results from the exponential absorption of both the incident UV light (excitation) and emitted fluorescent light (emission).

Kalle (1963) summarized much of his pioneering work in various aqueous systems. He used an excitation of $\lambda = 387$ nm and an emission of $\lambda = 420$ nm. The fluorescence method (Kalle, 1963) was used successfully as a natural tracer for river water in nearshore oceanic waters (Willey and Atkinson, 1982). Additional work has been reported using fluorescence in the marine environment (Duursma, 1961; Otto, 1967; Zimmerman and Rommets, 1974) that indicated conservative mixing of fresh and saline water.

Fluorescence has also been used to measure lignin sulfonates in natural waters (Christman and Minear, 1967; Thruston, 1970; Baumgartner et al., 1971; Almgren and Josefsson, 1973; Wilander et al., 1974). Christman and Minear (1967) showed that three excitation wavelengths could be used; 253, 293, and 340 nm. All three wavelengths resulted in an emission maximum at 400 ± 2 nm; however, the 340-nm excitation resulted in the most sensitive determination. Chloride ion concentration through 20,000 mg/L did not affect the location of the maximum excitation or emission wavelength. Highly acidic or alkaline solutions decreased the intensity of the fluorescence and was thought to result from the change in ionization of the functional groups. The pH range, from 4 to 7, resulted in no change in intensity, thereby allowing an easy analysis. The effect of pH on fluorescence was also noted by Hall and Lee (1974) who confirmed that at either high or low pH extremes, the fluorescence of lake water decreased, but that in the range 4–7 no pH effect was observed. However, Wilander et al. (1974) noted that fluorescence of natural compounds can present a significant interference when using fluorescence to estimate organic carbon. The pH effect on fluorescence is not the same as on color. Color appears to increase in intensity from pH 2 to 13 (Hall and Lee, 1974), confirming the earlier work of Shapiro (1957) and Christman and Ghassemi (1966).

The use of fluorescence has been reported for monitoring drinking water and correlated to the carbon chloroform extract (CCE) (Sylvia, 1973; Sylvia et al., 1979) and to TOC (Sylvia and Donlon, 1980). The correlations obtained between fluorescence (λ_{exc} = 340 nm, λ_{emm} = 390 nm) and TOC were 0.969, 0.938, and 0.963 for distribution system water before and after GAC and GAC effluents, respectively. Thus, Sylvia and Donlon (1980) concluded that fluorescence could be used as an alternative to TOC.

Smart et al. (1976) reported correlations of >0.87 between fluorescence (λ_{exc} = 365 nm, λ_{emm} = 400–600 nm) and TOC levels in natural waters. In a sample group where nonfluorescent organic material was suspected, the correlation was not significant, and the correlation coefficient was 0.46. They concluded that fluorescence could be used as a predictor of TOC and that extension of the technique would most probably be site specific.

Brun and Milburn (1977) automated a fluorometric procedure for determining humic substances in freshwater samples. An excitation wavelength of 270 nm was used while the emission was monitored at 460 nm. Using a Technicon Auto Analyzer system, they were able to analyze 30 samples per hour and obtained a correlation with TOC.

Stewart and Wetzel (1981a) reported the use of fluorescence : absorbance ratios as molecular weight tracers of DOC in freshwater lakes. They concluded that the ratio of fluorescence to absorbance, when used with caution, could provide a general index of seasonal or vertical distribution of low and high molecular weight dissolved humic species.

V. ELECTROCHEMICAL METHODS

Inorganic conductivity has been used extensively in general water quality assessment. The extension of this concept to organic carbon would be helpful, provided that a low limit of detection were possible using inexpensive instrumentation. Several reports have appeared that discuss the organic "conductivity" measurement (Hauden and Richard, 1975; Davenport *et al.*, 1978, 1979; Cowen *et al.*, 1979, 1981).

When an electrode is immersed in an aqueous solution, an electrode–surface interface is set up. The interface acts as a capacitor; the interfacial capacitance is determined by the potential of the electrode, the electrode material, the aqueous phase, and the composition of the aqueous phase.

Organic compounds can absorb to the electrode surface in aqueous solution. The fractional coverage of the electrode surface is represented by ϕ. The value of ϕ is a function of the potential of the electrode and, at maximum absorption potential, the logarithm of concentration of the organic compound. Thus the capacitance of the electrode–solution interface C is related to ϕ as

$$\frac{C_0 - C_\phi}{C_0 - C_t} = \phi$$

where C_0 is the interfacial capacitance for pure water, C_t is the interfacial capacitance for total coverage of the electrode by the organic compound, and C_ϕ is the interfacial capacitance of the test sample. C_0 and C_t are constants for a given electrode–compound pair and therefore

$$C_0 - C_\phi \propto \phi$$

and the interfacial capacitance following absorption of the organic compound(s) is a measurement of the concentration of the compound.

Very limited data have been reported using the interfacial capacitance measure of organic compounds in aqueous solution. Hauden and Richard (1975) reported that compounds such as pyridine, parathion, napthalene, and phenol gave responses at concentrations as low as 0.01 mg/L; an anionic detergent, tetrapropylene benzosulphonate, produced a response to 0.1 mg/L. Davenport *et al.* (1981) reported a significant signal at 0.1 mg/L for Triton X-100 and phenol and indicated that a lower concentration would probably produce adequate capacitance to be measured. On the other hand, compounds such as acetone and methanol did not produce measurable capacitance below $\sim$10 mg/L. Davenport *et al.* (1981) reported that by incorporating an oxidation–reduction cycle followed by absorption and capacitance measurements, the cycle could be repeated in a flow-through cell.

As yet no instruments are available commercially that measure the interfacial capacitance of organic compounds in water. There is the possibility that this measurement could provide a simple, potentially automated measure of organic compounds in aqueous solution. However, additional research is necessary.

VI. CLASS DETERMINATIONS

A. Amines: Fluorescamine

The use of ninhydrin for the determination of amino acids and peptides has been known for some time. It was shown that the reaction of phenylalanine with ninhydrin yielded a highly fluorescent product (Samejima *et al.*, 1971a,b) the structure of which was elucidated by Weigele *et al.* (1972a) and then synthesized (Weigele *et al.*, 1972b). The reagent, 4-phenylspiro[furan-2(3*H*),1′-phtalan]3,3′-dione, has been given the trivial name fluorescamine. At pH 9, primary amines react with fluorescamine with a $t_{1/2}$ of 100 to 500 ms while the hydrolysis of fluorescamine has a $t_{1/2}$ of 5 to 10 s (Udenfriend *et al.*, 1972). The reaction is shown in Fig. 7.

This method has been used to determine primary amines in seawater (North, 1975; Packard and Dortch, 1975; Zika, 1977) and river water (Conroy *et al.*, 1981). The reaction of fluorescamine and amines is carried out in pH 7-buffered samples. The reaction products are fluorescent, while the hydrolysis products are nonfluorescent. A significant advantage of this reaction is the negligible fluorescence from ammonia or urea (Stein *et al.*, 1973; Bohlen *et al.*, 1973).

The intensity of the fluorescence of the products with different primary amines is variable and thus difficult to quantitate in complex mixtures (Zika, 1977). However, because of its speed this procedure can serve as a screening method for primary amines and small peptides in natural water systems, with a detection limit of 50 picomol.

B. Amines: o-Phthalaldehyde-Reactive Substances

The determination of amino acids using *o*-phthalaldehydes as a fluorometric reagent has been described (Roth, 1971; Roth and Hampai, 1973).

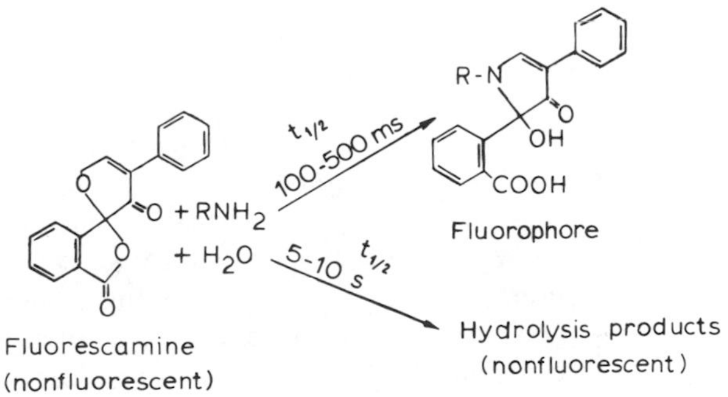

Fig. 7. Reactions of fluorescamine reagent. [Reprinted with permission from "Chemistry in Water Reuse" (W. J. Cooper, ed.). Ann Arbor Science Publ., Inc., 1981.]

The use of this reagent in seawater was automated by Josefsson *et al.* (1977). Unlike fluorescamine, the reagent is stable in aqueous solution. There is no salt dependence on the reaction; however, ammonia can interfere.

The method used by Josefsson *et al.* (1977) to determine *o*-phthalaldehyde-reactive substances (ORS) has been tested in many waters. The reagent is 100 mg of *o*-phthalaldehyde, dissolved in 10 mL of absoluted ethanol, diluted to 1 L with 0.4 *M* boric acid, and adjusted to pH 10.5 with NaOH; it is activated by the addition of 0.5 mL of mercaptoethanol. Once prepared, the reagent is stable for several days provided no ammonia is allowed to contact it. The seawater sample is pumped at 0.8 mL/min and the reagent is pumped at 0.42 mL/min. A reaction time of 2 min is provided. The excitation wavelength is 340 nm, and the fluorescence is measured at 455 nm.

The limit of sensitivity is determined by the purity of the water and therefore the water should be stored in a vessel with a 50% sulphuric acid trap to reduce ammonia (Dawson and Liebezei, 1981). Determination of picomolar levels is possible with this procedure. Extension of this procedure to freshwater systems has not been reported.

C. Carbohydrates

There have been several reagent systems used in the determination of carbohydrates in aqueous solution, and these were reviewed by Josefsson *et al.* (1972) and Burney and Sieburth (1977). Johnson and Sieburth (1977) combined several techniques to give a test for monosaccharides. The spectrophotometric assay uses sodium borohydride to reduce free monosaccharides to sugar alcohols, followed by oxidation to formaldehyde (using periodate) and determination of formaldehyde with 3-methyl-2-benzothiazolinone hydrazone hydrochloride. There is no salinity dependence of this method and few appreciable interferences (Dawson and Liebezeit, 1981).

Another absorbance method for monosaccharides, polysaccharides, and glycosides uses the anthrone test (Dreywood, 1946; Morse, 1947; Viles and Silverman, 1949; Black, 1951; Samsel and Delap, 1951). Jermyn (1975) found that by adding formic acid and hydrochloric acid the rate of formation of color increased, enhancing the sensitivity of the test.

Conroy *et al.* (1981) used this method to determine carbohydrates in river waters. The procedure involved dissolving 20 mg of anthrone in 100 mL of 80% (v/v) sulfuric acid. One mL of concentrated hydrochloric acid and 0.1 mL of 88% formic acid were added to a 1-mL riverwater sample. Eight mL of the anthrone reagent were then added; the mixture was heated for 12 min in boiling water and quenched in a cold-water bath. After 5 min (to disperse bubbles), the absorbance was read at 627.5 nm. These authors reported that their methods resulted in a linear plot from 1 to 20 mg/L, 5×10^{-6}, to 1×10^{-4} M, as glucose.

Fluorometric procedures for pentoses and hexoses using anthrone and 5-hydroxy-1-tetralone, respectively, have been reported (Hirayama, 1974). The procedure is affected by sodium chloride and may require a desalting step.

VII. ON-LINE MONITORS

Carbon and TOD analyzers are adaptable to on-line monitoring of process or wastewater streams. This involves using a sample injection system so that samples are metered to the analyzers as a series of injections or in a continuous stream. Often the injection interval or metering rate can be varied by the analyst.

Automatic sampling valves or injectors are of two types, the slider or the rotary valve. The slider type consists of a plunger fitted into a precision bore. The plunger is activated by a timed signal and operated with air pressure. Rotary sampling valves usually consist of a syringe mounted on a rotating head that actuates the syringe plunger to pick up samples from a trough. As rotation continues, the syringe tip is brought over an injection hole above the catalyst bed and the rotary motion of the head causes the sample to be injected. Either the slide or rotary injector reportedly can accept samples containing particulates up to 200 μm in diameter.

Continuous injection devices meter sample into the analyzer at flow rates of 0.25–4 mL/min. This sample is picked up by a carbon dioxide-free carrier gas stream that sweeps the sample into the catalytic combustion zone.

One potential problem with on-line monitors is that the continuous or intermittent injectors are not always reliable or require frequent cleaning or maintenance. Although test data specifically relating to these injectors are not available, potential problems include plugging of small orifices, plungers, or valves. There is little doubt that these injection devices must be cleaned or maintained frequently except with samples containing essentially no suspended solids.

Ideal use of on-line monitors is in industrial processes where an increase or decrease in feed stock or final product signifies a change in the desired reaction. Usually these streams have a fairly constant constituency so that variations in type of compound do not cause problems with data interpretation. Once a change in magnitude of organic content is detected with on-line instruments, specific means, such as gas chromatography or chemical tests, may be used to determine the type of compound causing the change.

VIII. COMPARISON OF NONSPECIFIC PARAMETERS

One of the first questions asked about a new or revised analytical test for measuring pollution parameters is "How do the results compare to standard or other accepted tests?" Much of this interest comes from the desire to find substitutes for the traditional and standardized BOD_5 or COD tests, which are time consuming and cannot be easily automated. The ODI test, for example, was developed specifically to produce estimates of the BOD_5 of a wastewater sample. A rapid substitute for the BOD_5 test has for many years been a desirable but unrealized goal. Because of this interest, published literature abounds with articles making comparisons between various water pollution test measurements.

The principal difficulty with obtaining a good correlation between various tests is that each measures a different parameter: carbon analyzers measure carbon, chemical and TOD tests measure oxidation of most organic compounds, and biochemical tests measure oxygen uptake associated only with decomposition of biodegradable organic materials. Even theoretically these test parameters have too little in common to expect test results to provide a one-to-one correspondence.

The first attempts to correlate chemical and biological tests involved a comparison of COD and BOD. Generally, the COD would be expected to approximate the BOD_u of wastewaters. However, this seldom occurs because

1. The COD test does not distinguish between biodegradable and non-biodegradable organic material.
2. Some inorganic compounds, such as reduced sulfur compounds, nitrites and iron, are oxidized by dichromate or by catalytic oxidation to a different extent than might be oxidized in the BOD test.
3. Seed differences, lack of acclimation of seed organisms, and presence of toxic substance, among other variables, introduce variations in the BOD test that do not occur in the COD test.

Because TOC measures only carbon, there is a different theoretical correlation of TOC to BOD_u or COD for each compound in a wastewater sample. In most cases, the mixture of organic materials is such that the ratio of COD to TOC approximates the molecular ratio of oxygen to carbon ($32:12 = 2.67$), but actual values may range from less than half to more than twice this value (Table IX). Total oxygen demand more closely equals COD because of the similarity of reactions for the two tests.

Specific correlations between test results often can be developed with a high degree of precision indicating that the relationship between test measurements is meaningful (Table X). However, correlations of measurements between plants and from point to point within a plant can be quite different (Shriver and Young, 1972). A common method of comparing two parameters is to use a simple ratio between them. Theoretically this may be correct, but actual measurements may show a definite bias at high or low measurements. In some cases, multiple straight lines or a nonlinear equation may provide better fit of the data. Analysts should review their data carefully to determine which method best suits their needs.

Thus, although good correlations between test parameters can be obtained, their value is limited to the specific waste being tested and to the point of sampling. Collecting the large amount of data needed to establish such a correlation requires a great deal of time and expense. Analysts must weigh the direct costs and assess the practical value of the correlation before investing time in establishing a correlation between any two or more given tests. In many cases, it is better to develop a direct relationship between process performance and a new parameter than to use the new parameter to estimate, through correlation, a value of the parameter normally used to evaluate performance.

TABLE IX

Total Organic Carbon (TOC) and Chemical Oxygen Demand (COD) per Unit of Weight of Pure Compounds

Compound	TOC (mg/mg)	COD (compound)	COD/TOC (mg/mg)
Formic acid (CH_2O_2)	0.27	0.36	1.35
Glucose ($C_6H_{12}O_6$)	0.40	1.07	2.67
Glutamic acid ($C_5H_9NO_4$)	0.41	0.98	2.39
Acetic acid ($C_2H_4O_2$)	0.40	1.07	2.67
Propionic acid ($C_3H_6O_2$)	0.49	1.51	3.08
Butyric acid ($C_4H_8O_2$)	0.54	1.82	3.37
Oleic acid ($C_{18}H_{34}O_2$)	0.77	2.89	3.76
Ethyl alcohol (C_2H_6O)	0.52	2.09	4.02
Phenol (C_6H_6O)	0.77	2.38	3.10
Methyl alcohol (CH_4O)	0.38	1.50	4.00
Methane (CH_4)	0.75	4.00	5.33

TABLE X

Representative Correlations Between Waste-Strength and Pollution Parameters

Comparison (Y:X)	Wastewater or compound	$Y = a + bX$ a	b	Correlation coefficient	Number of points	Reference
COD:TC	Domestic sewage	132.3	2.76	0.93	—	Stenger and Van Hall, 1968
COD:TOC	Domestic sewage	12.16	3.48	0.92	—	
COD:TOC	Papermill waste	3.16	3.21	0.99	—	U.S.E.P.A., 1971a
COD:TOC	Papermill waste	13.03	2.54	0.99	—	
COD:TOD	Domestic sewage	—	1.00	0.96	—	Wood et al., 1970
COD:TOD	Secondary effl.	—	1.05	0.97	—	Wood et al., 1970
COD:TOD	Secondary effl.	—	1.03	0.97	—	Wood et al., 1970
COD:TOD	Pesticide mfg. waste	—	0.95	0.96	—	Wood et al., 1970
COD:TOD	Petrochemical	—	0.98	0.43	—	Wood et al., 1970
COD:TOD	Petrochemical	—	1.20	0.92	—	Wood et al., 1970
COD:TOD	Petrochemical	—	1.12	0.96	—	Wood et al., 1970
COD:TOD	Plastic mfg.	—	1.25	0.99	—	Wood et al., 1970

COD:TOD	Cryogenics plant	—	1.04	0.99	—	Wood et al., 1970
BOD$_5$:TOC	Papermill waste	-0.35	0.0725	0.99	5	Hwang et al., 1974
BOD$_5$:TOD	Domestic sewage	—	0.54	0.91	—	Wood et al., 1970
BOD$_5$:TOD	Secondary effluent	—	0.35	0.89	—	Wood et al., 1970
BOD$_5$:TOD	Secondary effluent	—	0.42	0.20	—	Wood et al., 1970
BOD$_5$:TOD	Pesticide manufacturing	—	0.52	0.96	—	Wood et al., 1970
BOD$_5$:TOD	Petrochemical plant	—	0.52	0.94	—	Wood et al., 1970
BOD$_5$:TOD	Petrochemical plant	—	0.62	0.91	—	Wood et al., 1970
BOD$_5$:ODI	Domestic sewage	-43.15	0.98	0.89	360	Reynolds and Goellner, 1974
BOD$_5$:ODI	Primary effluent	-27.63	1.00	0.88	360	Reynolds and Goellner, 1974
BOD$_5$:ODI	Final effluent	-20.12	0.57	0.90	100	Reynolds and Goellner, 1974
BOD$_5$:ODI	Domestic sewage	52.71	0.96	0.86	93	Shriver and Young, 1972
BOD$_5$:ODI	Primary effluent	23.59	0.80	0.89	59	Shriver and Young, 1972
BOD$_5$:ODI	Final effluent	19.17	0.53	0.71	90	Shriver and Young, 1972
BOD$_5$:COD$_{rapid}$	Domestic sewage	-51.65	0.59	0.96	50	Wells, 1970
BOD$_5$:COD$_{rapid}$	Final effluent	-35.3	0.81	0.97	50	Wells, 1970
BOD$_5$:COD$_{rapid}$	Final effluent	6.82	0.19	0.50	30	Wells, 1970
BOD$_5$:COD$_{rapid}$	Domestic sewage	109.1	0.07	0.075	—	Wells, 1970
BOD$_5$:COD	Papermill waste	-0.73	0.0285	0.99	5	Hwang et al., 1974
COD:COD$_{rapid}$	Industrial	96.7	0.80	0.98	21	McLean and Spicher, 1973
COD:COD$_{rapid}$	Industrial	49.8	0.97	0.99	61	McLean and Spicher, 1973

REFERENCES

Ahnoff, M., Josefsson, B., Tunde, G., and Anderson, G. (1979). *Water Res.* **13**, 1233–1237.
Almgren, T., and Josefsson, B. (1973). *Sven. Papperstidn.* **76**, 19.
American Public Health Association (APHA) (1981). "Standard Methods for the Examination of Water and Wastewaters," 15th. ed. APHA, New York.
Arends, J. M., Cerfontain, H., Herschberg, I. S., Preusen, A. J., and Wanders, A. C. M. (1964). *Anal. Chem.* **36**, 1802.
Arin, M. L. (1974). *Environ. Sci. Technol.* **8**, 898–902.
Armstrong, F. A., and Tibbets, S. (1968). *J. Mar. Biol. Assoc. U.K.* **48**, 143–152.
American Society for Testing and Materials (ASTM) (1980). *Annu. Book ASTM Stand.* **31**, 665–669.
Baker, C. D., Bartlett, P. D., Farr, I. S., and Williams, G. I. (1974). *Freshwater Biol.* **4**, 467–481.
Baldwin, J. M., and McAtee, R. E. (1974). *Microchem. J.* **19**, 179–190.
Ballinger, D. G., and Lishka, R. J. (1962). *J. Water Poll. Cont. Fed.* **34**, 470–474.
Baumgartner, D. J., Feldman, M. A., and Gibbons, C. L. (1971). *Water Res.* **5**, 533.
Black, H. C. (1951). *Anal. Chem.* **23**, 1792.
Bohlen, P., Stein, S., Dairmon, W., and Udenfriend, S. (1973). *Arch. Biochem. Biophys.* **155**, 213.
Bramer, H. C., Walsh, M. J., and Caruso, S. C. (1966). *Water Sewage Works* **113**, 275–278.
Briggs, R., and Melbourne, K. V. (1968). *Water Treat. Exam.* **17**, 107–120.
Briggs, R., Schofield, J. W., and Gorton, P. A. (1976). *Water Pollut. Control* **75**, 46–57.
Brun, G. L., and Milburn, D. L. D. (1977). *Anal. Lett.* **10**, 1209.
Burney, C. M., and Sieburth, J. McN. (1977). *Mar. Chem.* **5**, 15–28.
Butzelaar, P. F., and Hoogeveen, L. P. J. (1975). *Water Serv.* **34**, 50–54.
Christman, R. F., and Minear, R. A. (1967). *Trend Eng. Univ. Wash.* **19**, 3.
Christman, R. F., and Ghassemi, M. (1966). *J. Am. Water Works Assoc.* **58**, 723–741.
Christman, R. F. and Ghassemi, M., (1968). *Limnol. Oceanogr.* **13**, 583.
Clifford, D. A. (1968). *Proc. Ind. Waste Conf.* **23**, 772–779.
Collins, K. J., and Williams, P. J. LeB. (1977). *Mar. Chem.* **5**, 123–141.
Conroy, L. E., Maier, W. J., and Shih, Y. T. (1981). *In* "Chemistry in Water Reuse" (W. J. Cooper, ed.), Vol. 1, pp. 65–84. Ann Arbor Sci. Publ., Ann Arbor, Michigan.
Cooper, W. J., and Suffet, I. H. (1981). "Water Reuse Symposium II," Vol. 3, pp. 2299–2324. Water Works Assoc., Denver, Colorado.
Cotruvo, J. A. (1978). *In* "Water Chlorination: Environmental Impacts and Health Effects" (R. L. Jolley, H. Gorchev, and D. H. J. Hamilton, eds.), Vol. 2, pp. 817–821. Ann Arbor Sci. Publ., Ann Arbor, Michigan.
Cowen, W., Peterman, B. W., and Davenport, R. J. (1979). "Water Reuse Symposium," Vol. 3, pp. 1975–1989. Am. Water Works Assoc., Denver, Colorado.
Davenport, R. J., Cooper, W. J., and Wynveen, R. A. (1978). "Monitoring Organic Impurity Concentrations in Water Using Differential Capacitance Measurements." Presented at the Division of Environmental Chemistry, Am. Chem. Soc., Anaheim, California.
Davenport, R. J., Cooper, W. J., Huang, J. C., and Wynveen, R. A. (1979). "Development of an Instrument for Continuous, Automated and Low-Cost Monitoring of the Organic Loading in Water." Presented at the Pittsbourgh Conference on Analytical Chemistry and Applied Spectroscopy; Symposium, New Techniques on the Horizon, Cleveland, Ohio.
Davenport, R. J., Wynveen, R. A., and Cooper, W. J. (1981). *J. Am. Water Works Assoc.* **73**, 555–558.

Dawson, R., and Liebezeit, G. (1981). *In* "Marine Organic Chemistry, Evaluation, Composition, Interactions and Chemistry of Organic Matter in Seawater" (E. K. Duursman and R. Dawson, eds.), pp. 445–496. Am. Elsevier, New York.

Dobbs, R. A., Wise, R. H., and Dean R. B. (1972). *Water Res.* **16,** 1173–1180.

Dressman, R. C., McFarren, E. F., and Symons, J. M. (1977). *Proc.—AWWA Water Qual. Technol. Conf.* Paper 3A-5.

Dressman, R. C., Najar, B. A., and Redzikowski, R. (1979). *Proc.—AWWA Water Qual. Technol. Conf.* pp. 69–92.

Dreywood, R. (1946). *Ind. Eng. Chem., Anal. Ed.* **18,** 449.

Duursma, E. K. (1961). *Neth. J. Sea Res.* **1,** 1–148.

Duursma, E. K., and Rommets, J. W. (1961). *Neth. J. Sea Res.* **1,** 391.

Ehrhardt, M. (1969). *Deep-Sea Res.* **16,** 393–397.

El-Dib, M. A., and Ramada, F. M. (1966). *J. Sanit. Eng. Div., Am. Soc. Civ. Eng.* **92,** 97–101.

Fleet, B., Ho, A. Y. W., and Tengyl, J. (1972). *Analyst* **97,** 321–333.

Ford, D. L. (1968). *Public Works* **99,** 89–92.

Ford, D. L., Eller, J. M., and Gloyna, E. F. (1971). *J. Water Pollut. Control Fed.* **43,** 1712–1723.

Foulds, J. M., and Lunsford, J. V. (1968). *Water Sewage Works* **115,** 112–115.

Games, L. M., and Hayes, J. M. (1976). *Anal. Chem.* **48,** 130–135.

Gaudy, A. F., Jr., and Gaudy, E. T. (1972). *Ind. Waste Eng.* **9,** 30–34.

Gershey, R. M., MacKinnon, M. D., and Williams, P. J. LeB. (1979). *Mar. Chem.* **7,** 289–306.

Gether, J., Lunde, G., and Steinnes, E. (1979). *Anal. Chim. Acta* **108,** 137–147.

Glaze, W. H., Peyton, G. R., and Rawley, R. (1977). *Environ. Sci. Technol.* **11,** 685–690.

Goldstein, A. L., Katz, W. E., Melle, F. H., and Murdock, D. M. (1968). "Total Oxygen Demand—A New Automated Instrumental Method for Measuring Pollution and Loading on Oxidation Processes" Ionics Inc., Watertown, Massachusetts (unpublished paper).

Gordon, D. C., Jr., and Sutcliffe, W. H., Jr. (1973). *Mar. Chem.* **1,** 231–244.

Goulden, P. D., and Brooksbank, P. (1975). *Anal. Chem.* **47,** 1943–1946.

Hall, K. J., and Lee, G. F. (1974). *Water Res.* **8,** 239.

Harkness, N., Hey, A. E., and Willets, D. G. (1972). *Water Pollut. Control* **73,** 261–277.

Hauden, D., and Richard, Y. (1975). *Prog. Water Technol.* **7,** 41–55.

Helfgott, T. B. (1977). "Parameters for Measuring Organics in Waters." Tahal—Water Planning for Israel Ltd., Tel Aviv.

Hirayama, H. (1974). *Anal. Chim. Acta* **70,** 141–148.

Hwang, C. P., Good, G., and Davis E. (1974). *Water Pollut. Control* **112,** 28–34.

Jekel, M., and Roberts, P. V. (1980). *Environ. Sci. Technol.* **14,** 970–975.

Jenkins, S. H., Harkness, N., Hewitts, P. J., Snadden, X. V. M., Ellerker, R., Divitto, B., and Dee, H. L. (1965). *J. Proc. Inst. Sewage Purif.* **6,** 533–565.

Jeris, J. S. (1969). *Water Sewage Works* **116,** 89–91.

Jermyn, M. A. (1975). *Anal. Biochem.* **68,** 332.

Jirka, A., and Carter, M. (1975). *Anal. Chem.* **47,** 1397–1402.

Johnson, K. M., and Sieburth, J. McN. (1977). *Mar. Chem.* **5,** 1–13.

Jones, R. H., and Degaford, A. F. (1968). *ISA Trans.* **7,** 267–272.

Josefsson, B. O., Upsstrom, L., and Ostling, G. (1972). *Deep-Sea Res.* **19,** 385–395.

Josefsson, B. O., Lindroth, P., and Ostling, G. (1977). *Anal. Chim. Acta* **89,** 21–28.

Kalle, V. K. (1949). *Deutsche Hydrographischen Zeitschrift* **16,** 117–124.

Kalle, V. K. (1963). *Deutsche Hydrographischen Zeitschrift* **16,** 154–166.

Kankare, J. J. (1970). *Anal. Chem.* **42**, 1322–1325.

Kay, H. (1954). *Kiel. Meeresforsch.* **10**, 26–35.

Kehoe, T. J. (1975). "Critical Influences Relating to the Analysis of Total Organic Carbon in Waters." Paper presented at the AWWA Water Qual. Technol. Conf., Atlanta, Georgia.

Keith, L. H., ed. (1982). "Advances in the Identification and Analysis of Organic Pollutants in Water," Vol. 1. Ann Arbor Sci. Publ., Ann Arbor, Michigan.

Kolle, W., and Sontheimer, H. (1973). "Experience with Activated Carbon in West Germany," Pap. 15. Univ. of Karlsruhe, Karlsruhe, FRG.

Kuhn, W., and Sontheimer, H. (1973). *Vom Wasser* **41**, 65–79.

Kuhn, W., and Sontheimer, H. (1974). *Vom Wasser* **43**, 327–341.

Lantz, J. B., Davenport, R. J., Wynveen, R. A., and Cooper, W. J. (1981). *In* "Chemistry in Water Reuse" (W. J. Cooper, ed.), Vol. 1, pp. 147–163. Ann Arbor Sci. Publ., Ann Arbor, Michigan.

Lunde, G., Gether, J., and Josefsson, B. (1977). *Bull. Environ. Contam. Toxicol.* **13**, 656–661.

McCahill, M. P., Conroy, L. E., and Maier, W. J. (1980). *Environ. Sci. Technol.* **14**, 201–203.

McCahill, M. P., Conroy, L. E., and Maier, W. J. (1981). *In* "Chemistry in Water Reuse" (W. J. Cooper, ed.), Vol. 2, pp. 119–126. Ann Arbor Sci. Publ., Ann Arbor, Michigan.

MacKinnon, M. D. (1978). *Mar. Chem.* **7**, 17–37.

MacKinnon, M. D. (1981). *Elsevier Oceanogr. Ser. (Amsterdam)* **31**, 415–443.

McLean, D. A., and Spicher, R. G. (1973). *Proc. Ind. Waste Conf.* **28**, 1017–1024.

Maier, W. J., and Conroy, L. E. (1981). *In* "Chemistry in Water Reuse" (W. J. Cooper, ed.), Vol. 1, pp. 85–145. Ann Arbor Sci. Publ., Ann Arbor, Michigan.

Malcolm, R. L., and Leenheer, J. A. (1973). *Inst. Environ. Sci.* **19**, 336–341.

Mattson, J. S., Smith, C. A., Jones, T. T., Gerchakov, S. M., and Epstein, B. D. (1974). *Limnol. Oceanogr.* **19**, 530–535.

Menzel, D. W., and Vaccaro, R. F. (1964). *Limnol. Oceanogr.* **9**, 138–142.

Metzler, D. C., Harris, C. M., Reaves, R. L., Lawton, W. H., and Maggio, M. S. (1977). *Anal. Chem.* **49**, 864–874.

Michail, M., and Idelovitch, E. (1981). *In* "Chemistry in Water Reuse" (W. J. Cooper, ed.), Vol. 1, pp. 35–64. Ann Arbor Sci. Publ., Ann Arbor, Michigan.

Mitchell, D. G., Aldous, K. M., and Canelli, E. (1977). *Anal. Chem.* **49**, 1235–1238.

Moore, W. A., Kroner, W. A., and Ruchhoft, S. C. (1949). *Anal. Chem.* **21**, 593–595.

Morse, E. E. (1947). *Anal. Chem.* **19**, 1012.

Mrkva, M. (1969). *J. Water Pollut. Control Fed.* **41**, 1923–1931.

Mrkva, M. (1975). *Water Res.* **9**, 587–589.

Nasr, M. S., and MacDonald, D. G. (1975). *Pulp Pap. Can.* **76**, 91–93.

National Academy of Sciences (1977). *Fed. Regist.* **42**, 35764.

North, B. B. (1975). *Limnol. Oceanogr.* **20**, 20–27.

Otto, L. (1967). *Neth. J. Sea Res.* **3**, 532.

Packard, T. T., and Dortch, Q. (1975). *Mar. Biol.* **33**, 347–354.

Reynolds, J. F., and Goellner, K. A. (1974). *Water Sewage Works* **121**, 31–34.

Roth, M. (1971). *Anal. Chem.* **43**, 880–882.

Roth, M., and Hampai, A. (1973). *J. Chromatogr.* **83**, 353–356.

Samejima, K., Dairman, W., and Udenfriend, S. (1971a). *Anal. Biochem.* **42**, 222–236.

Samejima, K., Dairman, W., Stone, J., and Udenfriend, S. (1971b). *Anal. Biochem.* **42**, 237.

Samsel, E. P., and Delap, R. A. (1951). *Anal. Chem.* **23**, 1795.

Schalekamp, M., Dietlicher, K., Valenta, J., Wattenhofer, R., and Zimmerman, U. (1978). *Z. Wasser Abwasser Forsch.* **11**, 150–157.

Scheuch, L. E., and Edzwald, J. K. (1981). *J. Am. Water Works Assoc.* **73**, 497–502.

Shapiro, J. (1957). *Limnol. Oceanogr.* **2**, 161.

Shriver, L. E., and Young, J. C. (1972). *J. Water Pollut. Control Fed.* **44**, 2140–2147.

Skopintsev, B. A. (1960). *Mar. Hydrophys. Inst.* **19**, 1–14.

Sylvia, A. E. (1973). *J. N. Engl. Water Works Assoc.* p. 87.

Sylvia, A. E., and Donlon, R. J. (1980). *In* "Activated Carbon Adsorption of Organics from the Aqueous Phase" (M. J. McGuire and I. H. Suffet, eds.), Vol. 2, pp. 559–565. Ann Arbor Sci. Publ., Ann Arbor, Michigan.

Sylvia, A. E., Bancroft, D. A., and Miller, J. D. (1979). *Proc.—AWWA Water Qual. Technol. Conf.* p. 87.

Smart, P. L., Finlayson, B. L., Rylands, W. D., and Ball, C. M. (1976). *Water Res.* **10**, 805.

Stein, S., Bohlen, P., Stone, J., Dairman, W., and Udenfriend, S. (1973). *Arch. Biochem. Biophys.* **155**, 202–212.

Stenger, V. A., and Van Hall, C. E. (1968). *J. Water Pollut. Control Fed.* **41**, 1755–1762.

Stewart, A. J., and Wetzel, R. G. (1981a). *Limnol. Oceanogr.* **26**, 590–597.

Stewart, A. J., and Wetzel, R. G. (1981b). *Arch. Hydrobiol.* **92**, 265–286.

Suffet, I. H. (1980). *In* "Activated Carbon Absorption of Organics from the Aqueous Phase" (M. J. McGuire and I. H. Suffet, eds.), Vol. 2, Ann Arbor Sci. Publ., Ann Arbor, Michigan.

Takahashi, Y. (1979). *Proc. EPA/NATO-CCMS Conf. Absorption Tech. 1979* (in press).

Takahashi, Y., Moore, R. T., and Joyce, R. J. (1973). *Am. Lab.* **5**, 31–38.

Takahashi, Y., Moore, R. T., and Joyce, R. J. (1981). *In* "Chemistry in Water Reuse" (W. J. Cooper, ed.), Vol. 2, pp. 127–146. Ann Arbor Sci. Publ., Ann Arbor, Michigan.

Thruston, A. D., Jr. (1970) *J. Water Pollut. Control Fed.* **42**, 1551.

Udenfriend, S., Stein, S., Bohlen, P., and Dairman, W. (1972). *Science* **178**, 871.

U.S. Environmental Protection Agency (U.S.E.P.A.) (1971a). "AQCL Newsletter No. 11." Analytical Quality Control Laboratory, Cincinnati, Ohio.

U.S. Environmental Protection Agency (U.S.E.P.A.) (1971b). "Method Research Study 3, Demand Analysis," U.S.E.P.A., Analytical Quality Control Laboratory, Cincinnati, Ohio.

U.S. Environmental Protection Agency (U.S.E.P.A.) (1979). "Methods for Chemical Analysis of Water and Wastes," Rep. No. EPA-600/4-79-020. U.S.E.P.A., Washington, D.C.

Van Hall, C. E., and Stenger, V. A. (1967). *Anal. Chem.* **39**, 503–507.

Van Hall, C. E., Safranko, J., and Stenger, V. A. (1963). *Anal. Chem.* **35**, 315–319.

Van Rensburg, J. F. J., Hasset, A. J., and Theron, S. J. (1981). *In* "Chemistry in Water Reuse" (W. J. Cooper, ed.), Vol. 1, pp. 17–34. Ann Arbor Sci. Publ., Ann Arbor, Michigan.

Van Steenderen, R. A. (1976a). *Water SA* **2**, 126–130.

Van Steenderen, R. A. (1976b). *Water SA* **2**, 156–159.

Van Steenderen, R. A. (1979). *J. Autom. Chem.* **1**, 88–92.

Van Steenderen, R. A. (1981). *Water SA* **7**, 28–34.

Van Steenderen, R. A., and Lin, J.-S. (1981). *Anal. Chem.* **53**, 2157–2158.

Van Steenderen, R. A., Basson, W. D., and Van Duuren, F. A. (1979). *Water Res.* **13**, 539–543.

Viles, F. J., and Silverman, L. (1949). *Ind. Eng. Chem., Anal. Ed.* **21**, 950.

Weigele, M., Blount, J. F., Tengi, J. P., Czaikjowski, R. C., and Leimgruber, W. (1972a). *J. Am. Chem. Soc.* **94**, 4052.

Weigele, M., DeBernardo, S. L., Tengi, J. P., and Leimgruber, W. (1972b). *J. Am. Chem. Soc.* **94,** 5927–5928.

Wells, W. N. (1970). *Water Sewage Works* **117,** 123–129.

West, D. L. (1964). *Anal. Chem.* **36,** 2194–2195.

West, S. J., Frant, M. S., and Ross, J. W., Jr. (1976). *Intersoc. Conf. Environ. Syst., 1976.*

West, S. J., Frant, M. S., and Franks, S. H. (1977). *Intersoc. Conf. Environ. Syst., 1977.*

Westerhold, A. F. (1965). *Digester* **117,** 123–129.

Wetzel, R. G. (1974). *Arch. Hydrobiol.* **73,** 31–56.

Wilander, A., Kvarnos, H., and Lindell, T. (1974). *Water Res.* **8,** 1037.

Willey, J. D., and Atkinson, L. P. (1982). *Estuarine, Coastal Shelf Sci.* **14,** 49.

Wilson, R. F. (1961). *Limnol. Oceanogr.* **6,** 259–261.

Wolfel, P., and Sontheimer, H. (1974). *Vom Wasser* **43,** 315–325.

Wolff, C. J. M. (1975). *Water Res.* **22,** 4–6.

Wood, E. D., Perry, A. E., and Hitchcock, M. G. (1970). Edna Woods Laboratory, Houston, Texas (unpublished paper).

Zika, R. G. (1977). Ph.D. Thesis, Dalhousie University, Halifax, Nova Scotia.

Zimmerman, J. T. F., and Rommets, J. W. (1974). *Neth. J. Sea Res.* **8,** 117.

3

CONCENTRATION, PARTITIONING, AND ISOLATION TECHNIQUES

Jerry A. Leenheer

Water Resources Division
U.S. Geological Survey
Denver, Colorado

WATER ANALYSIS, VOL. III
Copyright © 1984 by Academic Press, Inc.
All rights of reproduction in any form reserved.
ISBN 0-12-498303-0

I. INTRODUCTION

Analysis of organic materials in water frequently involves concentrating the sample to improve sensitivity of the analytical method, partitioning complex organic mixtures into more homogeneous fractions for compound class analysis or as a preparatory step for more specific organic analysis, and isolating the organic material of interest from the solvent and solute matrix, which may interfere with its analysis. Range of concentrations for organic constituent mixtures in various types of water spans more than five orders of magnitude, as shown in Table I. Certain organic constituents in wastewaters are often sufficiently concentrated that direct analysis by gas chromatography (Ho *et al.*, 1976) or liquid chromatography (Felix *et al.*, 1977) is possible. However, most organic analyses of water involve natural-water or drinking-water samples, whose total organic carbon (TOC) concentration is usually <10 mg/L and whose specific organic compound concentrations are measured as micrograms per liter (parts per billion) or even nanograms per liter (parts per trillion). These samples usually must be concentrated by two to three orders of magnitude before analysis.

The first approaches (1950–1960) to concentration and isolation of organic materials from water were tied to classical organic analysis schemes such as Shriner and Fuson (1948), which required milligram-sized quantities of the organic mixture. The carbon filter (Braus *et al.*, 1951; Middleton *et al.*, 1962) and solvent extraction (Hoak, 1962) were scaled up so hundreds to thousands of gallons of water were processed to obtain necessary amounts of organic materials. The advent (1952 to present) of first gas-chromatographic and later liquid-chromatographic analysis with sensitive detectors, such as the mass spectrometer, enabled the sample size to be reduced to a few liters when these analyses were coupled with more efficient concentration and isolation techniques. Recent developments in analysis of organic volatiles in water, such as the "sparge-and-trap" tech-

TABLE I

Representative Concentrations of Organic Carbon in Various Types of Water

Water type	Total organic carbon (mg/L)	Dissolved organic carbon (mg/L)	Particulate organic carbon (mg/L)
Groundwater[a]	0.7	0.7	—
Seawater[b]	1.1	1.0	0.1
Drinking water[c]	2.0	—	—
Surface water (lakes)[d]	7.7	7.0	0.7
Surface water (rivers)[e]	8.0	5.0	3.0
Untreated domestic sewage[f]	200	80	120
Ammoniacal liquors from coal and oil-shale conversion processes[g]	10,000	—	—

[a] Leenheer et al. (1974). [e] Malcolm and Durum (1976).
[b] Williams (1971). [f] Besik (1971) and Hunter (1971).
[c] Symons et al. (1975). [g] Poulson (1978).
[d] Wetzel (1975).

niques in which use of organic solvents can be omitted, illustrate simplification of sample preparation procedures. Therefore, the trend in organic analysis of water is toward smaller samples with a minimum of sample preparation. The goal towards which analysis of organics in water is proceeding is direct analysis with minimum concentration and isolation procedures.

A report on drinking water and health issued by the National Academy of Sciences (1977) concluded that 90% of volatile and semivolatile organic compounds and only 5–10% of nonvolatile organic compounds in drinking water have been identified and quantitated. As volatile and semivolatile organic compounds represent only ~10% of the total organic material in drinking water, ~90% of the material remains to be identified at the compound level of analysis. An even smaller percentage of organic material in natural waters has been identified because natural waters have not been analyzed as thoroughly as drinking water. The objective of this chapter is to provide a critical review of current methods of water sample preparation for organic analysis to assist the analyst in the goal of identification and quantitation of both known and unknown organic constituents of water. The scope of water sample preparation procedures extends from sampling to compound class separations and extract cleanup procedures, although most of the emphasis is on concentration, partitioning, and isolation techniques.

The following texts provide a useful introduction to the subjects of this chapter. A discussion of concentration and separation techniques of both organic and inorganic constituents in water is given by Andelman and Caruso (1971). It is oriented to readers with relatively little experience in concentration and separation methods for water. Fundamental principals of methods of separation are presented by Karger *et al.* (1973). A reference guide to current methodology for organic compounds in water (Baker and Yates, 1975) includes sections on water sampling, sample preparation and preservation, concentration and isolation techniques, and analysis of specific organic constituents. Finally, many examples of applications of concentration and separation of methods used to analyze organic compounds in water are given in a series of papers edited by Keith (1976).

A report by Kaiser (1971) concluded that although many organic compounds are present at concentrations in water at which they should be easily detected by current analytical methods, these compounds are lost during sampling, storage, or enrichment. He listed the following sources of systematic error that can occur throughout the scope of organic analysis:

1. The sample source
   a. Wrong place (nonhomogeneity of the source or sample flow)
   b. Wrong time and duration (strong fluctuations of the concentration over a given period)
2. Sample handling
   a. Method of drawing the sample (environmental influence, O_2, H_2O, light)
   b. Sample drawing duct
   c. Size of vessel
   d. Temperature, pressure of intermediate systems, and flow
   e. Temperature, pressure, and duration of storage
   f. Walls of the vessel (contamination, unintended selective adsorption, catalytic acceleration of transformation)
3. Sample preparation and final sampling
   a. Chemical reaction (intended selective adsorption or transformation that is incomplete or not sufficiently selective)
   b. Enrichment (concentration, temperature, and auxiliary substances that may cause a reaction of the substances with each other, a combination with memory of the auxiliary substance, or a reaction with the auxiliary substance. Auxiliary substance equals stationary phase)

   c. Nonrepresentative secondary sampling out of the sampling
      vessel
4. Separation process
   a. Incomplete desorption
   b. Chemisorption, transformation
   c. Incomplete separation (partial or total overlapping)
   d. Incomplete elution from the connecting ducts

The remaining sources of error listed are beyond the scope of this chapter. Many of these errors, which formerly went unnoticed, can now be detected and quantitated by total and dissolved organic carbon analysis of water in an analytical scheme (Malcolm and Leenheer, 1973). This chapter emphasizes sources of systematic errors for each technique and procedure discussed.

The organizational framework of this chapter is based on organic compound physical classifications as major headings, rather than on various sampling, concentration, and separation techniques, because each technique must be discussed as it applies to a particular class of compounds. For example, methodology for solvent extraction and concentration of volatile halomethane compounds from water is different from solvent extraction and concentration of polynuclear aromatic hydrocarbons because of volatility differences between these two physical types of compounds. The four major compound divisions for this chapter, classified by properties and physical structure, are volatiles, semivolatiles, low molecular weight nonvolatiles, and high molecular weight compounds. Volatile compounds are generally defined as those with sufficient vapor pressure to be easily sparged from water with an inert gas; semivolatile compounds cannot be efficiently sparged from water, but possess sufficient volatility when isolated from water to be analyzed by gas chromatography. Low molecular weight nonvolatile compounds cannot be analyzed by gas chromatography without derivatization, but their low molecular weight (<500) facilitates identification because they are usually monomers or simple polymers. High molecular weight compounds (>500) are usually complex polymeric organic materials in water that are very difficult to identify structurally.

It is recognized that there is considerable overlap in this compound classification; it is presented only for use in this chapter and is not intended to serve as a standard classification. Specialized techniques for sampling, preserving, concentrating, partitioning, and isolating each of the four types of compounds are presented in this chapter as a preamble to the following chapters, which discuss separation and analysis of specific compounds.

II. EQUIPMENT AND REAGENT SELECTION AND CLEANUP

As the limit of detection of many trace organic analyses of water has decreased to parts per billion and parts per trillion levels, the need for "contaminant-free" equipment and reagents has increased accordingly. Equipment and reagents must not cause loss of constituents of interest by contamination, sorption, precipitation, volatilization, or reaction. Most contaminant introductions or analyte losses occur during sample collection and sample preparation procedures.

A. Equipment Selection

Plastic tubing, containers, and filter membranes are the most common sources of organic contaminants. Organic contaminants introduced into water flowing through polyethylene, polypropylene, black latex, and polyvinyl chloride tubing were determined by Junk *et al.* (1974a). Organic contaminant concentrations ranged from 1 to 5000 ppb. The most common contaminants were various phenol stabilizers and phthalate plasticizers. High flow rates and large flushing volumes did not eliminate organic contaminant bleed from tubing. Organic filter membranes are also a common source of organic contaminants when used to filter samples for organic analysis. These filters usually contain organic detergents and wetting agents that leach into the filtrate. Many of these compounds are also published as common contaminants found in water; therefore, distinguishing the source of certain organic compounds may be impossible if plastics are used in sampling and analysis of organic compounds in water.

Loss of the analyte through sorption on plastic tubing, containers, and filter membranes is also a common occurrence. Hydrophobic organic constituents are most likely to be lost through sorption, but soluble organic acid and base constituents can be lost if they react with base or acid residues sorbed on walls of sampling vessels. To avoid this potential problem, sample containers should be rinsed with dilute acid (HCl) if organic acid compounds are to be analyzed, and dilute base (NaOH) if organic base compounds are collected. Sorption of organic constituents on organic filter membranes is often the greatest source of loss, because of intimate contact of organic solutes and suspended sediment with the filter membrane. Again, hydrophobic organic constituents show greatest losses, but certain compound classes such as phenols may be selectively sorbed on filter membranes such as cellulose acetate.

Equipment and containers for water sample collection, storage, and processing should be constructed of glass or metal, if possible. Borosilicate glass is preferable because it can be baked at high temperature to free it of organic contaminants; stainless steel is usually the metal of choice because of its inertness, resistance to corrosion, and strength. Metals such as zinc and copper may cause catalytic changes of certain organic molecules, and aluminum reacts with alkaline water samples. Teflon can be used for flexible tubing and seals in bottle caps, but hydrophobic organic constituents may be lost from solution through sorption.

A number of sampling devices have been designed that are suitable for collection of organic constituents in water. In flowing rivers and streams, depth-integrating samplers developed for collection of representative suspended-sediment samples (Guy and Norman, 1970) are suitable for collection of trace organic constituents such as pesticide residues (Feltz and Culbertson, 1972). These samplers consist of streamline aluminum or brass castings that enclose glass sample containers. The sample is collected through a metal intake nozzle at stream velocity. For sampling of organic constituents, it is recommended that rubber gaskets and seals be replaced with Teflon seals. For collection of dip samples and weighted-bottle samples, only cleanliness of the glass sampling bottle needs to be assured; for Kemmerer-type samplers (Welch, 1948), possible contamination from organic stoppers at both ends of the cylinder barrel may interfere with organic analysis.

Sampling of groundwater poses special problems in obtaining samples free of organic contaminants introduced by drilling a well, from the well casing, or from the pump. Whenever possible, groundwater should be obtained at springs or artesian wells that have been flowing at high rates for long periods of time. When pumps must be used, a large production well equipped with a water-lubricated submersible pump that has already pumped several million gallons of water is the best source of groundwater. Wells equipped with oil-lubricated, shaft-turbine pumps frequently have a layer of oil on top of the water column. For small observation wells, a thief-type sampler is often used to obtain small water volumes from a point source in the well, but these samplers suffer from the same limitations as the Kemmerer-type sampler just discussed. An ultraclean well pump designed for sampling organic constituents in water from small-diameter observation wells was designed by M. C. Goldberg and L. L. DeLong (unpublished data, 1971). A diagram of this pump is shown in Fig. 1. Stainless steel is the only material that the water sample contacts, and helium, purified by passage through a molecular sieve, is the contaminant-free pumping gas. A limitation of the ultraclean pump is its low pumping capacity, which may not be sufficient to "dewater" a well to obtain a fresh groundwater sample.

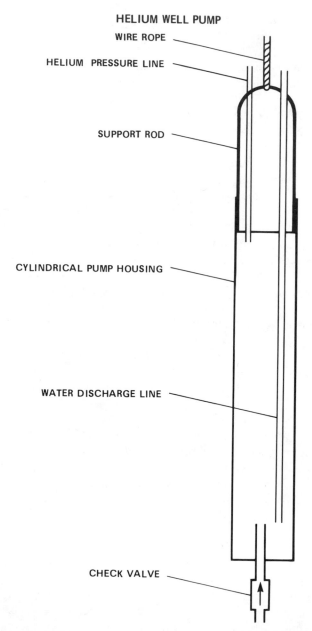

HELIUM WELL PUMP

WIRE ROPE

HELIUM PRESSURE LINE

SUPPORT ROD

CYLINDRICAL PUMP HOUSING

WATER DISCHARGE LINE

CHECK VALVE

Fig. 1. Ultraclean well pump for sampling trace organic compounds in groundwater.

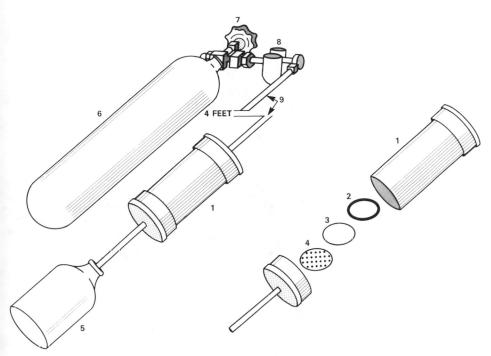

Fig. 2. Apparatus for pressure filtration of organic constituents in water: (1) Filter barrel; (2) Teflon O ring; (3) silver-membrane filter; (4) stainless steel filter support; (5) glass sample bottle; (6) pressure cylinder; (7) pressure cylinder valve; (8) pressure regulator; (9) high-pressure flexible tubing. (From Malcolm and Leenheer, 1973; published with permission of the Institute of Environmental Sciences.)

After a water sample is collected, suspended sediment is often separated by filtration or centrifugation. Although it is desirable to separate suspended sediment on site to minimize sample changes, in general filtration has proven to be the only feasible methodology for use in the field. The only filter materials that are largely free of organic contaminants are glass-fiber filters (without an organic binder) and silver-membrane filters. Both of these filters can be used in suction-filtration apparatus made from metal or glass, or in a pressure-filtration apparatus (Malcolm and Leenheer, 1973) as shown in Fig. 2. Pressure filtration is frequently preferable to suction filtration because loss of volatile organic compounds is minimized, but care must be taken to use a nonreactive, contaminant-free pressurizing gas such as nitrogen or helium, and high pressures must not be used because of the possibility of rupture of cellular constituents.

B. Equipment Cleanup

Giam and Wong (1972) detected contaminants from glassware and
Teflon equipment, a stainless steel high-speed blender, filter paper, glass
wool, aluminum foil, and Teflon–rubber laminated disc sample-bottle
seals in a gas-chromatographic procedure for analyses of chlorinated hydrocarbons in seawater. The most effective procedure for removing these
contaminants from equipment was heating to 300–350°C overnight when
applicable and washing with detergent and solvent for heat-labile materials.

A good general procedure to follow for equipment and sample-con-
tainer cleanup is specified in the American Society for Testing and Materials (ASTM) tentative test method for chlorinated phenoxy acid herbicides
in water (ASTM, 1978c). Glassware and equipment should first be cleaned
with detergent, followed by a dilute hydrochloric acid rinse, and lastly a
distilled-water rinse. To remove the last organic contaminants, glassware
should be heated in a muffle furnace to 400°C for 15–30 min. Glassware
and equipment not compatible with heating should be rinsed with redis-
tilled acetone and finally with pesticide-quality hexane. Glass-fiber filters
and silver membranes should only be rinsed with dilute acid, distilled
water, acetone, and hexane. Any necessary deviations from this cleanup
procedure are covered in the following discussions on sampling and sample preparation for each division of organic compounds.

C. Preparation of Reagent Water

Preparation of reagent water free of organic contaminants is frequently
one of the major challenges for recovery studies designed to test analytical procedures. ASTM specifies use of their Type II-grade reagent water
for analytical procedures that require trace concentrations of organic
compounds (ASTM, 1978a). Type II-grade water is prepared by distillation, using a still designed to produce a distillate having a conductivity
less than 1.0 μmho/cm at 25°C. Ion exchange, distillation, or reverse
osmosis may be required as pretreatment prior to final distillation if purity
cannot be obtained in a single distillation. However, reverse osmosis and
ion exchange frequently release steam-volatile organic contaminants that
are difficult to remove by distillation. Multiple distillations in all-glass
stills in the presence of potassium permanganate is frequently used to
obtain reagent water of acceptable quality (Giam and Wong, 1972). Distilled reagent water can be purified of traces of sorbable organic contaminants by passage through columns containing activated carbon and/or

XAD-2 resin (Junk *et al.*, 1974b). Both activated-carbon and resin adsorbents should be recently cleaned and not used over long periods of time to avoid "breakthrough" of previously sorbed contaminants from the resin and the establishment of bacterial populations on the adsorbent.

A spray-vaporization technique was developed by Chriswell (1977) to remove gas-chromatographable organic compounds from reagent water. This technique consists of atomizing water with a nebulizer into a high-velocity gas stream and impacting the mist produced onto a glass surface, where it condenses and drains into the reagent water reservoir. Gas-chromatographable impurities are removed from the system in the gas stream. This technique has particular utility for testing effects of natural-water matrices on recoveries of model compounds added to "real" waters. Gas-chromatographable compounds in "real" water, which may interfere with analysis of the test compound, are removed without removing nonvolatile constituents that are responsible for matrix effects.

If the equipment is available, "organic-free" reagent water can also be produced by controlling freezing of water (Shapiro, 1967). Freezing was found to be very effective in removing volatile organic substances that distillation did not remove, and reagent water was then obtained simply by melting the ice.

D. Selection and Preparation of Organic Solvents and Chemical Reagents

In most cases, organic solvents suitable for trace organic analysis can be commercially obtained simply by specifying the grade of the solvent. However, contaminants found in even the most highly purified organic solvents may interfere with certain analyses. Solvent extraction and concentration techniques of polynuclear aromatic hydrocarbons (PAH) from water should include correction for PAH blanks from the solvent because even the purest solvents may contain traces of PAH (Fedonin *et al.*, 1970). This finding emphasizes the importance of conducting solvent and reagent blank tests to determine whether interfering contaminants are present. If additional solvent purification is necessary in the laboratory, a wide variety of distillation, fractional crystallization, adsorptive filtration, and drying techniques are available (Gordon and Ford, 1970). The analyst should be aware of the toxicity and flammability of the solvent used for analysis, the possibility of formation of explosive peroxides in ether solvents, and stabilizers and denaturants that many commercial solvents contain. Purified organic solvents should be stored in a solvent storage cabinet or safety refrigerator in sealed glass bottles and under an inert atmosphere.

Chemical reagents are almost as common as organic solvents for introducing organic contaminants to a water sample. Mercuric chloride, sodium hydroxide, hydrochloric acid, and sulfuric acid are commonly used as water-preservation agents; sodium hydroxide, sodium carbonate, hydrochloric acid, and sulfuric acid are commonly used to adjust sample pH during analysis; sodium chloride and sodium sulfate are used to salt out solutes for solvent extraction; anhydrous sodium sulfate is used to dry solvent extracts. All these reagents should be tested by running blanks; a contaminated reagent or reagent solution may be purified by solvent extraction (Giam and Wong, 1972). However, inorganic base reagents have a very high affinity for organic acids, and inorganic acids have a similar affinity for organic bases. As acids and bases form salts, solvent extraction is relatively ineffective for removing these contaminants. A hydroxide-saturated, strong-base, anion-exchange resin can be used to sorb organic acids from sodium hydroxide solutions, and a hydrogen-saturated, strong-acid, cation-exchange resin can be used to sorb organic bases from mineral acids (Khym, 1974).

E. Preparation of Sorbents and Membranes

Activated carbon, synthetic resins, and semipermeable membranes used to concentrate and separate organic materials from water frequently contain preservation agents, wetting agents, unpolymerized material, and other potential contaminants that must be removed before use. Because activated carbon has been heated to a high temperature for its preparation, it usually contains very little volatile material that interferes with gas-chromatographic analysis. It has not been common practice to preclean activated carbon before use, because of concern that precleaning impairs its adsorptive power. However, Van Rossum and Webb (1978), noting that activated carbon gave unacceptable blanks for certain analysis, found that precleaning with chloroform reduced blanks to acceptable levels. They recommend in-column washing with acetone, chloroform, and methanol.

Synthetic nonionic resins, such as the XAD series, frequently are supplied with salt preservatives in the resin beads, which must be removed by in-column washing with water before they can be used (Simpson, 1972). Junk *et al.* (1974b) also recommended that XAD resins be Soxhlet extracted with methanol, acetonitrile, and diethyl ether. On-column washing of XAD resins with 0.1 N NaOH and 0.1 N HCl was found to be necessary if pH adjustment is used during adsorption and desorption steps (Thurman *et al.*, 1978a).

Ion-exchange resins usually must be pretreated before use. Directions for ionic saturation and solvent precleaning are usually available from the manufacturer. Most organic ion-exchange resins are slightly unstable and give rise to organic bleeds if they are not used immediately after purification. Basic amines bleed from anion exchangers and organic acids leach from cation exchangers. A good rule to follow in precleaning resins is to use the same reagents and solvents for precleaning as are to be used in the analysis.

Semipermeable membranes are used mainly for organic solute separations in aqueous systems; therefore, washing with organic solvents is not necessary. Many organic solvents can damage certain membranes by solubilization (Klein *et al.*, 1975). As with resins, membranes are cleaned by washing with both reagent water and with water solutions of reagents to be used in the analytical procedure.

III. TECHNIQUES FOR VOLATILES

This section discusses many recently developed techniques for isolation and concentration of volatile organic substances in water. Volatile-organics analysis of water is presently a rapidly developing field, because various techniques have enabled rapid analysis of small-volume samples at ultratrace levels. Volatiles are defined as sparingly soluble substances whose mole fraction in the gaseous phase above water is one to two orders higher than in the liquid phase (Voznakova *et al.*, 1978).

A. Sampling and Sample Preservation

Because of their high vapor pressure and low water solubility, volatile constituents present difficult sampling, sample preservation, and sample transfer problems. In deep groundwater under great hydrostatic pressure, volatile dissolved hydrocarbons are dissolved in water at concentrations near 1000 mg/L. A survey of the amount and type of dissolved hydrocarbons in deep subsurface brines of the Gulf Coast Region of the United States used a cylindrical stainless steel sampling device that could be lowered down the well to obtain a sample at specified depth and pressure and be retrieved while maintaining the sample under hydrostatic pressure (Buckley *et al.*, 1958). Hydrocarbon analysis was performed by coupling the sampling device to the inlet of a gas chromatograph and slowly releasing the pressure so the gases that effervesced from the water were introduced into the gas chromatograph. Gas that effervesces from groundwater at the wellhead may also be collected by connecting a glass gas-collector tube to the well outlet, displacing the air with groundwater, and trapping the effervescing gas in the tube as groundwater flows through the tube (Leenheer et al., 1976).

In surface waters, volatile constituents can be sampled with less difficulty at a given point source because pressure changes encountered during sampling are less than in groundwater. Sample-collection vessels consist of glass serum vials that are filled with water to overflowing and capped with Teflon-faced septa that are held in place with crimped-on aluminum seals (Kopfler et al., 1976). Sample transfer from sealed serum bottle to laboratory apparatus for volatile analysis can be accomplished by carefully pouring the sample into the purging chamber without excessive agitation of the sample, which might cause loss of volatiles.

Since the discovery of organic haloform formation in waters treated with chlorine (Rook, 1974), it has been shown that haloform formation continues in sample bottles unless the sample is treated (Kopfler et al., 1976; Kissinger and Fritz, 1976). Water samples for haloform analysis can be preserved by adding ascorbic acid (Kissinger and Fritz, 1976), potassium ferricyanide (Kopfler et al., 1976) or sodium thiosulfate to samples to destroy free-chlorine reactant. Chilling the sample on ice and analyzing within a few days after collection was also found to limit increases in haloform content (Kopfler et al., 1976).

B. Headspace and Sparging Techniques

The first report of a "sparge-and-trap" technique was given by Swinnerton and Linnenbom (1967), who stripped dissolved hydrocarbons from seawater by purging with helium and trapping the hydrocarbons in cold traps filled with activated alumina and activated carbon. Kaiser (1971) advocated use of sparging and trapping techniques as analytical procedures with the lowest possible systematic error for vapors in liquids, gases, and solids. Water samples were purged with nitrogen, which was then passed through a gas-chromatographic column cooled with liquid nitrogen; volatile constituents were trapped at the head of the column.

Shortly after these early developments in sparge-and-trap techniques, analysis of volatile organic components in headspace gas above a water sample was developed by McAuliffe (1971) and Rook (1972), who sampled static headgas after equilibration with water at a certain temperature. McAuliffe analyzed the gas phase directly, but Rook employed an activated-silica trap. Mieure and Dietrich (1973) continuously swept the headspace over the water sample with air, which was then passed into a porous polymer trap. Zlatkis *et al.* (1973a) developed a similar technique that heated the water sample almost to boiling to volatilize the organic solutes into the headspace gas. They applied this technique to profiling volatile metabolites in urine (Zlatkis *et al.*, 1973b), and Dowty and Laseter (1975) used this method of headspace analysis to determine volatiles in drinking water.

In more recent developments in headspace techniques, Stepanenko designed all-glass concentration cells in which volatiles were exchanged into a carrier-gas phase when the sample formed a thin liquid film on the heated walls of the concentration cell (Stepanenko and Golovina, 1975; Stepanenko, 1977). The carrier gas was sufficiently concentrated with volatiles that no trap was required for additional concentration prior to chromatographic analysis. Sample volumes ranged from 10 mL to 1 L; concentration times ranged from a few minutes to 1 h; and the lowest concentration determinable with a gas chromatograph equipped with a flame-ionization detector was 5×10^{-10} M. Cowen *et al.* (1975) developed an evacuated gas sampling valve for sampling static headspace gases over a heated water sample. The valve is coupled directly to the inlet of a gas chromatograph, and organic volatiles are enriched in the reduced pressure of the sampling-valve tubing without the aid of any sorbent or cold traps. The simplest device yet designed for isolation of organic volatiles from water and introduction into the analyzer is a hollow fiber probe developed by Westover *et al.* (1974). Sealed hollow silicone-rubber probes containing the water sample are directly inserted into the inlet of a mass spectrometer and exposed to high vacuum. Volatile solutes in water permeate the membrane and are determined by the mass spectrometer. The detection limit for chloroform in water was 10 ppb and 1 ppm for methanol in water.

In spite of developments in direct headspace analysis of volatiles, Novak *et al.* (1973) claimed that sparging organic volatiles from water was several times more efficient than any other extraction method. They enhanced sparging efficiency by adding a salting-out agent to samples. Recent developments in volatiles analysis have centered on sparging, coupled with various types of traps for concentration prior to analysis. Grob (1973) designed a closed-loop gas-sparging system that avoided high volumes of gas needed to remove compounds of low vapor pressure from water. A charcoal trap was used in the system. Advantages of this closed system are limitations on the amount of water vapor removed by sparging and a reduction in contaminants from the carrier gas. Organic compounds up to C_{24} were recovered by this system. Using this system coupled to a capillary-column gas chromatograph with flame-ionization detection, certain organic compounds at the part per trillion concentration level were analyzed in several waters (Grob and Grob, 1974). The most recent refinements in equipment developed for the closed-loop stripping procedure are given in a report by Grob and Zurcher (1976).

The sparge-and-trap procedure that has been most widely applied was developed by Bellar and Lichtenberg (1974) and has since been adapted for use in a commercial instrument (Grote, 1975). The original procedure used only 5.0 mL of water, which was purged for 11 min with nitrogen without heating. The trap was filled with a porous polymer (Tenax-GC) that trapped organic compounds with a retention index greater than 500. The trap, after interfacing with an inlet of the gas chromatograph, was desorbed by heating to 130°C and backflushing with nitrogen. This method was limited to organic compounds that are less than 2% soluble in water and boil below 200°C. For the 5-mL sample, the useful concentration range was 1–2500 μg/L for most gas chromatographs. Because of the sensitivity, reproducibility, and relative simplicity of the method, it was selected for use in a National Organics Reconnaissance Survey for halogenated organics in drinking water in 80 cities in the United States (Symons *et al.*, 1975).

The Bellar and Lichtenberg sparge-and-trap technique was modified and scaled up for a survey of all possible identifiable volatile organic constituents in drinking water of five United States cities. Modifications are described in detail by Kopfler *et al.* (1976), and results of the survey are given by Coleman *et al.* (1976). The modified apparatus shown in Fig. 3 provided a completely sealed system of sample transfer to the sparging apparatus, which was constructed for both 140- and 500-mL sample chambers. Volatiles in water samples were fractionated by conducting three successive 30-min purges of the sample. The first purge at 6°C removed highly volatile compounds of low solubility. The haloforms are included in this fraction. A stoichiometric amount of potassium ferrocyanide was added before the second purge to remove free chlorine that reacts with organic solutes at higher temperature. A second purge conducted at 95°C removed low-solubility compounds of lesser volatility. A third purge at 95°C detected the presence of volatile, water-soluble compounds. This modified technique detected approximately four times the compounds in drinking water as the original technique, and the limit of detection was reduced to 0.1 μg/L.

Various types of adsorbents for volatile traps have been investigated. Grob (1973) favored a charcoal trap because of its low water affinity and thermal stability. The merits of graphatized carbon black as a trapping material were described by Raymond and Guiochon (1975). The material was stable at 400°C where compounds such as polynuclear aromatic hy-

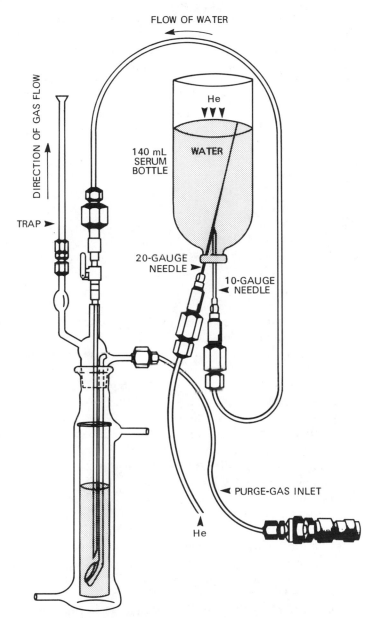

Fig. 3. Sparge-and-trap apparatus for volatile-organic analysis of water. (From Kopfler *et al.*, 1976; published with permission of Ann Arbor Science Publishers, Inc.)

drocarbons can be desorbed. Silica gel was found to be generally unsuitable because some aromatics above C_9 were irreversibly adsorbed and certain olefins were stereochemically rearranged (Bellar and Lichtenberg, 1974). Many styrene–divinylbenzene polymers outgas contaminants during thermal desorption. Chromosorb 103 and Tenax-GC are acceptable porous polymers for volatile traps because of thermal stability (Bellar and Lichtenberg, 1974); these polymers have subsequently received the most use as trapping materials. The upper limit for desorption of paraffin hydrocarbons from Tenax-GC at 300°C was found to be C_{25} by Ligon and Johnson (1976), who also designed a device for rapid thermal desorption of volatiles from the trap into gas chromatographs.

A trap consisting of two-thirds Tenax-GC and one-third silica gel is being used in the analysis of volatile priority pollutants in water (U.S. Environmental Protection Agency, 1977). However, Keith *et al.* (1979) found that silica gel degenerated after repeated usage and substituted chromosorb 102, which gave acceptable performance in combination with Tenax-GC.

Cryogenic trapping techniques have also been employed to fractionate and recover organic volatiles in a gas stream. Khare and Dondero (1977) used five low-temperature traps in series at successively lower temperatures to fractionate organic volatiles in a vacuum distillation of water. A temperature-gradient tube based on the reversion gas chromatograph was used by Bergert *et al.* (1974) to concentrate organic volatiles after they were sparged from water. The main disadvantage of cryogenic traps is that organic volatiles with vapor pressures similar to those of water condense with water; whereas these same compounds are often isolated from water when sorbents are used as traps because of differences in affinity of water and volatile compounds for the sorbent.

The range of application of the sparge-and-trap technique has been rapidly expanding. Grob (1973) stated that he did not see any serious boiling-point limitation to the sparge-and-trap technique, and the method, without fundamental limitations, could be extended beyond the C_{24} substances he detected. He may have been slightly optimistic; however, a method based on the sparge-and-trap technique for semivolatile aromatic hydrocarbons in water has been published (Voznakova *et al.*, 1978). The range of compounds isolated from water by the sparge-and-trap technique has been extended beyond highly volatile substances detected by original methods to many semivolatile materials as well. This wide range of application, good precision of the technique, and elimination of an organic solvent step in the procedure insures continuing development and wide application of the sparge-and-trap technique. Its chief disadvantage is the specialized equipment required, especially the trap, which is not available in most laboratories.

C. Solvent Extraction Techniques

Solvent extraction techniques for the analysis of volatiles in water have found limited application, because the volatile analyte is usually lost during the solvent concentration step that follows extraction. One approach is to extract with a low-boiling solvent in a pressurized extraction apparatus. A technique called pressure distillation (Gjavotchanoff, 1972) was developed that used *n*-butane as the extractant. After extraction, *n*-butane was evaporated at temperatures below 0°C to reduce loss of extracted volatiles. However, Grob (1973) and Novak *et al.* (1973) found that the limit of detection of this technique for alkanes dissolved in water was ~1 ppm.

A second approach to solvent extraction of volatiles is to eliminate the solvent evaporation step by extracting with a small volume of solvent so extracted volatiles are sufficiently concentrated for chromatographic analysis. Grob *et al.* (1975) developed a rapid extraction technique in which 1 L of water was extracted with only 200 μL of pentane. They concluded that sensitivity and overall recovery of this extraction procedure for volatiles was much less than the closed-loop stripping procedure they developed. The extraction procedure was superior for heavier hydrocarbons, above eicosane, dimethylnaphthalene, etc. The solvent peak of the gas chromatogram obscured many highly volatile constituents of the pentane extract when flame-ionization detection was used.

When a specific detector is used, such as the electron-capture detector which does not respond to the solvent peak, solvent extraction can be used to determine volatile organic haloforms in water. Mieure (1977) and Richard and Junk (1977) used gas chromatographs equipped with electron-capture detectors to measure volatile organohalides in solvent extracts, and Henderson *et al.* (1976) used a halogen-specific electrolytic conductivity detector for halomethane determination in a pentane extract. Extremely simple techniques for solvent extraction of halomethanes are possible by injecting a small volume of solvent into the water through the septum seal of the sample-collection bottle (Henderson *et al.*, 1976). Detection limits of 0.1 μg/L obtained for solvent extraction with halogen-specific detectors are equivalent to sparge-and-trap methods, and the number of analyses per unit time can be greatly increased.

D. Techniques Based on Sorption from Water

Isolation of volatiles by sorption from aqueous solution has not met with much success, because the volatiles were invariably lost when the sorbent was dried to remove water, or when the solvent eluate from a sorbent column was concentrated by evaporation. Renberg (1978) employed adsorption of volatile halogenated hydrocarbons in water on XAD-4 resin for preconcentration, elution with ethanol, and determination of volatile haloforms by gas chromatography with electron-capture detection. The sensitivity and specificity of the electron-capture detector enables direct determination of these organic volatiles in the ethanol eluate without additional concentrations of the eluate.

The development of techniques for thermal desorption of volatile traps with the sparge-and-trap techniques just discussed led Mieure and Dietrich (1973) to pack a column of Chromosorb 102 for sorption of organic volatiles and semivolatiles from water. After organic standards at part per million levels in water were passed through the column, the column was coupled to the inlet of a gas chromatograph and the standards were thermally desorbed. The main limitation of this procedure was large amounts of water that were introduced from the moist sorption column into the gas chromatograph.

The water problem was circumvented by Chang (1976), who first used XAD-2 resin for sorption from water, then thermally desorbed volatiles and water from XAD-2 to a Tenax-GC trap that retained only volatiles and allowed water to pass. The Tenax-GC trap was thermally desorbed in the usual manner at the inlet of a gas chromatograph. Good recoveries at the low part per billion level for solutes as volatile as chloroform and as nonvolatile as 2-methylnaphthalene were obtained. Ryan and Fritz (1978) improved Chang's technique by using XAD-4 resin instead of XAD-2 and by designing a specialized thermal desorption apparatus for transfer of volatiles from XAD-4 to Tenax-GC and then to the gas chromatograph.

This resin extraction–thermal desorption procedure has great potential because it provides adequate sensitivity over a wide volatility range without the use of organic solvents. Sampling and resin extraction can be accomplished simultaneously in the field by passing a 15-mL water sample in a glass syringe through a small column packed with XAD-2 or XAD-4 resin. It may be possible to use this technique for semivolatiles that need to be derivatized prior to gas chromatographic analysis, if a method for derivatizing organic solutes sorbed on substrates such as XAD-2 or XAD-4 resin can be devised.

IV. TECHNIQUES FOR SEMIVOLATILES

Recently developed techniques for concentration, partitioning, and isolation of semivolatile organic compounds in water are discussed in this section. The majority of analytical techniques for analysis of specific organic compounds in water have involved semivolatile compounds, because most phenols, organic pesticides, and most recently, polychlorinated biphenyls (PCB) fall within this compound class. Most of the 10% of identified organic material in drinking water in a National Academy of Sciences report (1977) falls in the semivolatile class of compounds. The widespread use of the gas chromatograph for semivolatile analysis has resulted in tailormade sample preparation techniques for gas chromatography. Semivolatiles in this chapter are defined to be soluble organic compounds of moderate-to-high volatility and sparingly soluble organic compounds of low-to-moderate volatility. Although many of these compounds in the gas phase can be isolated by headspace and sparging techniques for volatiles, isolation and separation techniques, such as solvent extraction or resin sorption that are based on solute forms of semivolatile species, are generally more efficient. As a class, semivolatiles are amenable to analysis by gas chromatography without derivatization, although derivatization may be used to either decrease or increase volatility for gas-chromatographic analysis. As discussed previously, there is considerable overlap of semivolatiles with both the volatile class at one extreme, and the low molecular weight nonvolatiles at the other extreme. However, this method-dependent definition for semivolatiles is useful when used in the context of concentration, partitioning, and isolation techniques.

A. Sampling and Sample Preservation

Semivolatiles in water are generally sampled in glass containers and preserved by refrigeration. Because of their lower volatility compared to volatiles, gas-tight sample containers and sample transfer devices are not required, although these devices should be sealed to avoid loss of water or introduction of contaminants. For general water sampling, the reader is directed to the discussion on selection of sampling equipment in Section II,A.

Specialized equipment for sampling of surface-active semivolatiles in surface films on water has been developed. A 16-mesh stainless steel screen dipped through the surface film was used by Duce et al. (1972) to sample for fatty acids and hydrocarbons found in the surface film of seawater. Miget et al. (1974) developed a Teflon disk that retained fatty acids and hydrocarbons when contacted with a surface film. A comparison of these two sampling devices by Ledet and Laseter (1974) found the Teflon disk more efficient in recovering saturated free fatty acids and the steel screen more efficient in recovering n-alkane hydrocarbons.

The most common sources of loss for semivolatiles during sampling and storage are by sorption from solution and microbial transformations.

Filtration through an organic-membrane filter is to be especially avoided for organic solutes with high affinities for organic substrates. For example, an analytical method for isolation of PCBs and DDT dissolved in water was based on quantitative retention of these compounds on the common cellulose triacetate-membrane filter after filtration of a water sample (Kurtz, 1977). Sediment should be separated by centrifugation or use of glass-fiber or silver-membrane filters. Sulfur-containing organic solutes may react with the silver-membrane filter. If the semivolatile is already sorbed on suspended sediment, use of all glass and metal samplers is not always necessary. Parr *et al.* (1974) claimed that plastic tubing components of the PS-69 automatic pumping sampler for suspended sediment did not adversely affect analysis of several organic pesticides sorbed on sediment.

Changes of semivolatiles during sample storage caused by microbiological processes can usually be minimized by chilling the sample on ice and analyzing as quickly as possible. Freezing is sometimes used, but freeze-concentration effects and formation of insoluble precipitates may interfere with recovery of the analyte. Adjusting the pH of the sample to low values with acid or high values with alkali was effective in killing *Pseudomonas* bacteria responsible for phenol degradation (Afgan *et al.*, 1974); however, high pH increases the rate of oxidation for many phenols, and low pH increases their tendency for sorption and precipitation. Various salt preservatives such as mercuric chloride, copper sulfate, and sodium azide are often used; however, aside from possible introduction of reagent contaminants and their hazards as poisons, mercuric and cupric ions often form undesirable organic precipitates, and azide may form explosive compounds in drains when samples are disposed. As always, the best sampling and preservation technique is to isolate the organic analyte at the sampling site whenever possible, or as soon after sampling as practical.

B. Solvent Extraction

Solvent extraction is the most common technique used to isolate sparingly soluble semivolatiles of low-to-moderate volatility from water. It is not the purpose of this section to review the basic concepts (Andelman and Caruso, 1971; Karger *et al.*, 1973) of solvent extraction or to present routine applications (Andelman and Caruso, 1971; Irving, 1974) that are more adequately presented in the referenced reports. Recent concepts, new techniques, and new applications of old techniques are presented. Many old techniques of solvent extraction developed for processing large volumes of water (Hoak, 1962; Kahn and Wayman, 1964) have been superseded by techniques that require water samples of small to moderate volume and use small quantities of solvent (Grob *et al.*, 1975; Garrison *et al.*, 1976).

Solvent extracts are frequently partitioned into acid, base, and neutral fractions according to the procedure first presented by Shriner and Fuson (1948) and later given in a standardized procedure by ASTM (1978b). The acid, base, and neutral partitioning can also be carried out during solvent extraction if sample pH is adjusted accordingly. In their analytical procedure guidelines for screening of industrial effluents for priority pollutants, the U.S. Environmental Protection Agency (1977) recommends the water sample be first extracted at pH 11 with methylene chloride for extraction of basic and neutral compounds, and then extracted at pH 2 for isolation of organic acids and phenols. An extraction scheme presented by Garrison *et al.* (1976) showed acids and neutrals extracted first on a pH 2 water sample, followed by extraction of the bases at pH 11. It is often more advantageous to extract the acid fraction first, because surface-active organic acids common in many wastewaters often form stable emulsions when the sample is first extracted at high pH. Acid, base, and neutral solvent partitioning also does not give clean compound class fractionations, particularly when the water contains high concentrations of extractable solutes. Jackson *et al.* (1975) noted significant carryover of acids into base fractions and bases in acid fractions when oil shale retort water was partitioned by solvent extraction.

A useful parameter for determining extraction efficiency of a solvent system for a particular solute is the extraction *p*-value (fraction of solute partitioning into the nonpolar phase of an equal-volume, two-phase solvent system). A simple device based on the Mohr pipette was devised by Bowman and Beroza (1966) to determine extraction *p*-values. In a series of three reports, Suffet used the *p*-value approach to determine the most efficient solvents and pH conditions for serial batch extractions of organophosphate pesticides from water (Suffet and Faust, 1972); determination of optimum water: solvent ratios and number of extractions necessary for quantitative recoveries of certain pesticides and herbicides from water (Suffet, 1973a); and determination of optimal conditions for serial extraction of phenoxy acid herbicides from natural water (Suffet, 1973b). Organic solutes whose *p*-value equals or exceeds 0.9 can be extracted from 1 L of water with 95% recovery in five or fewer successive serial extracts of 50 or more mL of nonpolar phase with a total volume of 500 mL of solvent (Suffet, 1973a).

Although batch extractions only require a separatory funnel for equipment, they are time consuming, can only handle a limited volume of water, and cannot be easily automated. Goldberg *et al.* (1973) designed a flow-through system whereby 8 L/h of water were extracted by a series of lighter-than-water and heavier-than-water continuous extractors of conventional design. An unlimited number of different extraction solvents could be used, an unlimited volume of water could be extracted, and concentration factors of 10^5 were claimed. Continuous liquid–liquid extraction was coupled with continuous evaporative concentration in a system designed by Wu and Suffet (1977). It featured a Teflon helix mixing coil, a flow rate of 900 mL/min, and a water-to-solvent ratio of 10:1. Greater than 80% efficiency compared to batch extraction was demonstrated. Autoanalyzer systems with solvent extraction as part of the procedure have also been built (Carter and Nickless, 1970). Afgan *et al.* (1974) used a manifold employing chloroform extraction of phenol for automated analysis of phenol in natural waters.

Continuous solvent extraction can even be conducted *in situ* at the sampling point at desired depth. This apparatus, invented by Ahnoff and Josefsson (1974), consists of two mixing chambers in series designed such that solvent extraction occurs in the upper part of the chamber, and phase separation occurs in the lower portion. The device is operated at a flow rate of 3 L/h with 100–300 mL of solvent in each chamber. It can be powered externally by a 220-volt source or internally by a 12-volt DC battery. Extraction efficiencies for PCBs and chlorinated hydrocarbon pesticides were comparable with serial batch extractions of these compounds. Advantages include extraction of a time-integrated sample in the field, extraction of large volumes without collection of large volumes, and no losses or contaminants introduced during sample collection and storage. Unfortunately, only very insoluble solvents can be used because slightly soluble solvents such as diethyl ether are lost to the water passing through the extractor.

One of the biggest operational advantages of solvent extraction techniques versus carbon- or resin-filter techniques is that moderate-to-high suspended-sediment concentrations can be managed with solvent extraction, whereas sorbent filters tend to become plugged. However, in whole-water samples, the analyte frequently is not a solute, and extraction may have to be tailored to an analyte sorbed on suspended matter. Easty and Wabers (1977) found that a paper-mill effluent containing cellulose fibers gave low recoveries for PCBs in a batch extraction with hexane–methylene chloride. The fibers had to be filtered from the sample, refluxed with alcoholic potassium hydroxide to remove sorbed PCBs, and extracted with hexane to obtain complete recovery.

New extraction solvents and solvents containing various extracting agents are continually being developed. Previously mentioned in the section on volatiles was the pressure extraction system of Gjavotchanoff (1972), which used *n*-butane as the solvent. Austern *et al.* (1975) used highly volatile Freon at atmospheric pressure for extraction of wastewater. Because of Freon's volatility, solute losses during solvent concentration were minimal.

Soluble fatty acids and hydroxy acids are not normally efficiently extracted because of hydrogen bond formation with water solvent; however, Bethge and Lindstrom (1974) demonstrated that when these acids were titrated with tetra-*n*-butylammonium hydroxide, lipophilic ion pairs were formed that were extractable into aprotic organic solvents. Korenman and Bortnikova (1977) found that camphor dissolved in butyl acetate increased the distribution coefficient for various phenols in water an average of 10 times.

Membrane separations are sometimes used in conjunction with solvent extraction for solute separations. A polyethylene membrane placed between water and an organic solvent in a specially constructed dialysis cell allowed efficient extraction of PCB mixtures from the water. Mieure *et al.* (1976) claimed the following advantages for this solvent dialysis system:

1. Dialysis using a water miscible solvent
2. Dialysis using membrane selectivity to prevent removal of otherwise extractable components
3. Analysis of solutions with high emulsion potential
4. Automated sampling and analysis

Liquid-membrane separation is a good example of a separation technique in search of analytical applications. Cahn and Li (1974) developed a liquid-membrane technique for separating phenol from wastewater, and Fig. 4 illustrates the principles of the liquid-membrane separation. Cussler and Evans (1974) state that liquid-membrane separations are similar to separation techniques involving a two-stage solvent extraction. Advantages of the liquid-membrane technique are speed and selectivity, which can be designed into liquid membranes. High concentration factors and high capacities (compared to ion-exchange resins) make this technique attractive for solute concentration. Disadvantages include the high level of technical expertise required, and handling problems associated with viscous organic membranes.

C. Sorption Techniques

Techniques based on sorption of semivolatile solutes from water can be used to concentrate solutes on the sorbent, desalt organic-solute fractions, partition the organic solutes into compound classes by selective sorption and desorption, and isolate solutes from water by desorbing with an organic solvent. Additional concentration can be performed by evapo-

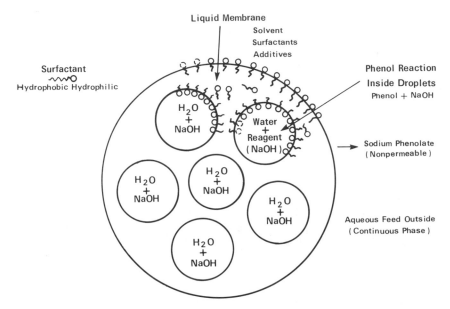

Fig. 4. Liquid-membrane system designed for phenol removal. (Reprinted from Cahn and Li, 1974, p. 509, by courtesy of Marcel Dekker, Inc.)

rating the organic solvent eluate. Sorption is a comprehensive term including both adsorption (solute bound at substrate surface) and absorption (solute solubilized within substrate). This section discusses mainly nonchromatographic techniques (as defined by Khym, 1974), which need relatively few theoretical plates to effect a concentration or separation. Nonchromatographic techniques can use simple equipment and small sorbent columns because solutes are sorbed with capacity factors (k') of 100 or greater.

A wide variety of sorbents are available that can bind solutes by physical and chemical sorption mechanisms in various systems. For a discussion of physical and chemical sorption mechanisms in aqueous systems, the reader is referred to Kipling (1965). Matrix composition, functional group composition, and possible solute-bonding mechanisms for various sorbents commonly used for organic solutes in water are listed in Table II. In general, sorbents with most bonding mechanisms are the most efficient sorbents; however, they are often unsuited for analytical applications because sorbed solutes are difficult to desorb. For example, a study by Suffet *et al.* (1978) found that granular activated carbon (four sorption mechanisms) was a much more efficient sorbent for removing volatile and semivolatile trace organic solutes from drinking water than was the

TABLE II

Matrix Composition, Functional Group Composition, and Possible Solute Bonding
Mechanisms for Various Sorbents

Sorbent	Matrix composition	Functional groups	Possible bonding mechanisms[a]
Activated carbon	Amorphous carbon and graphite microcrystallites	Phenolic hydroxyl, quinone, carboxyl, lactone	a,b,d,c
Graphatized carbon	Graphite	None	a,b
Styrene–divinylbenzene copolymer	Aromatic and aliphatic hydrocarbon	None	a,b
Phenol–formaldehyde copolymer	Aromatic and aliphatic hydrocarbon	Phenolic hydroxyl	a,c,b
Methylmethacrylate ester polymer	Aliphatic hydrocarbon	Ester	a,c
Polyamide (Nylon)	Aliphatic hydrocarbon	Amide	c,a
Polyurethane foam	Aromatic and aliphatic hydrocarbon	Urethane	a,c,b
C_{18}-Bonded Silica	Silica and aliphatic hydrocarbon	None	a
Dextran	Carbohydrate	Hydroxyl and ether	c
Cellulose triacetate	Carbohydrate	Ether and acetate ester	c,a
Strong-acid cation-exchange resin	Styrene–divinylbenzene copolymer	Sulfonic acid	d,a,b
Strong-base anion-exchange resin	Styrene–divinylbenzene copolymer	Quarternary amine	e,a,b
Weak-acid cation-exchange resin	Methylmethacrylate polymer	Carboxyl and ester	d,c,a
Weak-base anion-exchange resin	Phenol–formaldehyde copolymer	Secondary and tertiary amines, phenolic hydroxyl	e,c,a,b
CM cation-exchange cellulose	Carbohydrate	Carboxyl, hydroxyl, ether	d,c
DEAE anion-exchange cellulose	Carbohydrate	Tertiary amine, hydroxyl ether	e,c

[a] a, Hydrophobic effect; b, π-electron interactions; c, hydrogen bonding; d, cation exchange; e, anion exchange. Bonding mechanisms are listed in order of estimated importance.

styrene–divinylbenzene XAD-2 resin (two sorption mechanisms). However, studies by Chriswell *et al.* (1977) and Van Rossum and Webb (1978) indicated that XAD-2 gave better recoveries for similar solutes at trace levels in analytical procedures requiring solute desorption than did granular activated carbon because of irreversible sorption on carbon.

To successfully apply sorption techniques, knowledge of factors affecting sorption kinetics is as essential as knowledge of sorption mechanisms. In a study of the kinetics of solute sorption on carbon, Weber and Morris (1963) found that diffusion of the solute within pores of carbon was the rate-limiting step for solute sorption. Temperature, affinity of sorbent surface for water, size of sorbent pore, and the size of the solute all affect diffusion rates and sorption kinetics. With these considerations in mind, a series of copolymer resins with macroporous or macroreticular structure were developed (Kun and Kunin, 1964). Macroporous resins usually are large 20–50 mesh beads, each of which is composed of a large number of microspheres. Controlled pore size, rapid rates of solute diffusion within the macrobead, high internal surface areas, and low-pressure, high flow-rate column packings are major advantages of macroporous resins for nonchromatographic applications.

The XAD series of sorbent resins have macroporous structures; physical properties of commonly used members of the series are shown in Table III. Because of its high surface area, XAD-4 frequently has the highest equilibrium capacity factors (Pietrzyk and Chu, 1977) for small-sized solutes, but it may not be suitable if solute size is too large for its small pore diameter. Gustafson *et al.* (1968) recommended that average pore diameter should be six times the sorbate diameter to maintain rapid sorption kinetics on XAD resins. Very large pores that allow high flow rates and rapid sorption kinetics were cited as reasons for using polyure-

TABLE III

Physical Properties of XAD Copolymers[a]

Sorbent	Copolymer	Average surface area (m^2/g)	Average pore diameter (Å)
XAD-1	Styrene divinylbenzene	100	200
XAD-2	Styrene divinylbenzene	330	90
XAD-4	Styrene divinylbenzene	750	50
XAD-7	Methylmethacrylate	450	80
XAD-8	Methylmethacrylate	140	250

[a] Kunin (1977). Published with permission of *Polymer Engineering and Science.*

thane foam as a sorbent for polynuclear aromatic hydrocarbons from large quantities of water (Saxena *et al.*, 1977).

Sorption kinetics is directly related to temperature and sorbent pore size, and is inversely related to solute size. Sorbent capacities at equilibrium are inversely related to temperature and directly related to surface area; pore size can be a factor if its dimensions exclude the solute. Column operations represent a combination of all these factors, and sorbent columns can be optimized by correct selection of the right surface chemical and physical properties, use of flow rates that allow equilibrium sorption, and reasonable control of temperature.

Activated carbon has been the most widely used sorbent for semivolatile solutes in water. The activated-carbon filter developed by Braus *et al.* (1951) was designed to give an organic fraction, the carbon–chloroform extract (CCE), which served as a gross guideline to determine efficiency of drinking-water treatment plants. The CCE method has been scaled up for processing large volumes of water (Middleton *et al.*, 1962), and modified by adding an alcohol extract (CAE) and developing different filters for high and low flow rates (Buelow *et al.*, 1973a,b). A standard CCE method has been adopted (American Public Health Association *et al.*, 1976a), and the CCE was used as a preconcentration and isolation technique for the National Organics Reconnaissance Survey of drinking water in 13 U.S. cities (Keith *et al.*, 1976). The CCE has been responsible for identification of many organic contaminants at trace levels in drinking water, and it is one of the simplest and most convenient preconcentration techniques ever developed. Activated carbon has a complex surface chemistry (Snoeyink and Weber, 1967), which sorbs organic solutes very efficiently from water by multiple sorption mechanisms (Table II). Use of the carbon filter for analytical preconcentration is decreasing because certain solutes sorbed on carbon have been shown to be oxidized and bacterially degraded during the carbon drying step of the CCE procedure (Hoak, 1962); certain sorbed solutes cannot be desorbed by chloroform or other solvents (Van Rossum and Webb, 1978; Chriswell *et al.*, 1977); more sensitive detectors have lessened the need for processing large volumes of water through a carbon filter; and better sorbents have been developed.

The XAD series of resins listed in Table III, particularly XAD-2, has been found to be good sorbents for semivolatile solutes with lipophilic or hydrophobic properties. Burnham *et al.* (1972) demonstrated the utility of XAD-2 and XAD-7 for isolating trace semivolatile solutes in drinking water, and Junk *et al.* (1974b) developed a comprehensive procedure for isolating several classes of organic compounds at the part per billion level

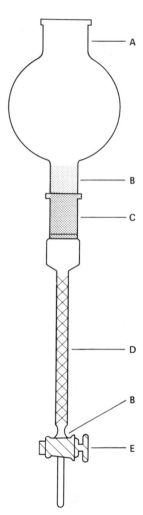

Fig. 5. Sample reservoir and XAD-2 resin column: (A) 5-L reservoir; (B) glass-wool plugs; (C) 24/40 ground-glass joint with Teflon sleeve; (D) 40–60 mesh XAD-2 resin; (E) Teflon stopcock. (From Junk *et al.*, 1976; published with permission of Ann Arbor Science Publishers, Inc.)

from water using XAD-2 resin. The sample reservoir and sorption column used for the XAD-2 resin sorption method is shown in Fig. 5.

The following factors have been found important in sorption of semi-volatile organic solutes on low-polarity XAD resins:

1. Polarity of solute. For monofunctional compounds, a ratio of five or more carbon atoms to one polar, functional group was needed for solute capacity factors k' of 100 or more for nonchromatographic separations (Thurman *et al.*, 1978b). Solute capacity factors are dependent on the

"hydrophobic effect" physical adsorption mechanism described by Tanford (1973).

2. Polarity of the solvent. Reducing the polarity of water by addition of a lower polarity solvent such as alcohol reduced the capacity factor for a sorbed solute (Grieser and Pietrzyk, 1973). Reduction of solvent polarity is the basis for solute desorption.

3. Ionization of solute. To optimize capacity factors for organic acids and bases, solute sorption should be conducted at a pH two units below pK_a for acids and two pH units above pK_b for organic bases (Thurman et al., 1978b; Pietrzyk et al., 1978).

4. Ionic strength. Solute capacity factors increase with ionic strength, but effects are significant only above 0.1 N salt concentration (Grieser and Pietrzyk, 1973).

5. Flow rates. Flow rates greater than two bed volumes per minute for XAD-2 and XAD-7 decreased recoveries (Stepan and Smith, 1977), especially for marginally held solutes with capacity factors near 100. Lower flow rates are needed for small-pore XAD-4, and higher flow rates can be used with large-pore XAD-1 and XAD-8.

6. Solvent miscibility. More efficient desorption was obtained when a "bridging solvent" such as methanol or acetone was used for desorption of water-moist resin before final desorption with an immiscible, nonpolar solvent such as chloroform or carbon tetrachloride (Van Rossum and Webb, 1978).

7. Secondary sorption mechanisms. π-electron interactions can occur between the π-electron-donating aromatic matrix of XAD-1, 2, and 4, and π-electron-accepting solutes such as picric acid (Stepan and Smith, 1977). π-bonding is highly specific and enhanced in many nonpolar solvents, thus causing desorption problems. Higher affinities of phenol compounds on acrylate ester XAD-7 and XAD-8 resins (Burnham et al., 1972; Pietrzyk et al., 1978) are probably the result of hydrogen bonding of solute phenolic hydrogen with ester oxygen of the resin.

The XAD-resin sorption–solvent elution method for concentrating and isolating semivolatile solutes from water has been evaluated versus the CCE method and solvent extraction method. XAD resins gave generally better recoveries than the carbon filter for most compounds tested (Van Rossum and Webb, 1978; Chriswell et al., 1977), and Webb (1975) found that an equal mixture of XAD-4 and XAD-8 was the most efficient sorbent. XAD resins also compared favorably with solvent extraction procedures with the exception of aliphatic hydrocarbons, which were not efficiently sorbed on the resins (Webb, 1975). Long-chain aliphatic chlorides (Glaze et al., 1977) and the pesticide Mirex, a chlorinated alicyclic hydrocarbon (Coburn *et al.*, 1977) were not efficiently sorbed on XAD-2 or XAD-4 resins. Thus, solvent extraction is the method of choice for aliphatic hydrocarbons and chlorinated aliphatic hydrocarbons greater than C_{10}.

XAD-2 was advocated by Burnham *et al.* (1973) as the sorbent of choice for concentration and isolation of a broad spectrum of neutral semivolatile organic solutes from drinking water, river water, and groundwater. Richard and Fritz (1974) found that an XAD-2 resin sorption procedure for isolating chlorinated hydrocarbon pesticides from river water gave better recoveries than a solvent extract procedure. Musty and Nickless (1974) claimed that XAD-4 gave better recoveries of PCBs and chlorinated pesticides from tap water than XAD-2. These differing recovery results may be the result of the varied water matrices in which the standards were dissolved. Van Rossum and Webb (1978) and Coburn *et al.* (1977) noted significant recovery decreases going from distilled water to tap water to natural water.

By acidifying a water sample to pH 2, sorbing on XAD-2 resin, and eluting with acetone, Niederschulte and Ballschmiter (1974) efficiently recovered chlorophenoxyacid herbicides dissolved at the 25-ppb level in water. Rogers and Keith (1974) used XAD-8 to isolate and identify chlorinated guaiacols in kraft bleaching wastes. Osterroht (1974) developed an XAD-2 extraction apparatus that featured six parallel sorption columns through which high volumes of seawater were pumped. The apparatus was operated on a ship, and good recoveries at the low-ppb level were demonstrated for chlorinated pesticides, PCBs, PAH, and natural hydrocarbons. The final example of an application of XAD resins is provided by Glaze *et al.* (1977), who developed a total organic halogen (TOX) water quality parameter based on the isolation of chlorinated organics from water using XAD-2 and XAD-4 resins.

A number of sorbents other than XAD resins have been used for concentration of semivolatile solutes from water by the physical sorption mechanism. Tenax [poly(*p*-2,6-diphenylphenylene oxide)] was used by Leoni *et al.* (1976) to extract various pesticides and PCBs from polluted natural waters at the 1-ppb level. Recoveries equivalent to solvent extraction were obtained for waters of low suspended-sediment concentrations; high suspended sediment decreased recoveries on Tenax. The main advantage of Tenax is its purity; no preliminary cleanup procedure is necessary. A disadvantage is its high price.

Matrix-affinity chromatography on cation-exchange resins was used to preconcentrate PAH (Walton, 1975) and PCB (Hanai and Walton, 1977) from water at the head of a liquid-chromatographic column. Calcium-form styrene–divinylbenzene matrix exchange resins with 4% cross-linkages were used. Two advantages were cited: (1) ionic exclusion of natural organic acids from the resin; these acids frequently must be removed in a subsequent extract cleanup; and (2) rapid sorption–desorption kinetics because of the polarity sulfonic acid groups impart to the resin.

Ahling and Jensen (1970) coated a diatomacious earth support with a mixture of n-decane and Carbowax-4000 monosterate. This adsorbent enabled higher flow rates and higher recoveries in petroleum ether eluate at ultratrace levels (parts per quadrillion) of chlorinated pesticides and PCBs than was obtainable on XAD-2. Kummert et al. (1978) used an octadecylsilica high-pressure liquid chromatographic (HPLC) column to preconcentrate tetrachloroethylene in natural water prior to its desorption and separation by a water–methanol solvent gradient.

Polyurethane foams have been very useful sorbents for PCBs and PAH in water. Their first reported use was by Gesser et al. (1971) who concentrated PCBs on a foam plug in a sorption column and extracted the plug with hexane. By coating polyurethane foams with a GC liquid phase such as DC-200, a number of chlorinated-hydrocarbon pesticides were extracted from water with high efficiency (Uthe et al., 1972). These coated-foam plugs were later used as indwelling monitors for organochlorine pesticides and PCBs in river water (Uthe et al., 1974). Only the surface of the foam plug serving as an indwelling monitor was utilized for extraction of water; however, adequate recoveries of PCBs from the plug monitor were obtained and could be related to time-integrated levels of PCBs in river water.

Saxena et al. (1977) cited the advantage of high flow-rate capabilities of foam adsorbents needed to process large volumes of drinking water that contained part per trillion levels of PAH. They used a flow rate of 250 mL/min to concentrate 100 ppt of benzo(a)pyrene (BAP) in water and found that a water temperature of 62°C significantly enhanced BAP recoveries. They hypothesized that enhanced recovery at high temperature was linked to desorption of BAP from particulate matter in water. Bedford (1974) also noted that turbid waters decreased PCB recoveries on polyurethane foam plugs because of the association of PCBs with suspended sediment.

Another limitation of polyurethane foam sorbents is that they efficiently sorb only semivolatile solutes whose water solubility is extremely low. Gough and Gesser (1975) noted that sparingly soluble dimethyl and diethyl phthalates were not sorbed nearly as efficiently as more insoluble di-n-butyl phthalate. However, higher phthalates, such as di-n-heptyl phthalate, were not sorbed efficiently because of steric factors. Webb (1975) evaluated polyurethane foam for its potential as a sorbent for a broad spectrum of organic compounds and found it unsuitable compared to XAD resins and solvent extraction, because of its selectivity for certain classes of compounds.

In summary, polyurethane foams are good sorbents for PCBs, PAH, and certain organochlorine pesticides in low-turbidity water. Their main advantages are high flow-rate applications, use as indwelling monitors, convenience of use, and low cost.

The final topic in this section concerns sorption procedures in which the solvent elution step is eliminated from or combined with the chromatographic separation stage of analysis. A technique called "trace enrichment" uses a HPLC column or precolumn in which solutes are sorbed from water at the head of the column under conditions in which there is little solute migration as the water sample is pumped through the column. Examples of this technique include the sorption of low-polarity semivolatile solutes on a reverse-phase octadecylsilica HPLC column (Gough and Gesser, 1975; Little and Fallick, 1975; Greed, 1976) or calcium-form cation-exchange resin packed in a HPLC column (Walton, 1975; Hanai and Walton, 1977). Chromatographic separation of the sorbed solutes is effected by lowering their capacity factors through pumping a water–solvent gradient or pH gradient through the column. A technique that completely eliminates solvent elution of the sorbent uses thermal desorption of solutes from the sorbent into a gas chromatograph. This technique, discussed in Section III (Chang, 1976; Ryan and Fritz, 1978), can be extended for many semivolatile, solutes as well.

D. Solvent Evaporation and Extract Cleanup

The solvent extraction and sorption techniques just discussed give a solvent extract that must be dried, cleaned of interfering coextracted contaminants, and evaporated to 1 mL or less before an extract aliquot can be injected into a gas chromatograph. Many steps can be involved in solvent evaporation and extract cleanup. For example, a solvent extraction procedure for chlorinated-hydrocarbon insecticides in suspended sediment and bottom material (Goerlitz and Brown, 1972) includes the following steps:

1. Wash hexane–acetone extract with water to remove acetone
2. Dry extract with anhydrous sodium sulfate
3. Evaporate hexane to 0.5 mL
4. Pass extract through an alumina cleanup column and elute with 8.5 mL of hexane
5. Evaporate hexane eluate to 1.0 mL

This sequence of steps constitutes a representative procedure for most semivolatile solutes; the discussion in this section follows that sequence. An excellent, up-to-date description of most solvent evaporation and cleanup procedures for pesticide extracts is given in a manual by Sherma (1976).

Solvent extracts of soils, sediments, and biological materials usually employ a water-miscible solvent such as acetone or acetonitrile to serve as a solvent bridge between water and an immiscible solvent such as hexane. This bridging solvent must be removed before the extract can be dried; back extraction of the bridging solvent into water is the procedure. However, many extracted polar organic solutes also partition into the water phase during the back-extraction procedure.

The most common drying agent is anhydrous sodium sulfate that has been precleaned by solvent extraction or baking at 400–500°C. If the extraction solvent is moderately polar, such as diethyl ether or chloroform, a volatile nonpolar solvent such as petroleum ether may be added to the extract to decrease water solubility and enhance water removal on the sodium sulfate. Alternate methods of drying include freezing out the water by immersing the solvent receiver vessel in liquid nitrogen (Junk *et al.*, 1974b), or by sorption of water on glass wool packed in a column and prewet with a nonpolar solvent (Webb, 1975). Webb (1975) found that drying of methylene chloride, chloroform, and other hydrophobic solvents was an unnecessary step for concentration purposes because semivolatile solutes tested were better recovered if the extract was not dried before solvent evaporation; however, drying is usually necessary for derivatization procedures.

The major source of semivolatile analyte loss is frequently the solvent evaporation stage of analysis, because of analyte volatilization or loss on walls of the concentration vessel. The most widely used evaporative concentration is the Kuderna–Danish apparatus shown in Fig. 6. Solvent extracts can be evaporated to 5 mL in the large apparatus, and final evaporation to volumes as low as 0.1 mL can be performed in the micro-Snyder column apparatus (Fig. 6). Water baths, steam baths, sand baths, hot plates, and a block tube heater have been used as heat sources with the Kuderna–Danish apparatus. The block tube heater originally described by Beroza and Bowman (1967) avoids evaporation to dryness by extending the tip of the concentrator tube beneath the heated zone. A similar block tube heater is used with an ebullator tube that evaporates the solvent by passing nitrogen or helium through the solvent extract (Beroza *et al.*, 1972). The standard rotary vacuum evaporator can be used for very low volatility analytes, and even the more volatile analytes can be recovered when a "keeper solution," such as paraffin oil, is added to the solvent extract (Sherma, 1976).

A comparison of efficiencies of various solvent evaporation procedures was performed by Webb (1975) and Junk *et al.* (1974b). They both found

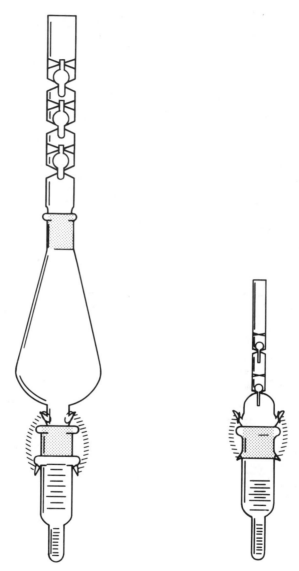

Fig. 6. Kuderna–Danish evaporative concentrator (left) and micro-Snyder column (right).

that the common practice of "blowing down" the 5-mL concentrate with nitrogen to the 1-mL or less final concentrate was the major source of analyte loss. Use of the micro-Snyder column evaporator for final concentration resulted in acceptable recoveries. Both studies found that the

Kuderna–Danish evaporator (Webb, 1975) and a similar evaporator (Junk *et al.,* 1974b) gave the best recoveries, although other devices gave acceptable recoveries within their design limitations. Whenever possible, walls of the solvent concentrator devices should be washed with a minimal amount of fresh solvent during sample transfer. When an exact volume of final concentrate is needed, a micro-Snyder column should be used to concentrate the extract below the required volume, and the extract can be diluted with solvent to the exact volume. The solvent extract should never be allowed to go dry during concentration.

Cleanup of concentrated solvent extract can be accomplished by a wide variety of techniques. The most common technique is classical column chromatography with a polar sorbent and a nonpolar mobile phase. Coextracted contaminants of natural origin are usually polar solutes that have high affinities for polar sorbents in nonpolar solvents, whereas semivolatile analytes isolated by solvent extraction or physical sorption have low affinities for polar sorbents. Silica gel, alumina, magnesia, and Florisil (a magnesia–silica gel) are the most common sorbents, and diatomaceous earth, cellulose, and even carbon have been used to lesser extents (Sherma, 1976). The oxide sorbents are heated (activated) to a specific temperature where their sorptive properties are the greatest and frequently are deactivated for a specific application by adding small amounts of water or ethylene glycol (McClure, 1972).

Compound class separations are also performed during the column cleanup procedure. The Rosen fractionation (Rosen and Middleton, 1955) uses silica gel to separate aliphatic hydrocarbons, aromatic hydrocarbons, and polar organic solutes by eluting the sorbent with an iso-octane, benzene, and chloroform–methanol solvent sequence. Very similar compounds such as PCBs and DDT and its metabolites can be separated on silica gel precisely deactivated with diethyl ether using hexane as the eluent (McClure, 1972).

Thin-layer chromatography (TLC) has been used as a cleanup procedure prior to gas chromatographic determination of pesticide residues (Smith and Eichelberger, 1965). Thin layer chromatography provides better resolution than classical column chromatography and cleaner fractions are obtained; however, solute capacities are much lower than with column chromatography.

The most promising development in cleanup procedures is the use of HPLC. Both the speed and resolving power of HPLC separations provide significant advantages over TLC and classical column chromatography. LaRose (1974) used HPLC to clean up lindane extracts in which coextractives are difficult to remove, and (Thruston, 1978) used HPLC to both clean up and provide compound class separations of many compounds found in a drinking water extract. A gas chromatogram of the unfractionated drinking water extract gave an envelope of unresolved compounds, but gas chromatograms of HPLC fractions gave baseline resolution for most of the compounds.

Other methods of cleanup include gel-permeation chromatography (Griffitt and Craun, 1974), which is most useful when the solvent extract contains high concentrations of lipids that must be separated from analytes. Stalling *et al.* (1979) combined the techniques of gel-permeation chromatography and adsorption chromatography on carbon in a microprocessor-controlled cleanup system for rapid analysis of pesticides and PCBs in complex environmental samples. Sweep codistillation is used to separate volatile analytes from nonvolatile coextractives (Sherma, 1976). Ion-exchange resins can be used to separate ionic analytes from nonionic coextractives or nonionic analytes from ionic coextractives (Calderbank, 1966). Solvent partitioning can be used to back-extract coextractives or analytes into various solvents (Sherma, 1976). Natusch and Tomkins (1978) used solvent partitioning between pentane and dimethyl sulfoxide to separate pentane-extractable solutes into (1) aliphatic hydrocarbons, (2) alcohols, phenols, and low molecular weight aliphatic and aromatic acids, and (3) polycyclic organic compounds, phthalates, aromatic bases, and high molecular weight aliphatic acids.

E. Water Distillation and Evaporation Techniques

The class of semivolatile solutes to which distillation techniques of separation and concentration can be applied are polar, monofunctional acids, bases, and neutrals of moderate-to-high volatility. These compounds are frequently called volatile polar organics (Kuo *et al.*, 1977), but they are classified as semivolatiles in this chapter because of their reduced volatility when dissolved in water. Their volatility is reduced because of moderate-to-strong hydrogen bonding of water with the polar functional group of the low molecular weight acids, bases, alcohols, aldehydes, ketones, and esters. Water evaporation or freeze-drying as a concentration method for semivolatile compounds should only be utilized when the semivolatile, such as an acid or base, can be converted to a nonvolatile salt during evaporation.

A recent study conducted by Kuo *et al.* (1977) of the water distillation technique to concentrate volatile polar organics determined that 80–90% of the mass of the compounds could be recovered in the first 10% of the distillate. By using a two-stage distillation process, concentration factors of 50 to 100 were possible, and volatile polar analytes could be determined at the high part per billion level by combining distillation preconcentration with direct aqueous-injection gas chromatography. Chian *et al.* (1977) reduced the detection limit of volatile polar organics to the low part per billion level by combining distillation preconcentration with gas-chromatographic analysis of headspace vapors above the distillate heated to 70°C.

The efficiency of the distillation method for volatile polar organics varied in the following manner (Kuo *et al.*, 1977):

1. For compounds of the same carbon number, concentration factors were aldehydes > ketone > alcohols
2. Within a compound class, long-chain compounds were more concentrated than short-chain compounds
3. Branched isomers were more concentrated than normal isomers

Highly polar organic acids and bases can be concentrated by distillation only when the pII of the water is adjusted so these solutes are in the nonionic state. Simple distillation of extremely polar acetic acid was not successful as a concentration technique (Kuo *et al.*, 1977); however, steam distillation of water containing formic and acetic acid coupled with simple evaporation of the alkaline distillate concentrated these solutes for gas-chromatographic analysis (White and Leenheer, 1975).

Even when high concentration factors are not obtained by distillation techniques, they may be useful because they isolate volatile polar organic solutes from inorganic and nonvolatile organic constituents. The vacuum distillation and cold-trap train used by Khare and Dondero (1977) isolated volatile polar organics in a landfill leachate sample from volatile and non-volatile organic and inorganic fractions.

F. Freeze Concentration

Freeze-concentration methods are applicable to the same volatile polar organic class of compounds discussed under distillation techniques. Although a recovery study of the suitability of freeze concentration for all classes of semivolatile compounds has not been performed, there is some evidence to indicate that enhanced sorption effects at low temperatures during freeze concentration causes poor recoveries for more insoluble nonpolar classes of semivolatile compounds (Schaumberg, 1974).

The apparatus most often used for freeze concentration was designed by Shapiro (1961); the commercial version of this apparatus is shown in Fig. 7. Other freeze-concentration devices include a rotating, round-bottomed flask immersed in a ice–salt bath (Baker, 1969), and a cold finger condenser containing freezing coolant immersed in the water sample held in a rotating Dewar flask (Kepner *et al.*, 1969).

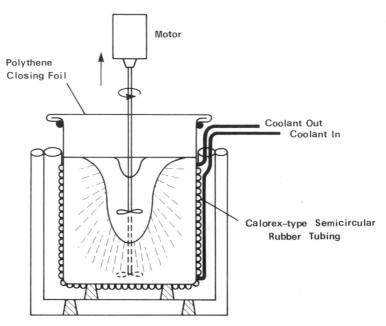

Fig. 7. Shapiro-type freeze concentrator. (Reprinted with permission from Kepner *et al.*, 1969. *J. Agric. Food Chem.*, **17**; copyright by the American Chemical Society.)

Baker (1969) performed a systematic series of studies on freeze concentration of phenols and volatile fatty acids dissolved in water. For these standards dissolved in distilled water, he found that

1. Variations in initial sample volume did not influence recovery, but recovery decreased if concentrate volume was too small
2. Recovery efficiency was not affected by mixing rate
3. There was no difference in recoveries for various compounds tested
4. Single stage or cascade freeze concentration was equally effective
5. Organic-solute mixtures were recovered with equal efficiency as single solutes

When these same standards were dissolved in solutions containing inorganic salts, he found that:

1. Organic solute recovery decreased as ionic strength increased
2. For a defined ionic strength, recovery increased with higher mixing rates
3. When significant losses of organic solutes occurred by ice entrapment, losses were approximately equal for each component of the

solute mixture; this finding suggested the use of an added internal standard for determining recoveries
4. No significant effect of pH on recovery was observed
5. Recovery was increased by washing after decanting the liquid concentrate

Therefore, the major variable affecting freeze concentration of organic solutes is inorganic salt content of the sample. For most natural-water samples, concentration factors of 10 to 100 are possible before high salt content and precipitate formation in the concentrate limit recoveries.

Freeze concentration was found to be especially useful for concentrating volatile esters in water that are easily lost by evaporation or hydrolysis when distillation techniques of preconcentration are used (Kepner *et al.*, 1969). Better than 90% recoveries for compounds as volatile as ethyl formate were obtained when a Shapiro-type freeze concentrator was used. It was important to cover the sample container with polyethylene foil during freeze concentration to limit evaporative losses of volatile constituents. After freeze concentration, either headspace analysis (Kepner *et al.*, 1969) or direct aqueous-injection gas chromatography (Baker, 1969) can be used to determine various volatile polar solutes in the concentrate.

Although freeze concentration is a very gentle preconcentration technique, it probably has not been extensively used for semivolatiles because it does not isolate the solute from the sample matrix, it is relatively slow, concentration factors are limited by ionic strength of the sample, and alternate preconcentration techniques are frequently available.

G. Membrane Techniques

Membrane techniques, such as osmosis, reverse osmosis, and dialysis, have not experienced much use for analytical preconcentration of semivolatile constituents in water because low molecular weights (<200) and rapid diffusion rates of many semivolatile solutes allow their passage through semipermeable membranes. Specific sorption interactions of semivolatile analytes with membranes also limit applicability of membrane techniques to certain classes of compounds. However, membrane concentration techniques have the ability to process much larger volumes of water per unit time than solvent extraction, sorption techniques, evaporative techniques, or freeze concentration. For an exhaustive discussion of the theory and technology of membrane separations of organic solutes in water, the reader is directed to the text edited by Sourirajan (1977).

A number of reverse-osmosis membranes were evaluated by Edwards and Schubert (1974) for removing the herbicide 2,4-D from water, but various membranes retained only an average of 50% of the 2,4-D. Cabasso *et al.* (1975a) concluded that separation and retention of semivolatile solutes by osmosis and reverse osmosis were rather poor compared to inorganic solute retention. Both osmosis and reverse osmosis gave comparable results when the same membranes and test solutes were used.

An evaluation of cellulose acetate, cellulose triacetate, and ethyl cellulose reverse-osmosis membranes for concentrating trace organic contaminants in drinking water was conducted by Klein *et al.* (1975). They devised a predictive model for organic solute permeation through these membranes based on solubility of the solute in membrane material. By plotting Hansen solubility parameters of solutes that result from hydrogen-bonding forces and polar forces, a solubility diagram was obtained that indicated the organic solutes that were soluble in the membrane. Low molecular weight solutes (<200) that fell within the solubility region of the diagram were not retained by the membrane, whereas solutes that fell outside the solubility region were retained. Membrane-soluble solutes greater than MW 200 are retained in varying degrees by the membrane because of their slower diffusion rate through the membrane relative to water. Hydrophobic solutes of low water solubility were often retained by physical sorption on the membrane surface; these solutes could not be recovered from the liquid concentrate. Ionizable solutes, such as the volatile fatty acids, were retained in the greatest quantity, whereas low molecular weight esters, aldehydes, ketones, phenols, and alcohols were retained in the least quantity (Klein *et al.,* 1975; Sourirajan, 1977).

Polyamide reverse osmosis membranes are much more stable in a wider pH range than cellulose-based membranes (McKinney, 1972); their solubility parameters and range of application are significantly different from cellulose-based membranes (Klein *et al.,* 1975), and they show better rejection characteristics than cellulose acetate for low molecular weight organic solutes (Cabasso *et al.,* 1975b).

A comprehensive study by Cabasso *et al.* (1975b), which evaluated a number of semipermeable membranes for concentration of organic contaminants in drinking water, recommended that cellulose acetate-derived membranes be used at the first stage of reverse osmosis concentration because of their excellent salt rejection, good water permeability, and resistance to chlorine. Organic solutes that pass through cellulose acetate membrane can then be concentrated by using reverse osmosis with polyamide membranes on the permeate from the cellulose acetate. Free chlorine that would damage polyamide membranes had been previously removed by the cellulose acetate membrane.

A reverse osmosis unit using cellulose acetate membranes was used to concentrate organic contaminants in 1600 L of Cincinnati drinking water (Deinzer *et al.*, 1975). Semivolatile solute retention followed the model of Klein *et al.* (1975), and total organic carbon (TOC) recoveries ranged from 60% for 10-fold concentration to 29% for 72-fold concentration.

Dialysis of semivolatile acids and bases through a hydrophobic membrane can be performed by sorption of the solute on the membrane in the nonionic state and desorption on the reverse side of the membrane in the ionic state. Separation of aniline by dialysis against acid (Klein *et al.*, 1972) and phenol by dialysis against base (Klein *et al.*, 1973) achieved removals up to 80% and 5- to 10-fold concentration factors for these solutes in acid and base dialysate.

H. Other Techniques

Semivolatiles are not usually concentrated and isolated from water by adsorptive bubble separation methods because of their low surface activity. However, these methods can be used with semivolatiles that form an adduct with an added collector surfactant. Poncha and Karger (1965) were able to foam fractionate 1-naphthylamine from water by adding sodium laural sulfate as the surfactant collector. Concentration factors of 100 were obtained for 1-naphthylamine in the foamate. Techniques of solvent extraction, solvent sublation, and foam fractionation were compared for isolation of phenol from water (Grieves *et al.*, 1974) after addition of a quaternary ammonium surfactant. Solvent sublation, which is a combination of foam fractionation and solvent extraction using amylacetate as the extraction solvent, required the lowest concentration of surfactant collector for efficient extraction of phenol.

A novel procedure based on vacuum sublimation was used to isolate volatile biogenic amines excreted by algae into water (Herrmann and Juttner, 1977). Algal suspension in water was first acidified and taken to dryness in a rotary flask evaporator. The residue was taken up in a small volume of water while frozen in a round-bottomed flask. This frozen amine hydrochloride solution was overlayered with frozen distilled water and frozen 2 N NaOH. The flask was then connected to the vacuum sublimation apparatus; the contents were thawed and mixed to convert hydrochlorides to free amines; the mixture was refrozen; vacuum sublimation was performed; and the water and amine sublimate was trapped with a liquid nitrogen-cooled trap. The isolated amines were highly purified of other contaminants found in the algal medium and were analyzed by gas chromatography after derivatization with trifluoroacetic anhydride.

V. TECHNIQUES FOR LOW MOLECULAR WEIGHT NONVOLATILES

Low molecular weight nonvolatile constituents in water are composed of amino acids, detergents and surfactants, monosaccharide carbohydrates, and other nonvolatile, water-soluble compounds of less than MW 500. These compounds generally are highly polar materials, which are not easily isolated from water by simple solvent extraction or physical sorption techniques. Their nonvolatility obviates gas chromatography as a separation procedure without derivatization, but liquid-chromatographic techniques can be readily applied.

Because of their water solubility and nonvolatility, low molecular weight nonvolatiles are the easiest compound class to sample. These solutes are not lost through evaporation and are rarely lost through sorption or precipitation in the sample container. The main consideration is sample preservation, as many of these solutes are readily metabolized by microbiological processes. If refrigeration is not sufficient to inhibit microbiological activity, a preservative such as mercuric chloride or sodium azide may have to be added to the sample. Otherwise, the usual considerations discussed in Section IV for sampling organic constituents in water apply.

A. Sorption Techniques

With the exception of detergents and surfactants, sorption techniques for low molecular weight nonvolatiles are based on ion exchange, ligand exchange, and hydrogen bonding. Techniques based on physical sorption usually cannot be used for nonchromatographic separations because these polar solutes have low capacity factors on hydrophobic sorbents. Physical-sorption capacity factors for weak diprotic acids and weak diprotic bases can be significantly enhanced by lowering or raising the system pH (Pietrzyk et al., 1978), but capacity factors for amphoteric amino acids cannot be elevated by pH adjustment above k' 100 needed for nonchromatographic separations, because these solutes have some ionic character at all pH values (Kroeff and Pietrzyk, 1978).

Amino acids are usually sorbed from water by cation exchange or ligand exchange. The cation-exchange technique is based on using a strong acid exchanger in hydrogen form, passing the water sample through at pH 4, and desorbing amino acids with 2 N NH$_4$OH (Degens and Reuter, 1964). Addition of a hydrolysis step to this procedure enables sorption of combined amino acids as well as free amino acids. This cation-exchange procedure was used to characterize dissolved and particulate, free and combined amino acids in surface waters of the Mackenzie River drainage basin (Peake *et al.*, 1972). This ion-exchange procedure is frequently less than satisfactory because of oxidative losses of sulfur-containing amino acids and irreversible adsorption of aromatic amino acids on the cation-exchange resin (Krutz, 1975).

In saline waters, the cation-exchange technique for isolating amino acids is unsatisfactory because of the high proportion of inorganic cations relative to cationic amino groups. Ligand exchange of amino acids with cupric ion immobilized on a chelating cation-exchange resin was successful in selectively concentrating trace quantities of amino acids in seawater (Siegel and Degens, 1966). Hydrophobic aromatic amino acids were retained to the greatest extent, and cystine was converted to cysteic acid. The sorbed amino acids were eluted with 3 N NH_4OH. Gardner and Lee (1973) used cation exchange followed by ligand exchange to isolate amino acids from lake water. Ligand exchange served as the cleanup step for the concentrate obtained by cation exchange.

Anion-exchange resins have been used to concentrate and isolate organic acids from water. A method for determining part per billion levels of citric and nitrilotriacetic acids in tap water and sewage effluents was based on a strong-base, formate-saturated, anion-exchange resin (Aue *et al.*, 1972). Citric acid was chromatographically separated from nitrilotriacetic acid when the resin was eluted with formic acid. Trace amounts of the herbicide glyphosate have been concentrated from natural waters by sorption on a chloride-form, strong-base, anion-exchange resin followed by elution with 0.1 N HCl (Bronstad and Friestad, 1976).

These few examples are among the many applications of sorption of organic compounds by ion-exchange resins. Excellent critical reviews of the factors affecting adsorption of various organic compounds by various exchange resins were presented by Abrams (1969) and Kim *et al.* (1976). These authors indicated that matrix-sorption effects are the most significant cause for loss of analyte on the resin and loss of resin capacity. Most sorbed aromatic carboxylic acids caused a 10–15% irreversible capacity loss for a styrene–divinylbenzene matrix, strong-base, anion-exchange resin because of physical sorption of the aromatic moiety of the solute on the aromatic matrix of the sorbent (Abrams, 1969). Dinitro-substituted aromatic acids caused a 50–60% capacity loss, probably because of formation of a π-bonded charge-transfer complex with the resin matrix. Physical sorption effects on the resin matrix can be lessened by mixing a water-miscible organic solvent with the water eluent during analyte elution. For example, the detergent ABS sorbed on a chloride-form, strong-base, anion-exchange resin can be eluted by a 50–50 mixture of organic solvent and mineral acid in water (Abrams and Lewon, 1962).

Weak-acid and weak-base exchange resins have some unique advantages for analytical preconcentration purposes. In contrast to strong-acid and strong-base exchangers, sorbed organic acids on weak-base resins can be eluted with dilute alkali because of conversion of the resin to a nonionic form, and organic bases sorbed on weak-acid resins can be eluted with dilute acid for the same reason. These resins act as organic solute sorbents in intermediate pH ranges through ion-exchange and hydrogen-bonding mechanisms, and desorption can be accomplished by ion-exchange replacement of the analyte with a higher affinity solute, or by conversion of the analyte or resin to nonionic forms by changing the system pH. Capacity factors may not be quite as high or sorption kinetics may not be as rapid as strong-acid and strong-base exchangers, but desorption efficiency is usually greater for weak-acid and weak-base resins. These exchangers also show a greater selectivity for weak-acid and weak-base solutes found in water.

In summary, ion-exchange resins that hold the most promise for analytical ion-exchange preconcentration applications are weak-acid and weak-base exchange resins on aliphatic hydrocarbon matrixes. The aliphatic matrix minimizes strong matrix-adsorption effects like π-bonding, and weak-acid and weak-base groups allow most options for analyte desorption.

B. Adsorptive Bubble Separation Techniques

Adsorptive bubble separation techniques are generally used for concentration and isolation of detergents and other surface-active materials from water, but can be used for nonsurface-active materials as well. The various techniques have been classified by Karger et al. (1967), as shown in Fig. 8.

Techniques most often used for concentration and isolation of organic materials in water are foam fractionation and solvent sublation; ion flotation and bubble fractionation are used to lesser extents. Foam fractionation is the removal of dissolved material by foaming. Solvent sublation is the removal of dissolved or particulate material by adsorption at the surface of rising bubbles, followed by extraction of the material by an immiscible solvent on top of the water sample. Ion flotation is the removal of a nonsurface-active ion by frothing, through the use of a surfactant that forms an insoluble product with the analyte ion. Bubble fractionation is the removal of dissolved or particulate materials by adsorption at the surface of rising bubbles, followed by redeposition of these materials just under the water surface. A comprehensive presentation of the theory, technology, and application of these techniques is found in Lemlich (1972).

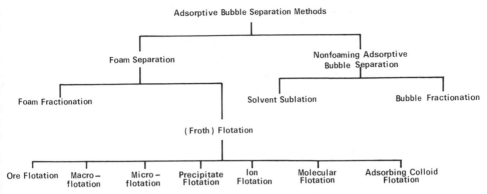

Fig. 8. Classification of adsorptive bubble separation techniques. (Reprinted from Karger *et al.*, 1967, p. 401, by courtesy of Marcel Dekker, Inc.)

Foam fractionation can be used for concentrating both naturally surface-active materials and organic ions associated with an added surfactant collector. Batch-type and continuous-feed foam-fractionation apparatus with various systems of foam recycle, collapse, and drainage are available (Lemlich, 1972). Efficiency of foam-fractionation concentration increases with decreasing bubble size, increasing gas flow, height, the amount of foam recycle, and decreasing surfactant concentration; it decreases with diameter of the fractionation column. Increasing efficiency of foam fractionation for decreasing surfactant concentrations is related to the formation of surfactant micelles at higher concentrations, which are not as efficiently sorbed at the gas–liquid interface as monodispersed surfactant molecules in dilute solution.

Increasing ionic strength by electrolyte addition sometimes aids efficiency by increasing surfactant sorption at the gas–liquid interface, but it frequently decreases efficiency by promoting micelle formation and by electrolyte competition with a surfactant collector for the analyte colligend. Foam stability and surfactant bubble adsorption are very sensitive to pH, and the optimum pH for foam fractionation for a particular analyte is usually established by experiment. By optimizing the variables just mentioned, Karger *et al.* (1966) demonstrated that enrichment factors as high as 200 could be obtained for foam fractionation of methyl orange dye and vegetable lecithin, when these nonfoaming solutes were paired with appropriate surfactant collectors.

Bubble fractionation is the simplest bubble separation technique. Although a surfactant-enriched foamate is not produced, surfactant concentration gradients as high as 50:1 can be formed between the top and bottom of the water column (Lemlich, 1972). Bubble fractionation with continuous sample feed can be used to raise the surfactant concentration to foaming threshold levels where foam fractionation can be used (Harper and Lemlich, 1965). Bubble fractionation is optimized by correct adjustment of the same variables as were discussed with foam fractionation.

The most widely used bubble separation technique is solvent sublation. It has been more widely accepted for analytical purposes than foam fractionation or bubble fractionation, because the sample is efficiently concentrated and isolated from water in an easily managed organic solvent. The solvent should be sufficiently nonvolatile that it does not evaporate during sublation; it should be sufficiently polar to extract the surfactant; and it should be sufficiently nonpolar to be immiscible with water. Solvents such as 2-octanol, anisole, and ethyl acetate have proved to be satisfactory for solvent sublation.

Solvent sublation differs from solvent extraction in that it is a dynamic, nonequilibrium process rather than an equilibrium process between two liquid phases in solvent extraction (Karger, 1972). In solvent sublation, liquid–liquid equilibrium exists only at the interface, and overall extraction is enhanced because of surfactant enrichment in the aqueous phase at the interface as a result of bubble fractionation in the column. Emulsion formation also tends to be less with solvent sublation than solvent extraction. A sublator column is shown in Fig. 9.

An analytical method for isolation of nonionic detergents based on solvent sublation into ethyl acetate was developed by Wickbold (1972). His procedure also isolated anionic and cationic surfactants that occasionally interfered with determination of nonionic surfactants. Boyer *et al.* (1977) developed a cleanup procedure in which ethyl acetate is evaporated from detergent sublate, the detergent residue dissolved in methanol, and ionic detergents removed by passing the methanolic solution through a column containing a hydrogen-form, cation-exchange resin and a hydroxide-form, anion-exchange resin.

Ion flotation is very similar to the bubble fractionation techniques just discussed, with the exception that a collector surfactant is always used that forms a colloid or precipitate with the ionic analyte. The development and potential applications of this technique were discussed by Sebba (1962). An example of an application is the determination of anionic and cationic surfactants in water by their reaction with a dye of the opposite charge (Lovell and Sebba, 1966). Excess dye is added to the surfactant solution and the reaction product containing all the surfactant is removed by ion flotation; decrease in dye color is a measure of surfactant concentration.

In conclusion, adsorptive bubble fractionation techniques hold considerable promise for analytical concentration and isolation of both surface-active and nonsurface-active materials, but most of their applications

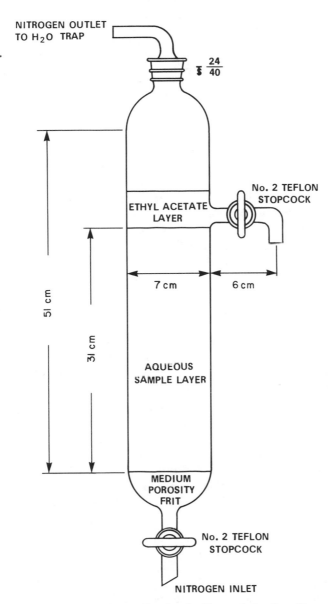

Fig. 9. Solvent sublation column. (Reprinted with permission from Boyer *et al.*, 1977. *Environ. Sci. Technol.* **11**; copyright by the American Chemical Society.)

have been in the areas of industrial and wastewater treatment processes. Their biggest advantage is that their efficiency increases as analyte concentration is reduced, in contrast to sorptive techniques and solvent extraction where extraction efficiency is directly proportional to analyte concentration.

C. Solvent Extraction

Low molecular weight nonvolatiles are not generally directly extractable by organic solvents. However, solvent extraction can be performed by addition of an extractant that forms an extractable complex in water with the low molecular weight nonvolatile analyte.

The most common example is the methylene blue method for estimating anionic detergents in water (American Public Health Association *et al.*, 1976b). Cationic methylene blue reacts with anionic detergents to form neutral ion pairs that are readily extracted into chloroform. The extracted complex is determined by absorbance measurement at 652 nm. Many other extractable and nonextractable organic and inorganic compounds can react with methylene blue giving positive interference to anionic detergent determination. A chloroform extraction of the sample prior to methylene blue addition can eliminate extractable compounds that interfere (Wang, 1975); however, chloride is not extracted and interferes, especially in brine samples.

A promising method for selective extraction and determination of anionic detergents was developed by Chrisp *et al.* (1975). The cationic metal complex, bis(ethylenediamine)copper(II) ion, selectively pairs with anionic detergents. The ion pair complex can be removed from water by a single extraction in chloroform, and the complex can be determined either colorimetrically or by atomic absorption photometry for copper. Sensitivity is equivalent to the methylene blue method, and the only interferences detected thus far are Fe^{3+}, Co^{2+}, and S^-, all of which can be eliminated by procedural modifications. Therefore, this method is applicable to fresh water, estuarine water, and saltwater.

A method was developed by Taylor *et al.* (1974) for selective extraction of anionic surfactants based on alkyl chain length of the surfactant. They determined that differences between chloroform–water extraction constants of compounds of adjacent homologous surfactants with a particular iron(II) chelate were sufficient to separate the surfactants. By using eight different iron(II) chelates, they developed selective solvent extractants that could separate C_5–C_{10} alkyl sulfates, C_8–C_{12} alkyl sulfonates, and C_4–C_8 alkylbenzene sulfonates.

Cationic surfactants are extracted and determined similarly to anionic surfactants. Chromate anion forms ion association complexes with primary, secondary, tertiary, and quarternary long-chain amines in strongly acidic water. By extracting with chloroform and determining chromium colorimetrically, amine detection limits as low as 0.01 ppm were obtained (Florence and Farrar, 1973).

Extraction of ion pair or association complexes is not limited to detergents and surfactants. Some examples of other applications include acetone extraction of hydroxy acids and volatile fatty acids paired with tetra-n-butylammonium cation (Bethge and Lindstrom, 1974). Various highly polar phenolic acids are efficiently extracted into nonpolar solvents, such as cyclohexane, when these acids are associated with tributylphosphate (Matysik and Soczewinski, 1977).

There are many more applications of solvent extraction of ion association species for concentrating and isolating trace metals from water than for organic ions found in water. Although considerable research is frequently needed to find the right complexation agent, the relative simplicity of the solvent extraction technique ensures that many future applications of solvent extraction for isolation of polar ionic organic solutes from water are forthcoming.

D. Water Evaporation Techniques

Evaporation techniques of concentration are best applied to low molecular weight nonvolatiles of all compound classes discussed in this chapter. Volatility losses are not a problem, and losses caused by precipitation and denaturation are not concerns, as they are for high molecular weight compounds.

Low-temperature are preferable to high-temperature evaporation techniques to minimize changes in heat-labile organic constituents. Low-temperature techniques include vacuum distillation, rotary evaporation, and freeze-drying. Freeze-drying provides maximum sample integrity, but it is relatively slow and not convenient to use for sample volumes more than 500 mL. Rotary evaporation under a partial vacuum is somewhat faster, but it is also limited to approximately the same sample volume. A technique that is applicable to large sample volumes is vacuum distillation, as discussed by Katz et al. (1972). The vacuum distillation system used is diagrammed in Fig. 10. They concentrated 3-L sewage effluent samples to 150 mL by vacuum distillation, and then used freeze-drying to take the sample to dryness. The residue was dissolved in 5–10 mL of acetate or borate buffer, and insoluble solids were separated by centrifugation. Organic compound recovery was greater than 95% (as determined by organic carbon analysis). Several carbohydrates and hydroxy acids were

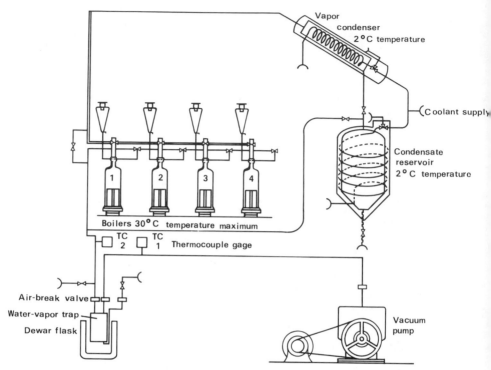

Fig. 10.   Vacuum distillation system for water removal. (From Katz *et al.*, 1972; published with permission of Pergamon Press, Inc.)

determined in the buffer concentrate by high-resolution ion exchange chromatography. A similar procedure was used by Pitt *et al.* (1975) to concentrate 10–100 L of sewage effluents and natural waters for trace organic analysis by high-resolution ion exchange chromatography. They passed the sample through a hydrogen-form, weak-acid, cation-exchange resin before vacuum distillation to reduce the amount of insoluble inorganic salts in the freeze-dried residue.

Sample cleanup and derivatization is frequently performed on the dry residue or liquid concentrate after water evaporation. Various aminopolycarboxylic acids and nitrilotriacetic acid in river water were concentrated and derivatized for gas-chromatographic analysis by freeze-drying the sample, esterifying organic acids in the residue with 3 N 1-butanol-HCl, evaporating the alcoholic HCl, extracting the 1-butanol esters from the residue with methylene chloride, and derivatizing the free amino groups of the extracted esters with trifluoroacetic anhydride (Warren and Malec, 1972).

E. Other Techniques

Reverse osmosis is probably superior as an analytical preconcentration method for low molecular weight nonvolatiles than for semivolatiles, although there have been few analytical applications on which to base a comparison. From Sourirajan's text (1977), ionic carboxylic acids, ionic aliphatic amines, and carbohydrates are effectively concentrated by reverse osmosis using cellulose acetate membranes, because these compounds do not sorb on and are not transported through the membranes. Physical sorption of low molecular weight nonvolatile solutes on polar reverse-osmosis membranes does not usually occur because of the solute's hydrophilic nature.

The weakly acidic phenolic hydroxyl group can hydrogen bond with cellulose acetate, and compounds with the phenolic hydroxyl group are frequently not effectively concentrated by reverse osmosis. Other low molecular weight nonvolatile compounds like urea have similar affinities for water and the reverse-osmosis membrane and pass through without concentration.

In a study where drinking water was concentrated by reverse osmosis (Deinzer *et al.*, 1975), hydrophilic organic solutes not extractable with methylene chloride gave an infrared spectrum very similar to the carbon–alcohol extract, which is composed of low and intermediate molecular weight, nonvolatile organic compounds. Therefore, reverse osmosis may be a viable concentration technique for nonvolatile organic compounds in drinking water.

Precipitation techniques are infrequently used with low molecular weight nonvolatiles because of their water solubility. One reported application of a precipitation technique is a field test for cationic surfactants in water (Geyer, 1974). Cationic surfactants are precipitated with methyl orange in alkaline solution. The precipitate is filtered, washed, and redissolved in 2 N HCl. The limit of detection is 2 mg/L of surfactant by colorimetric determination of redissolved methyl orange in hydrochloric acid.

Freeze concentration is an excellent but slow method for concentrating low molecular weight nonvolatiles in water. Kammerer and Lee (1969) demonstrated 90–100% recoveries for [^{14}C]glycine, -glucose, -citric acid, and -phenylalanine from distilled and lake water. Initial solute concentrations ranged from 0.01 to 1.0 mg/L, and concentration factors ranged from 3 to 22. However, availability of alternate techniques with higher concentration factors and higher concentration rates limit the usefulness of freeze concentration to concentrating unstable nonvolatile solutes that should not be taken to dryness or heated.

VI. TECHNIQUES FOR HIGH MOLECULAR WEIGHT CONSTITUENTS

Although high molecular weight materials in water are the least characterized at the structural and compound levels of analysis, they constitute most of the mass of organic carbon in dissolved and colloidal fractions. An example of molecular weight distributions of organic carbon in natural waters is given in a study by Smith (1976), who used various ultrafiltration membranes to determine molecular weight distribution of organic materials in the estuary of the Ogeechee River in Georgia. The results are presented in Fig. 11. By defining high molecular weight materials to be greater than MW 500, one can readily conclude that most organic carbon in the Ogeechee River is high molecular weight material. Other studies using ultrafiltration (Gjessing, 1970) and gel-permeation filtration (Gjessing and Lee, 1967) have also concluded that most organic constituents in water are high molecular weight materials. The upper limit for molecular weight and size determination depends on the method of filtration or centrifugation to remove suspended sediment. The most common particle-size break is established by membrane filtration using 0.45-μ pore size.

The most common and stable high molecular weight constituents in natural waters are humic and fulvic acids. The more soluble fulvic acid is found mostly in water (Christman and Ghassemi, 1966), whereas humic

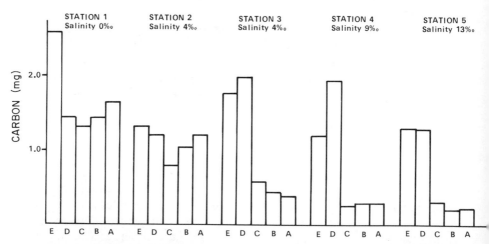

Fig. 11. Molecular weight distribution of Ogeechee River determined by ultrafiltration: (A) >100,000; (B) 50,000–100,000; (C) 10,000–50,000; (D) 1,000–10,000; (E) <1,000. (Reprinted with permission from Smith, 1976. *Anal. Chem.*; copyright by the American Chemical Society.)

acids are enriched in soils and sediment (Schnitzer and Khan, 1978). Polysaccharide carbohydrates and proteinaceous materials are also found in small amounts in natural waters (Leenheer and Malcolm, 1973) and in major amounts in sewage effluents (Manka et al., 1974).

High molecular weight constituents in water are easily sampled but can be difficult to preserve, because of the ease with which certain of these constituents aggregate and precipitate, denature and hydrolyze, and biodegrade. The aggregation of humic materials in water, determined by small-angle X-ray scattering, usually increases as pH decreases until in some cases a humic acid precipitate forms (Wershaw and Pinckney, 1973). At high pH, humic materials in water are liable to oxidation and photodegradation of phenolate and quinone functional groups (Chen et al., 1978). Therefore, adjustment of pH to either low or high values is not recommended as a sample preservation technique for humic materials in water.

Polysaccharides and polypeptides are easily biodegraded if the sample is not preserved, but addition of heavy metal preservation agents such as Hg^{2+} or Cu^{2+} can complex and precipitate these materials. Chilling, and especially freezing, also may cause precipitates to form as a result of decreased solubility and freeze-concentration effects.

A promising preservation technique that has been applied to samples for dissolved organic carbon (DOC) analysis is to field filter the sample through a silver-membrane filter (Malcolm and Leenheer, 1973). Filtration removes most microorganisms that cause sample biodegradation, and the filter releases 1–2 mg/L of silver, which serves as a preservation agent, into the sample. Chilling the sample to 4°C is also recommended as most large colloidal materials that may aggregate and precipitate on chilling have been removed by filtration. As always, the best preservation technique is to isolate these materials from water as soon as possible.

A. Adsorption

Analytical concentration and isolation of high molecular weight macromolecules in water by adsorption have been limited by difficulties in desorbing and recovering this class of analytes from the adsorbent. The reasons for irreversible adsorption are (1) multiple adsorption mechanisms, which bind polyfunctional sorbates to polyfunctional adsorbents and (2) large size of the sorbate molecules, which are trapped in the pores of the adsorbent and block sorption sites. A number of suitable adsorbents for macromolecular materials in water have been found by basing adsorbent selection on large pore size, few adsorption mechanisms, and reversibility of adsorption mechanism.

Humic substances have been recovered from water by adsorption on alumina, silica gel, magnesia, calcium carbonate, activated carbon, strong- and weak-base anion-exchange resins, nylon and polyamide resins, styrene–divinylbenzene resins, and methylmethyacrylate ester resins. Recoveries of humic materials from water using inorganic adsorbents tend to be less than 50%, mainly because of inefficiency of the adsorption stage (Jeffrey and Hood, 1958; Williams and Zirino, 1964). Organic adsorbents have greater capacity factors for humic substances; their possible sorption mechanisms are listed in Table II.

Activated carbon was found to adsorb 55–86% of DOC in seawater (Kerr and Quinn, 1975). However, only an average of 59% of adsorbed DOC could be recovered by sequentially eluting with 0.5 N NaOH and methanol. Chemical alteration of humic materials has also been reported to occur on carbon (Jeffrey and Hood, 1958). Humic materials are retained primarily by physical sorption on activated carbon, but secondary sorption mechanisms, such as ion exchange and π-bonding, are often responsible for low recoveries because these mechanisms are additive and often stronger in elution solvents.

Better recoveries of sorbed humic materials have been obtained using nonionic macroporous copolymers whose physical properties and matrix compositions were given in Table III. XAD-1 resin was found to be generally effective for analytical concentration of humic materials in river water and seawater, although exact recovery data were not given (Riley and Taylor, 1969). Optimum adsorption occurred at pH 2; desorption was by 0.2 N KOH. Mantoura and Riley (1975) developed a method for concentrating humic and fulvic acids from natural waters on XAD-2 resin by adsorbing at pH 2.2 and desorbing in 0.2 N NaOH. Adsorption efficiencies of 92% for humic acid and 75% for fulvic acid were achieved at flow rates of 35 bed volumes per hour. A 24-fold concentration factor was obtained when humic and fulvic acid were 95% desorbed in 0.2 N NaOH. Organic solvents such as methanol, ethanol, and acetone were less than 50% effective in desorbing humic acid and fulvic acid.

All XAD resins listed in Table III were evaluated by Aiken *et al.* (1979) as adsorbents for fulvic acid in water. They found that methylmethacrylate ester resins (XAD-7 and XAD-8) had higher capacity factors and gave better elution efficiency than polystyrene–divinylbenzene resins (XAD-1, XAD-2, and XAD-4). Higher column capacity factors of XAD-7 and XAD-8 were related to more rapid adsorption kinetics on the moderately polar resins, and greater desorption efficiencies were related to the absence of π-bonding, which was postulated for reduced recoveries on XAD-1, 2, and 4. Steric factors were also important. XAD-4 had the lowest capacity and the slowest adsorption kinetics because of steric exclusion of fulvic acid from the small pores. The resin that possessed the highest capacity was XAD-7, but continuing organic solute bleed from this resin negated its use. The best overall adsorbent for fulvic acid was XAD-8, which quantitatively adsorbed the fulvic acid at pH 2 as determined by color removal, and desorbed at 98% efficiency in 0.1 N NaOH. With proper cleaning, organic bleed from XAD-8 is <0.1 mg/L, and with multiple-stage concentration on XAD-8, concentration factors as high as 25,000 were obtained for fulvic acid in a groundwater sample (Thurman *et al.*, 1978a). XAD-8 was not a surprising choice for humic and fulvic acids as it was developed to remove similar lignosulfonic acids from Kraft paper-mill effluents (Kennedy and Minor, 1972).

Adsorption of humic materials from water on nylon and polyamide resins occurs by hydrogen bonding of phenolic hydroxyl groups in humic substances with the amide bond of the adsorbent (Endres and Hormann, 1963). A comparison of nylon and polyamide powders for isolating humic materials (*gelbstoff*) in terrestial and marine waters was performed by Sieburth and Jensen (1968). At pH 3.5, nylon consistently adsorbed 70% of the color measured at 450 nm, whereas various polyamide powders adsorbed 75–95% of yellow humic materials. However, humic color was quantitatively recovered from nylon by elution with 0.1 N NaOH, whereas 30% of the color was irreversibly adsorbed to the polyamide powders. Irreversible adsorption of quinones on free amino groups of polyamide adsorbents was observed by Endres and Hormann (1963). Single-stage concentration factors of 10,000 and recovery factors of 70% for humic materials on nylon adsorbents were obtained (Sieburth and Jensen, 1968).

Anion exchange, hydrogen bonding, physical sorption, and π-bonding are all possible mechanisms of humic and fulvic acid adsorption on anion-exchange resins. Packham (1964) used a chloride-form, strong-base, anion-exchange resin to adsorb humic material from water. The resin was regenerated by 2 M NaCl, and desorbed humic and fulvic acids were extracted from NaCl brine with n-butanol. This procedure frequently gives low recoveries caused by irreversible adsorption. In a test of four different strong-base, polystyrene-matrix, anion-exchange resins as adsorbents for removal of color in drinking water, Abrams and Breslin (1965) were only able to recover 1–10% of the sorbed humic and fulvic acids by regeneration with sodium chloride and sodium hydroxide. They also found that a number of weak-base, anion-exchange resins had higher capacity factors for color removal than strong-base resins, and essentially 100% desorption of humic material from weak-base resins could be obtained with sodium hydroxide, if loading was limited to less than half the resin capacity.

The mechanisms of humic and fulvic acid adsorption on weak-base anion exchangers is pH dependent. At and above neutrality, the secondary- and tertiary-amine functional groups are in the nonionic, free-base form. At this pH, adsorption of humic materials most likely occurs by hydrogen bonding of proton-donating phenolic hydroxyl groups in the adsorbate, with proton-accepting amino groups of the adsorbent (Kim *et al.*, 1976). At pH 11–12, ionization of the phenolic group occurs, and the sorbate is no longer bound by hydrogen bonding, which results in desorption. At acid pH values, the adsorbent acts as an anion exchanger because the amine groups are protonated and positively charged. Kim *et al.* (1976) found that both acid form and free-base form of weak-base anion exchangers could initially remove more organic material from drinking water and wastewater than activated carbon, and quantitative regeneration was achieved with sodium hydroxide solution.

Surprisingly, weak-base, anion-exchange resins have only been evaluated from the water treatment viewpoint, and analytical applications are lacking. Potential analytical advantages include (1) isolation of humic and fulvic acid on free-base resin without pH adjustment of the samples and quantitative elution in stoichiometric amounts of alkali; (2) single step isolation of all weak-acid solutes (both low and high molecular weight) without sample pH adjustment on the acid form of the weak-base resin, followed by quantitative elution in alkali; and (3) gradient elution and fractionation of organic acid sorbates by changing pH and displacing anion concentrations.

Weak-base and weak-acid ion-exchange celluloses have been used to concentrate and fractionate high molecular weight acids and bases in water. Sirotkina *et al.* (1974) used DEAE cellulose in free-base form to adsorb humic and fulvic acid from river water, and CM cellulose in hydrogen form to sorb proteins from river water. Adsorption occurred primarily through hydrogen bonding. Overall recovery of solute fractions was greater than 90%, as determined by standard additions to river water samples. Optimum pH of adsorption was near neutrality for each adsorbent, and fulvic acid equivalent capacity was about five times the calculated chemical equivalent exchange capacity of DEAE cellulose, suggesting that only one in five of the fulvic acid functional groups binds with the adsorbent's amino group.

Cation-exchange resins can be used to concentrate basic proteinaceous constituents from aqueous solution. Helfgott (1975) proposed cation-exchange resins to absorb positively charged polypeptides in sewage effluents that were not removed by activated carbon. MacCarthy and O'Cinneide (1974) isolated "cationic fulvic acid" by passing a fulvic acid extract from bog peat through a hydrogen-form, strong-acid cation-exchange resin, and desorbing the cationic fraction with 0.5 N NaOH. Weak-base anion-exchange resins can also be used to remove and concentrate proteins in water by hydrogen bonding and anion exchange of the carboxylate acid groups on the protein molecule (Foster *et al.*, 1977). High molecular weight, cationic organic bases are minor constituents in most natural waters, and analytical studies of adsorbents for their concentration and isolation are relatively rare.

Recommended adsorbents for analytical concentration of humic and fulvic acid from water are XAD-8, XAD-2, and DEAE-cellulose. Water treatment studies suggest that various weak-base anion-exchange resins may be suitable after additional research. Recommended sorbents for high molecular weight base constituents are CM-cellulose and possibly weak-acid cation-exchange resins. For high molecular weight solutes, large adsorbent-pore size (≥ 200-Å diameter) and aliphatic matrixes (to prevent π-bonding) are desirable.

B. Ultrafiltration and Gel Filtration

The techniques of ultrafiltration and gel filtration are primarily used to fractionate high molecular weight materials in water according to molecular size. They are also used to desalt organic concentrates, and ultrafiltration can be used to concentrate high molecular weight organic materials in a similar manner as reverse osmosis.

Ultrafiltration and gel-filtration techniques were developed around 1960 for separation and purification of high molecular weight biological materials, such as proteins, nucleic acids, and carbohydrates. Gel filtration was generally applied to natural waters and was accepted as a fractionation technique before ultrafiltration because of the work of Gjessing (1965) and Gjessing and Lee (1967). They related the fractionation obtained by gel filtration on Sephadex of natural organic matter in lakes and streams to estimated molecular weight, chemical oxygen demand, color, and organic nitrogen. Since this pioneering work, many other studies too numerous to mention have used gel filtration for molecular weight fractionation of high molecular weight organic materials in water.

The theory and practice of gel filtration was presented by Determann (1969). Ideally, chromatographic separation by gel filtration occurs by steric exclusion of large particles in solution from gel pores into which smaller sized particles can migrate by diffusion. Therefore, in a column arrangement, excluded particles will migrate faster than particles that can move into pores of the gel structure. Actually, this chromatographic separation, supposedly based on physical size differences, usually is also affected by solute adsorption and disaggregation effects.

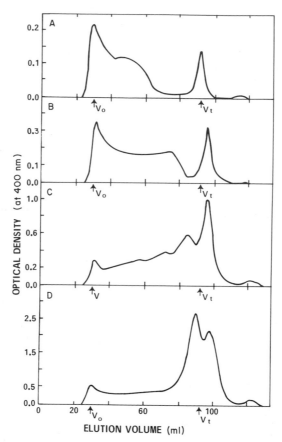

Fig. 12. Fractionation of humic acid (pyrophosphate extract) on Sephadex G-100 using distilled water as eluent. Sample concentrations are (A) 2 mg; (B) 4 mg; (C) 10 mg; and (D) 20 mg/2 mL. (From Swift and Posner, 1971. Published with permission of the Williams and Wilkins Co.)

It has been known for some time that humic materials are not clearly fractionated according to molecular size by gel filtration, but that adsorption and aggregation effects cause chromatographic separations unrelated to molecular size. Swift and Posner (1971) noted that Sephadex gel-filtration fractionation was dependent on the concentration of applied humic acid; this concentration dependence is illustrated in Fig. 12. These humic acid fractionations indicate that adsorption of humic acid on the Sephadex becomes more pronounced as sample concentration is increased. Other phenomena that affect adsorption and aggregation of humic materials during gel filtration are changes in ionic strength and pH, because of

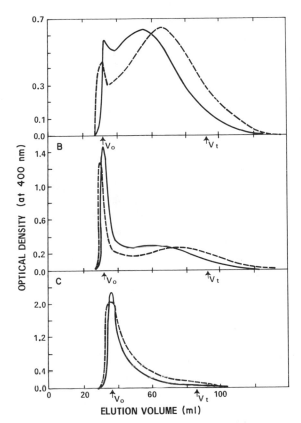

Fig. 13. Fractionation of (A) humic acid (pyrophosphate extract) on Sephadex G-100; (B) humic acid (NaOH extract) on Sephadex G-100; and (C) humic acid (pyrophosphate extract) on Sephadex G-50. Solid line is borate buffer eluent; broken line is tris buffer eluent. (From Swift and Posner, 1971. Published with permission of the Williams and Wilkins Co.)

desalting of humic materials during fractionation. Because most of the reactive functional groups in humic materials that cause adsorption and aggregation are acidic carboxyl and phenolic hydroxyl groups, Swift and Posner (1971) decided to reduce their reactivity by forming ion pairs with alkylamine and borate buffers. Humic acid fractionations obtained with tris(hydroxymethyl)aminomethane and borate buffers as eluents for gel filtration are shown in Fig. 13. These fractionations indicate that humic materials form a continuous molecular-size distribution broadly grouped about one or two mean sizes, rather than the discontinuous distribution grouped around several molecular weights as indicated by Fig. 12. The ultrafiltration fractionation shown in Fig. 11 also indicates a continuum of sizes for humic materials in water.

A direct comparison by Gjessing (1973) of the techniques of gel filtration and ultrafiltration for molecular-size fractionation of humic materials in water showed that the ultrafiltration method was preferable. In this study, concentrated humic substances from water were first fractionated by Sephadex gel filtration using distilled water as the eluent, and each fraction from the gel filtration was secondarily fractionated by ultrafiltration membranes. Molecular size results indicated a poor correspondence between the two methods. Using ultrafiltration of fractions from gel filtration as the criterion for size fractionation, the MW >50,000 fraction at the high molecular weight extreme had ~25% lower molecular weight material. At the low molecular weight extreme, the MW <700 fraction contained ~70% higher molecular weight material. Gjessing attributes these discrepancies to adsorption and disaggregation behavior of humic materials during gel filtration.

The general theory and practice of ultrafiltration is presented by Michaels (1968). Ultrafiltration is the term used to describe separations involving solutes of molecular dimensions greater than 10 solvent molecular diameters and below the resolution of the optical microscope; whereas, reverse osmosis involves solutes whose molecular dimensions are within one order of magnitude of those of the solvent. Ultrafiltration membranes are subject to the same adsorption and fouling problems as osmotic membranes. Polar neutral solutes pass through ultrafiltration membranes more readily than charged ionic solutes of comparable size. Adsorption effects are much less for ultrafiltration than gel filtration because of higher flow rates past the adsorbent surface and a smaller adsorbent surface area for ultrafiltration. Ultrafiltration membranes can be used as disks held in a stirred, pressurized ultrafiltration cell (Michaels, 1968), or can be used in the hollow-fiber configuration (Bio-Rad Laboratories, 1972), which provides a large effective filtration area.

An interesting application of ultrafiltration to water analysis was the characterization of soluble organic matter in a landfill leachate (Chian and DeWalle, 1977). Low molecular weight fatty acids, which permeated a MW 500 ultrafiltration membrane, were the largest organic fraction. Humic materials in the intermediate molecular weight range (500–100,000) were the second largest fraction, and significant amounts of proteins and carbohydrates were found in the highest molecular weight fractions (>100,000).

Dialysis membranes are similar to ultrafiltration membranes. Dialysis differs from ultrafiltration in that the driving force is a concentration gradient across the membrane instead of a pressure gradient. Dialysis is generally used to desalt high molecular weight organic concentrates, such as humic and fulvic acids, which are dialyzed against distilled water (Malcolm, 1976).

C. Precipitation and Coprecipitation

Most high molecular weight and certain low molecular weight organic materials in water can be removed and concentrated from water solution by coprecipitation with a carrier precipitate. Coprecipitation removes analytes from solution by adsorption upon or occlusion within a carrier precipitate without exceeding the analytes' solubility, whereas precipitation removes analytes from solution by adjusting solubility-determining parameters (pH, ionic strength, temperature) so that the analyte is no longer soluble. Coprecipitation of humic materials and suspended colloids with alum and various other metallic oxides and hydroxides is a common water-treatment technique; however, there are relatively few analytical applications.

One of the first reported applications was by Jeffrey and Hood (1958), who concentrated ^{14}C-labeled organic matter from an aged algae culture in seawater by coprecipitation with hydrated iron and magnesium oxides. Recoveries up to 79% and concentration factors of 10^4 were achieved. Williams and Zirino (1964) tested iron-, aluminum-, and copper-hydrated metal oxides as coprecipitate carriers for removal of organic matter in seawater. The most effective carrier was Fe^{3+} added to seawater at a concentration of 20 mg/L and a pH of 4 to 6. These conditions coprecipitated 40–60% of the dissolved organic carbon from seawater.

In fresh waters, Ziechmann (1976) coprecipitated humic and fulvic acid from the "black waters" of the Amazon region with aluminum hydroxide. Aluminum chloride was added to these acid waters, pH was adjusted to 6.5 with sodium hydroxide, and aluminum hydroxide coprecipitate was filtered from solution. Humic and fulvic acids were released from the precipitate by treatment with hydrochloric acid, which dissolved the aluminum hydroxide, but infrared spectra of the humic material showed the presence of residual aluminum oxides.

Various low molecular weight organic solutes may or may not be coprecipitated with humic materials. Sridharan and Lee (1972) found that natural organic matter in lake water increased recovery of phenol by coprecipitation with iron hydroxide, decreased recovery of citric acid, and had no effect on recovery of glycine.

Direct precipitation of humic materials as lead humates from bog waters was employed by Klocking and Mucke (1969). Lead-humate precipitates were formed by addition of lead(II) nitrate, and after filtration, humates were released from the precipitate by extraction with ammonium hydroxide.

Humic acids (not fulvic acids) can be directly precipitated from water (Martin and Pierce, 1971). Precipitation occurs by adding glacial acetic acid and isoamyl alcohol to the sample in a separatory funnel. After shaking this mixture, humic acid precipitates at the water–alcohol interface, and precipitation is complete after 5 h. The amount of humic acid in the sample can be determined by filtering and washing the precipitate and measuring it spectrophotometrically after dissolution of the precipitate in 0.5 N NaOH.

D. Water Evaporation

The low-temperature, reduced-pressure distillation techniques for water removal discussed in Section V for low molecular weight nonvolatiles are also applicable to high molecular weight constituents in water with this exception: freeze-drying should always be used when the sample is taken to dryness. When water is evaporated in the liquid state, solutes tend to migrate toward the surface layers in liquid concentrates, where high concentrations may damage high molecular weight solutes by denaturation; this process manifests itself as a case-hardened product that cannot be redissolved or redispersed in water (Broughton, 1956). In freeze-drying, the solutes are "locked" into their soluble and dispersed configurations during drying so that denaturation cannot occur.

Procedures for freeze-drying large volumes of water for concentrating humic materials are given by Malcolm (1968). He reported that the sample should be quick-frozen to avoid the possibility of precipitation of humic material through freeze concentration during slow freezing. Prior to freeze-drying, humic and fulvic acid preparations may be cleaned up by pressure filtration to remove sediment and large colloidal constituents, dialyzed to remove inorganic salts and low molecular weight nonhumic organics, and passed through hydrogen-form cation-exchange resins to remove complexed metals (Malcolm, 1976). Freeze-drying of humic materials can safely be carried out in the presence of neutral inorganic salts, volatile inorganic bases (NH_4OH), and volatile inorganic acids (HCl, H_2CO_3). However, freeze-drying of liquid concentrates that contain nonvolatile inorganic bases (NaOH) or acids (H_2SO_4, H_3PO_4) results in hydrolyzed and charred preparations. Freeze-drying is usually the final concentration step after the bulk of the water has been removed by a rotary evaporator under reduced pressure (Beck et al., 1974) or by vacuum distillation (Katz et al., 1972).

E. Other Techniques

High molecular weight natural polyelectrolytes are almost completely inextractable by nonpolar solvents and only partially extractable by polar solvents such as butanol. However, an elegant solvent extraction technique for humic acid, fulvic acid, and lignosulfonic acids in water was developed by Eberle and Schweer (1974). They used trioctylamine dissolved in chloroform to solvent-extract humic materials and lignosulfonates. Trioctylamines formed neutral ion pairs with organic acid polyelectrolytes, which were then extracted into chloroform. At pH 5, nearly 100% extraction efficiency was obtained. To recover the humic materials and lignosulfonates, the chloroform extract was back extracted with water at pH 10 or above. The trioctylamine remained in the water, and the humic materials and lignosulfonates partitioned into the water. A series of continuous extractors were set up, which first extracted low molecular weight volatiles and semivolatiles into chloroform, and then extracted polar nonvolatile acids and humic materials into trioctylamine–chloroform.

Freeze concentration has been successfully applied for an 80-fold concentration of fulvic acids in natural waters (Black and Christman, 1963). However, this procedure took 110 h of continuous operation, whereas rotary evaporation under partial vacuum of the same sample obtained a 40-fold concentration in only 6 h. Formation of humate precipitates at the low temperature of freeze concentration can also cause recovery problems as a result of precipitate entrapment in ice.

VII. COMPREHENSIVE ANALYTICAL SCHEMES

Individual techniques for concentrating, partitioning, and isolating organic constituents in water have been presented in this chapter. Comprehensive organic analysis of water requires analytical schemes that combine various techniques in a logical, systematic sequence. Christman and Hrutfiord (1973) stated that one of the most critical needs in water quality research was to develop comprehensive analytical schemes for isolation and identification of all classes of organic materials in aquatic environments. Some progress has been made in this direction; however, the National Academy of Sciences (1977) has recommended research to better characterize and more extensively determine the nature of TOC in drinking water, in which they estimated that an average of only 10% of the TOC was identified. Development of rapid and sensitive techniques to assay for TOC and DOC in water has given water chemists a valuable tool for devising quantitative and comprehensive analytical schemes for potentially determining all organic compounds in various waters.

Determination of organic constituents in water can be terminated at varying levels of specificity. Figure 14 (Leenheer and Huffman, 1976) illustrates a proposed organic solute classification that forms a pyramid with increasing specificity of analysis. Karger *et al.* (1973) cite the following considerations for designing multistep separation schemes for complex samples:

1. What methods (and how many) will be required?
2. In what sequence will these methods be used?
3. What scale of separation will be used; how much sample must be separated?
4. At what stage will separation be terminated?

For most complex water samples, comprehensive analysis requires more specific determinations than TOC or poorly defined groups of compounds; however, analysis at the specific compound level is presently impossible in some cases because of the complexity of polymeric high molecular weight constituents in water. Manpower, equipment, time, and funding considerations also tend to limit comprehensive organic analysis of water. The end result is always a compromise between the desired objective to analyze for all organic constituents in water and what is possible with the resources available.

A. Schemes Based on Solvent Extraction

Dry organic concentrate mixtures have classically been solvent partitioned into various acid, base, and neutral classes by the well-known method described in Shriner and Fuson (1948). This fractionation scheme was subsequently applied to solvent extracts from the carbon filter (Braus *et al.*, 1951) and then directly to water, as shown in Fig. 15 (Braus *et al.*, 1952). This basic scheme with various modifications has been widely used for organic analysis of water up to the present, and is partially responsible for the estimated 10% of organic constituents identified in drinking water. One reason 90% of organic constituents in drinking water remain unidentified is that they remain in the "water-soluble, nonvolatile" fraction of Fig. 15, and therefore are not easily amenable to gas chromatograph–mass spectrometric analysis.

One advantage of solvent extraction schemes compared to schemes that use sorbent filters is that solvents can extract sediments either suspended or previously separated from water. An example of a solvent extraction scheme for dry sediments separated from domestic sewage is

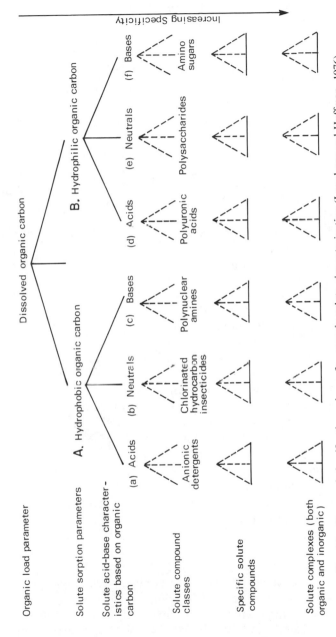

Fig. 14. Proposed classification scheme for organic solute characterization (Leenheer and Huffman, 1976).

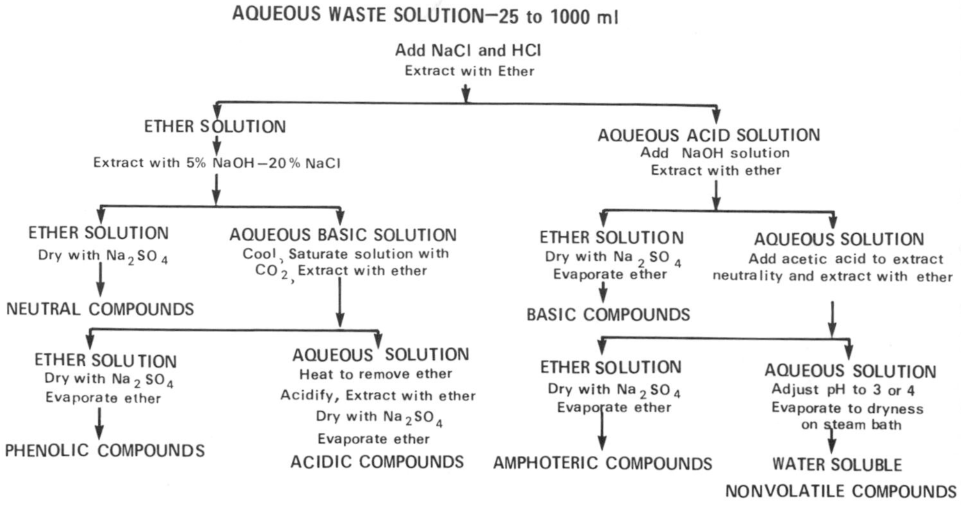

AQUEOUS WASTE SOLUTION—25 to 1000 ml

Add NaCl and HCl
Extract with Ether

ETHER SOLUTION
Extract with 5% NaOH—20% NaCl

AQUEOUS ACID SOLUTION
Add NaOH solution
Extract with ether

ETHER SOLUTION
Dry with Na_2SO_4

AQUEOUS BASIC SOLUTION
Cool, Saturate solution with
CO_2, Extract with ether

ETHER SOLUTION
Dry with Na_2SO_4
Evaporate ether

AQUEOUS SOLUTION
Add acetic acid to extract
neutrality and extract with ether

NEUTRAL COMPOUNDS

BASIC COMPOUNDS

ETHER SOLUTION
Dry with Na_2SO_4
Evaporate ether

AQUEOUS SOLUTION
Heat to remove ether
Acidify, Extract with ether
Dry with Na_2SO_4
Evaporate ether

ETHER SOLUTION
Dry with Na_2SO_4
Evaporate ether

AQUEOUS SOLUTION
Adjust pH to 3 or 4
Evaporate to dryness
on steam bath

PHENOLIC COMPOUNDS

ACIDIC COMPOUNDS

AMPHOTERIC COMPOUNDS

WATER SOLUBLE
NONVOLATILE COMPOUNDS

Fig. 15. Fractionation of organic compounds in water by solvent extraction. (Reprinted with permission from Braus *et al.*, 1952. *Anal. Chem.* Copyright by the American Chemical Society.)

shown in Fig. 16 (Hunter and Heukelekian, 1965). The total grease fraction and carbohydrate and lignin fraction were fractionated further by additional water–solvent partitioning. Organic solutes in the sewage were concentrated by freeze-drying the filtrate and fractionating the residue by solvent partitioning as described by Braus *et al.* (1951).

Solvent extraction was used by Rebhun and Manka (1971) to determine essentially 100% of COD in a secondary sewage effluent. Determinations made at the compound class level found that 40–50% of organic constituents were humic substances, ~22% proteins, ~14% anionic detergents, ~12% carbohydrates, 8% ether extractables, and ~2% tannins. With current analytical techniques, specific compound analyses of the humic fraction, proteins, tannins, and the high molecular weight carbohydrates are not feasible because of the structural complexity of these constituents, which together constitute ~80% of the COD of this wastewater. Specific compound analysis can be performed on the ether extractables, anionic detergents, and low molecular weight carbohydrates.

Classical techniques of solvent extraction for comprehensive organic characterization of water are well established but are relatively tedious and ineffective for high molecular weight fractions. The solvent extraction method of Eberle and Schweer (1974) for humic materials in water could be an important addition to solvent extraction schemes, and its utility should be tested.

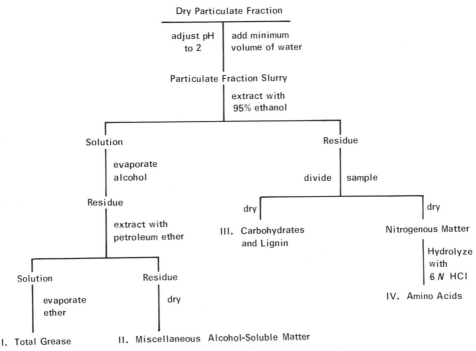

Fig. 16. Fractionation of particulates in domestic sewage by solvent extraction. (From Hunter and Heukelekian, 1965). Published with permission from the *Water Pollution Control Federation.*)

B. Schemes Based on Sorbents

The carbon filter, with its chloroform and alcohol extracts (Buelow *et al.,* 1973a,b), is the most common scheme used to concentrate and isolate organic constituents from water, but it is not comprehensive, as recoveries are usually <50% of the TOC in water.

Synthetic macroreticular-resin sorbents were used to separate all organic solutes in water according to the classification given in Fig. 14. The analytical scheme, termed DOC fractionation analysis, which is the basis for the classification, is shown in Fig. 17 (Leenheer and Huffman, 1979). In DOC fractionation analysis, hydrophobic solutes are first isolated from the sample by physical sorption on XAD-8 resin, then ion-exchange resins are used to isolate hydrophilic acids and bases. Irreversible sorbtion on strong-acid and strong-base exchange resins is minimized by prior removal of hydrophobic solutes, which are often irreversibly sorbed on polystyrene–divinylbenzene matrix exchange resins. The types of com-

TABLE IV

Compound Classes in Organic Solute Fractions

Hydrophobic bases	One- and two-ring aromatic amines except pyridine
Hydrophobic acids	Aliphatic carboxylic acids of five to nine carbons, one- and two-ring aromatic carboxylic acids, one- and two-ring phenols, fulvic acid
Hydrophobic neutrals	Hydrocarbons; aliphatic alcohols, amides, esters, ketones, and aldehydes of > five carbons; aliphatic carboxylic acids and aliphatic amines of > nine carbons; aromatic carboxylic acids and aromatic amines of three or more rings.
Hydrophilic bases	Aliphatic amines ≤ nine carbons, amino acids, pyridine
Hydrophilic acids	Aliphatic acids of ≤ five carbons, polyfunctional acids
Hydrophilic neutrals	Aliphatic amides, alcohols, aldehydes, esters, and ketones of ≤ five carbons; polyfunctional alcohols; carbohydrates

pounds found in the six fractions generated by DOC-fractionation analysis are listed in Table IV.

DOC fractionation analysis has been mainly used in a diagnostic manner, whereby only sixfold fractionation of DOC is determined. A standardized procedure for this analytical-scale methodology (Leenheer and Huffman, 1979) requires only 200 mL of a filtered sample and should be applied to samples whose DOC = 5–25 mg/L and whose specific conductance is <2000 μmho/cm at 25°C.

DOC fractionation analysis can also be scaled up and modified for use as a preparative fractionation procedure. Hydrophobic neutrals, hydrophilic bases, and hydrophilic acids are determined by DOC differences in the analytical procedure of Fig. 17; however, in the preparative procedure, hydrophobic neutrals are eluted from XAD-8 resin with methanol; hydrophilic bases are eluted from cation-exchange resin with 0.5 N NaOH; and hydrophilic acids are eluted from anion-exchange resin with 0.5 N HCl. The hydrophilic neutral fraction, because it is in deionized water, can be concentrated further by freeze concentration. Recoveries of hydrophobic solutes from XAD-8 are in the 90–100% range, whereas recoveries from ion-exchange resins are in the 60–90% range. Overall recoveries based on organic carbon analysis are 80–90%. An evaluation of the utility and applications of DOC fractionation analysis for natural waters was made by Stuber and Leenheer (1979), and for wastewaters from oil-shale processing by Leenheer and Farrier (1979).

Hydrophilic ion-exchange cellulose sorbents and Sephadex were used by Sirotkina *et al.* (1974) to systematically analyze for organic solutes in natural waters. The fractionation scheme is shown in Fig. 18. DEAE

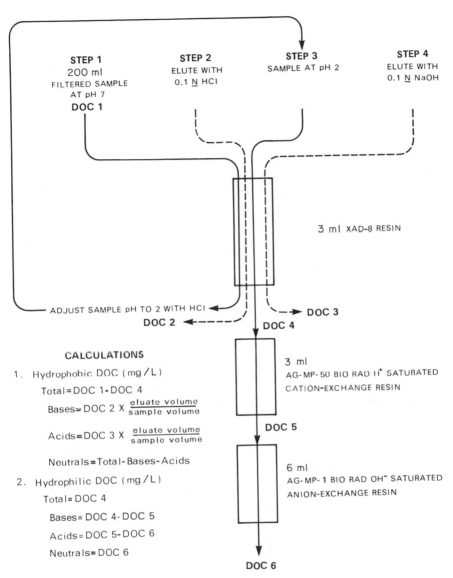

Fig. 17. Analytical scheme for DOC fraction analysis.

cellulose was used in the free-base form, and CM cellulose was used in the acid form. Sorption of natural organic acids on DEAE cellulose and organic bases on CM cellulose was apparently by hydrogen bonding as well as by ion exchange, because functional exchange groups were only

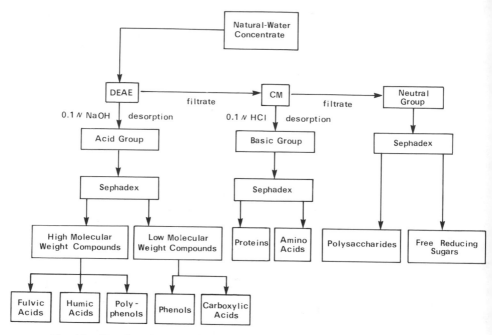

Fig. 18. Fractionation of organic solutes in water by ion-exchange celluloses and Sephadex. (From Sirotkina *et al.*, 1974. Published with permission from Plenum Publishing Corporation.)

slightly ionic at neutral pH where sorption occurred. Irreversible sorption of hydrophobic solutes was not a problem because of the hydrophilic nature of cellulose sorbents. Natural waters were first concentrated by freeze concentration and the sample was passed through these two sorbents in series without pH adjustment. Gel filtration on Sephadex was used on the acid, base, and neutral fractions to desalt the samples and broadly fractionate into low and high molecular weight components. Recovery studies based on standard additions determined that organic solute losses during fractionation did not exceed 10%, and the fractionation procedure was applied to five different river-water samples.

C. Schemes Using Various Technique Combinations

Reverse osmosis, ultrafiltration, and gel filtration were used in that order by DeWalle and Chian (1974) to size-fractionate organic solutes found in sewage effluents after their passage through activated-carbon columns. In this case, size-fractionations using membranes and gel filtration were especially successful, because those organic solutes that interfere through sorption interactions were no longer in the sample as a result of its passage through activated carbon.

The solvent extraction scheme of Rebhun and Manka (1971) was combined with gel filtration on Sephadex to size-fractionate humic and fulvic acid (Manka *et al.,* 1974). Most of the humic substances were found in the MW 1000–5000 range. To characterize humic materials in natural waters, Black and Christman (1963) first concentrated the water sample either by vacuum distillation or by freeze concentration. The liquid concentrate was then extracted with diethyl ether to isolate natural waxes, fats, and hydrocarbon. Humic acid was precipitated from the concentrate by acidification to pH 1, and after filtration, hymatomelanic acid was extracted from the humic acid with ethyl alcohol.

These examples of multistep separation schemes using various techniques are a few of many different schemes. However, most reported schemes do not attempt to analyze for all organic constituents in water and do not quantify compound class recoveries by organic carbon analysis. Many schemes have been developed only for industrial wastewaters and domestic sewage that contain high concentrations of organic constituents.

D. Proposed Comprehensive Isolation and Separation Scheme for Organic Solutes in Water

After evaluating current procedures presented in this chapter, a possible analytical scheme for concentration, isolation, and compound class separations is presented in Fig. 19. General advantages include (1) few steps, (2) no sample splitting, (3) no addition of acids, bases, solvents, or chemical reagents to the sample, (4) little change in pH during fractionation, and (5) no use of immiscible organic solvents.

The water sample should be pressure filtered on site through a 0.45-μm silver-membrane filter to remove sediment (Malcolm and Leenheer, 1973). The filtrate could be collected in the closed-loop stripping apparatus of Grob and Grob (1974), and the neutral volatiles and semivolatiles isolated by sparging and trapping on site to avoid volatility losses and chemical transformations. Neutral volatiles and most semivolatiles can be directly analyzed by thermal desorption of the trap into a gas chromatograph. DOC change in the sample before and after sparging is a direct measure of sparged organic constituents. These volatiles and semivolatiles should constitute the majority of organic compounds routinely identified in water.

Very volatile compounds, such as the halomethanes, are not effectively collected in the closed-loop stripping apparatus. If halomethane

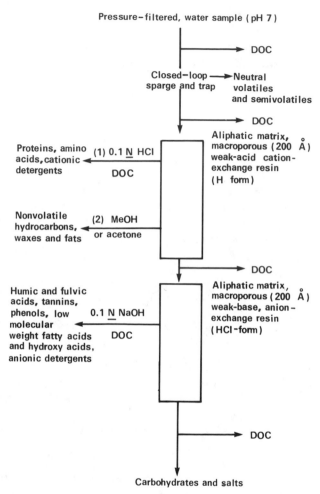

Fig. 19. Proposed comprehensive isolation and separation scheme for organic solutes in water.

analysis is desired, a separate sample should be collected and preserved, and shipped to the laboratory for analysis by the sparge-and-trap procedure specifically designed for halomethanes (Keith *et al.*, 1979).

After completing the sparge-and-trap procedure, the sample should be preserved by chilling on ice and shipped to the laboratory for passage through ion-exchange sorbents. Weak-acid and weak-base ion-exchange resins were chosen, based on the high capacity factor and recovery data given by Abrams (1969) and Kim *et al.* (1976). Weak-acid and weak-base ion-exchange resins should have higher capacities than ion-exchange celluloses used in a similar manner by Sirotkina *et al.* (1974). Aliphatic resin matrixes are desired to eliminate the possibility of irreversible sorption through π-bonding; average pore diameter should be 200 Å to allow for rapid diffusive transport of macromolecules within macroporous beads. Ion-exchange resins with the exact specifications desired are not yet available, although Bio-Rex 70 cation-exchange resin and AG3-X4A anion-exchange resin from Bio-Rad Laboratories have most of the desirable characteristics.

The cation-exchange resin was placed before the anion-exchange resin because amphoteric amino acids and proteins would be isolated with other organic acids in the reverse-column arrangement. The sample should be near pH 7 during passage through each column for optimization of hydrogen-bonding sorption. DOC determinations of sample and aqueous column eluates quantify the compound class separations and determine organic solute recoveries.

Proteins and amino acids in the 0.1 N HCl eluate can be determined by high-resolution, ion-exchange, and ligand-exchange chromatography. Cationic detergents can be determined by a number of solvent extraction and colorimetric procedures. Nonvolatile hydrocarbons and natural waxes and fats may be determined by a number of HPLC techniques.

The 0.1 N NaOH eluate should contain the largest assortment of organic acids and inorganic salts. Hydrophobic acid constituents (humic acid, fulvic acid, tannins, phenols, and fatty acids $>C_5$) could be isolated on XAD-8 resin after acidifying the sample to pH 2, and the remaining low molecular weight fatty acids and hydroxy acids in the eluate from XAD-8 could be directly determined by high-resolution, anion-exchange chromatography (Pitt *et al.*, 1975). Anionic detergents might be isolated by solvent sublation and determined colorimetrically.

After passage through both ion-exchange resins, the sample should only contain nonvolatile neutrals such as carbohydrates, and the sample can be evaporated by vacuum distillation. Gel filtration or ultrafiltration can then be used to desalt and fractionate the carbohydrates into low and high molecular weight fractions. The low molecular weight carbohydrates can be directly analyzed by liquid chromatography, but the high molecular weight carbohydrates (as well as proteins, humic materials, and tannins) require considerable research and analysis for structural characterization.

The proposed analytical scheme of Fig. 14 is only one combination of a large number of possibilities. As new techniques and refinements of existing techniques are presented, proven, and applied, the complex task of concentrating, partitioning, and isolating organic materials in water will undoubtedly be simplified.

VIII. SUMMARY

Since the early 1970s, there has been rapid development in concentration, partitioning, and isolation techniques for organic constituents in water. The most significant developments include the sparge-and-trap technique for volatile analyses, the development and use of synthetic resin adsorbents that replace techniques based on solvent extraction and carbon adsorption, and applications of high performance liquid chromatography to sample concentration, fractionation, and cleanup. These and other advancements have lowered the limit of detection for many organic constituents by orders of magnitude, have reduced the volume of water sample required, have reduced the number of steps for analytical procedures, and have reduced the time of analysis.

Future advancements depend largely upon continued research interest in organic analysis of water. Because considerable basic theory and technology for organic separations has been developed, new techniques can systematically be created from a theory and technology base rather than from the more inefficient trial-and-error approach that has often been used in the past. Frequently, separation techniques are developed for industrial applications, and adaptations for analytical applications follow. Presently, the techniques of reverse osmosis, liquid-membrane separations, and adsorption on weak-acid and weak-base, ion-exchange resins are applied primarily to industrial processes, and analytical applications of these techniques are waiting to be developed. Most of the attention in organic analysis of water has been given to low molecular weight volatile and semivolatile organic constituents, whereas analysis of high molecular weight constituents, which constitute the bulk of organic matter dissolved in water, demand new and different types of concentration, partitioning, and isolation procedures. Lastly, with the combination of analytical procedures into analytical schemes in which concentration, partitioning, and isolation procedures play an important part, comprehensive quantitative organic analysis of water is progressing from a dream into a reality.

REFERENCES

Abrams, I. M. (1969). *Chem. Eng. Prog., Symp. Ser.* **65**, 106–112.
Abrams, I. M., and Breslin, R. P. (1965). *26th Annu. Meet. Int. Water Conf. Eng. Soc. West. Pa., Pittsburg, 1965* pp. 1–7.
Abrams, I. M., and Lewon, S. M. (1962). *J. Am. Water Works Assoc.* **54**, 537–543.
Afgan, B. K., Belliveau, P. E., LaRose, R. H., and Ryan, J. F. (1974). *Anal. Chim. Acta* **71**, 355–366.
Ahling, B., and Jensen, S. (1970). *Anal. Chem.* **42**, 1483–1486.
Ahnoff, M., and Josefsson, B. (1974). *Anal. Chem.* **46**, 658–663.

Aiken, G., Thurman, E. M., Malcolm, R. L., and Walton, H. F. (1979). *Anal. Chem.* **51**, 1799–1803.

American Public Health Association, American Water Works Association, and Water Pollution Control Federation (1976a). "Standard Methods for the Examination of Water and Waste Water," 14th ed., pp. 535–543. Am. Public Health Assoc., Washington, D.C.

American Public Health Association, American Water Works Association, and Water Pollution Control Federation (1976b). "Standard Methods for the Examination of Water and Waste Water," 14th ed., pp. 600–603. Am. Public Health Assoc., Washington, D.C.

American Society for Testing Materials (ASTM) (1978a). "1978 Annual Book of ASTM Standards, Part 31–Water," D1193-77, pp. 20–22. Am. Soc. Test. Mater., Philadelphia, Pennsylvania.

American Society for Testing Materials (ASTM) (1978b). "1978 Annual Book of ASTM Standards, Part 31–Water," D2778-70, pp. 532–537. Am. Soc. Test. Mater., Philadelphia, Pennsylvania.

American Society for Testing Materials (ASTM) (1978c). "1978 Annual Book of ASTM Standards, Part 31–Water," D3478-75T, pp. 607–614. Am. Soc. Test. Mater., Philadelphia, Pennsylvania.

Andelman, J. B., and Caruso, S. C. (1971). *In* "Water and Water Pollution Handbook" (L. L. Ciacco, ed.). Vol. II, pp. 483–591. Dekker, New York.

Aue, W. A., Hastings, C. R., Gerhardt, K., Pierce, J. O., II, Hill, W. H., and Moseman, R. F. (1972). *J. Chromatogr.* **72**, 259–267.

Austern, B. M., Dobbs, R. A., and Cohen, J. M. (1975). *Environ. Sci. Technol.* **9**, 588–590.

Baker, R. A. (1969). Concentration of trace organic contaminants from aqueous solution by freezing. Sc. D. Hyg. Thesis. Available from University Microfilms Inc., Order No. 70-17,060, Ann Arbor, Michigan.

Baker, R. A., and Yates, M. L. (1975). *Geol. Sur. Open-File Rep. (U.S.)*

Beck, K. C., Reuter, J. H., and Perdue, E. M. (1974). *Geochim. Cosmochim. Acta* **38**, 341–364.

Bedford, J. W. (1974). *Bull. Environ. Contam. Toxicol.* **12**, 622–625.

Bellar, T. A., and Lichtenberg, J. J. (1974). *J. Am. Water Works Assoc.* **66**, 739–744.

Bergert, K. H., Betz, V., and Pruggmayer, D. (1974). *Chromatographia* **7**, 115–121.

Beroza, M., and Bowman, M. C. (1967). *Anal. Chem.* **39**, 1200–1203.

Beroza, M., Bowman, M. C., and Bierl, B. A. (1972). *Anal. Chem.* **44**, 2411–2413.

Besik, F. (1971). *Water Sewage Works* **118**, 213–219.

Bethge, P. O., and Lindstrom, K. (1974). *Analyst (London)* **99**, 137–142.

Bio-Rad Laboratories (1972). "The New Hollow Fiber Technique," Bull. 1004. Bio-Rad Laboratories, Richmond, California.

Black, A. P., and Christman, R. F. (1963). *J. Am. Water Works Assoc.* **55**, 897–912.

Bowman, M. C., and Beroza, M. (1966). *Anal. Chem.* **38**, 1427–1428.

Boyer, S. L., Guin, K. F., Kelley, R. M., Mausner, M. L., Robinson, H. F., Schmitt, T. M., Stahl, C. R., and Setzkorn, E. A. (1977). *Environ. Sci. Technol.* **11**, 1167–1171.

Braus, H., Middleton, F. M., and Walton, G. (1951). *Anal. Chem.* **23**, 1160–1164.

Braus, H., Middleton, F. M., and Ruchhoff, C. C. (1952). *Anal. Chem.* **24**, 1872–1876.

Bronstad, J. O., and Friestad, H. O. (1976). *Analyst (London)* **101**, 820–824.

Broughton, G. (1956). *In* "Techniques of Organic Chemistry" (A. Weissberger, ed.), 2nd ed., Vol. III, Part I, pp. 818–838. Wiley (Interscience), New York.

Buckley, S. E., Hocott, C. R., and Taggart, M. S., Jr. (1958). *In* "Habitat of Oil," pp. 850–852. Am. Assoc. Pet. Geol., Tulsa, Oklahoma.

Buelow, R. W., Carswell, J. K., and Symons, J. M. (1973a). *J. Am. Water Works Assoc.* **65**, 57–72.

Buelow, R. W., Carswell, J. K., and Symons, J. M. (1973b). *J. Am. Water Works Assoc.* **65,** 195–199.

Burnham, A. K., Calder, G. V., Fritz, J. S., Junk, G. A., Svec, H. J., and Willis, R. (1972). *Anal. Chem.* **44,** 139–142.

Burnham, A. K., Calder, G. V., Fritz, J. S., Junk, G. A., Svec, H. J., and Vick, R. (1973). *J. Am. Water Works Assoc.* **65,** 722–7245.

Cabasso, I., Klein, E., and Smith, J. K. (1975a). *Coat. Plast. Prepr. Pap. Meet. Am. Chem. Soc., Div. Org. Coat. Plast. Chem.* **35,** 498–502.

Cabasso, I., Eyer, C. S., Klein, E., and Smith, J. K. (1975b). "Evaluation of Semipermeable Membranes for Concentration of Organic Contaminants in Drinking Water," EPA Rep. No. 670/1-75-001 (NTIS PB-243-245). Environ Prot. Agency, Washington, D.C.

Cahn, R. P., and Li, N. N. (1974). *Sep. Sci.* **9,** 505–519.

Calderbank, A. (1966). *Residue Rev.* **12,** 14–34.

Carter, J. M., and Nickless, G. (1970). *Analyst (London)* **95,** 148–152.

Chang, R. C. Y. (1976). "Concentration and Determination of Trace Organic Pollutants in Water," NTIS Rep. No. IS-T-730. U.S. Energy and Development Admin.

Chen, Y., Khan, S. W., and Schnitzer, M. (1978). *Soil Sci. Soc. Am. J.* **42,** 292–296.

Chian, E. S. K., and DeWalle, F. B. (1977). *Environ. Sci. Technol.* **11,** 158–163.

Chian, E. S. K., Kuo, P. P. K., Cooper, W. J., Cowen, W. F., and Fuentes, R. C. (1977). *Environ. Sci. Technol.* **11,** 282–285.

Chrisp, P. T., Eckert, J. M., and Gibson, N. A. (1975). *Anal. Chim. Acta* **78,** 391–396.

Christman, R. F., and Ghassemi, M. (1966). *J. Am. Water Works Assoc.* **58,** 723–741.

Christman, R. F., and Hrutfiord, B. F. (1973). *Proc. Water Qual. Conf.* **15,** 58–72.

Chriswell, C. D. (1977). *J. Chromatogr.* **132,** 537–542.

Chriswell, C. D., Ericson, R. L., Junk, G. A., Lee, K. W., Fritz, J. S., and Svec, H. J. (1977). *J. Am. Water Works Assoc.* **70,** 669–674.

Coburn, J. A., Valdmanis, I. A., and Chau, A. S. Y. (1977). *J. Assoc. Off. Anal. Chem.* **60,** 225–228.

Coleman, W. E., Lingg, R. D., Melton, R. G., and Kopfler, F. C. (1976). *In* "Identification and Analysis of Organic Pollutants in Water" (L. H. Keith, ed.), pp. 305–327. Ann Arbor Sci. Publ., Ann Arbor, Michigan.

Cowen, W. F., Cooper, W. J., and Highfill, J. W. (1975). *Anal. Chem.* **47,** 2483–2485.

Cussler, E. L., and Evans, D. F. (1974). *Sep. Purif. Methods* **3,** 399–421.

Degens, E. T., and Reuter, J. H. (1964). *Adv. Org. Geochem., Proc. Int. Meet., 1st, 1962 Int. Ser. Monogr. Earth Sci.,* Monogr., Vol. 15, pp. 377–402.

Deinzer, M., Melton, R., and Mitchell, D. (1975). *Water Res.* **9,** 799–805.

Determann, H. (1969). "Gel Chromatography," 2nd ed. Springer-Verlag, Berlin and New York.

DeWalle, F. B., and Chian, E. S. K. (1974). *J. Environ. Eng. Div. [Am. Soc. Civ. Eng.]* **100,** 1089–1104.

Dowty, B., and Laseter, J. L. (1975). *Anal. Lett.* **8,** 25–32.

Duce, R. A., Quinn, J. G., Olney, C. E., Piotrowicz, D. R., Ray, B. J., and Wade, T. L. (1972). *Science* **176,** 161–163.

Easty, D. B., and Wabers, B. A. (1977). *Anal. Lett.* **10,** 857–867.

Eberle, S. H., and Schweer, K. H. (1974). *Vom Wasser* **41,** 27–44.

Edwards, V. H., and Schubert, P. F. (1974). *J. Am. Water Works Assoc.* **66,** 610–615.

Endres, H., and Hormann, H. (1963). *Angew. Chem., Int. Ed. Engl.* **2,** 254–260.

Fedonin, V. F., Guil'dis, V. Y., Malinouskaya, M. N., and Medvedev, A. M. (1970). *Gig. Sanit.* **35,** 78–80.

Felix, W. D., Farrier, D. S., and Poulson, R. E. (1977). *Pac. Chem. Eng. Congr.* [*Proc.*] **2**, 480–485.

Feltz, H. R., and Culbertson, J. K. (1972). *Pestic. Monit. J.* **6**, 171–178.

Florence, T. M., and Farrar, Y. J. (1973). *Anal. Chim. Acta* **63**, 255–261.

Foster, D. H., Engelbrecht, R. S., and Snoeyink, V. L. (1977). *Environ. Sci. Technol.* **11**, 55–61.

Gardner, W. S., and Lee, G. F. (1973). *Environ. Sci. Technol.* **7**, 719–724.

Garrison, A. W., Pope, J. W., and Allen, F. R. (1976). *In* "Identification and Analysis of Organic Pollutants in Water" (L. H. Keith, ed.), pp. 517–556. Ann Arbor Sci. Publ., Ann Arbor, Michigan.

Gesser, H. D., Chow, A., Davis, F. C., Uthe, J. F., and Reinke, J. (1971). *Anal. Lett.* **4**, 883–886.

Geyer, W. (1974). *Anal. Chem.* **46**, 1359.

Giam, C. S., and Wong, M. K. (1972). *J. Chromatogr.* **72**, 283–292.

Gjavotchanoff, S. (1972). *Vom Wasser* **39**, 37–40.

Gjessing, E. T. (1965). *Nature (London)* **208**, 1091–1092.

Gjessing, E. T. (1970). *Environ. Sci. Technol.* **4**, 437–438.

Gjessing, E. T. (1973). *Schweiz. Z. Hydrol.* **35**, 286–294.

Gjessing, E. T., and Lee, G. F. (1967). *Environ. Sci. Technol.* **1**, 631–638.

Glaze, W. H., Peyton, G. R., and Rawley, R. (1977). *Environ. Sci. Technol.* **11**, 685–690.

Goerlitz, D. F., and Brown, E., eds. (1972). *In* "Techniques of Water-Resources Investigations of the U.S. Geological Survey, Methods for Analysis of Organic Substances in Water," Book 5, Chapter A3, pp. 30–40. U.S. Govt. Printing Office, Washington, D.C.

Goldberg, M. C., DeLong, L. L., and Sinclair, M. (1973). *Anal. Chem.* **45**, 89–93.

Gordon, A. J., and Ford, R. A. (1970). "The Chemist's Companion," pp. 429–436. Wiley, New York.

Gough, K. M., and Gesser, H. D. (1975). *J. Chromatogr.* **115**, 383–390.

Greed, C. G. (1976). *Res. Dev.* **27**, 40–44.

Grieser, M. D., and Pietrzyk, D. J. (1973). *Anal. Chem.* **45**, 1348–1353.

Grieves, R. B., Charewicz, W., and Brien, S. M. (1974). *Anal. Chim. Acta* **73**, 293–300.

Griffitt, K. R., and Craun, J. A. (1974). *J. Assoc. Off. Anal. Chem.* **57**, 168–172.

Grob, K. (1973). *J. Chromatogr.* **84**, 255–273.

Grob, K., and Grob, G. (1974). *J. Chromatogr.* **90**, 303–313.

Grob, K., and Zurcher, F. (1976). *J. Chromatogr.* **117**, 285–294.

Grob, K., Grob, K., Jr., and Grob, G. (1975). *J. Chromatogr.* **106**, 299–315.

Grote, J. O. (1975). *Am. Lab.* **1**, 47–49.

Gustafson, R. L., Albright, R. L., Heisler, J., Lirio, J. A., and Reid, O. T., Jr. (1968). *Ind. Eng. Chem. Prod. Res. Dev.* **7**, 107–115.

Guy, H. P., and Norman, V. W. (1970). "Techniques of Water-Resources Investigations of the U.S. Geological Survey, Methods for Measurement of Fluvial Sediments," Book 3, Chapter C2. U.S. Govt. Printing Office, Washington, D.C.

Hanai, T., and Walton, H. F. (1977). *Anal. Chem.* **49**, 764–766.

Harper, D. O., and Lemlich, R. (1965). *Ind. Eng. Chem. Process Des. Dev.* **4**, 13–16.

Helfgott, T. B. (1975). "Chemical Analysis and Process Classification of Constituents of Effluents (Organic Nitrogen in Activated Carbon Effluents)," NTIS Rep. PB-249-816. Natl. Tech. Inf. Serv., Springfield, Virginia.

Henderson, J. E., Peyton, G. R., and Glaze, W. H. (1976). *In* "Identification and Analysis of Organic Pollutants in Water" (L. H. Keith, ed.), pp. 105–111. Ann Arbor Sci. Publ., Ann Arbor, Michigan.

Herrmann, V., and Juttner, F. (1977). *Anal. Biochem.* **78**, 365–373.

Ho, C. H., Clark, B. R., and Guérin, M. R. (1976). *J. Environ. Sci. Health, Part A* **A11**, 481–489.

Hoak, R. D. (1962). *Int. J. Air Water Pollut.* **6**, 521–538.

Hunter, J. V. (1971). *In* "Organic Compounds in Aquatic Environments" (S. D. Faust and J. V. Hunter, eds.), pp. 51–94. Dekker, New York.

Hunter, J. V., and Heukelekian, H. (1965). *J. Water Pollut. Control Fed.* **37**, 1142–1163.

Irving, H. (1974). *Chem. Ind. (London)* **17**, 639–644.

Jackson, L. P., Poulson, R. E., Spedding, T. J., Phillips, T. E., and Jensen, H. B. (1975). *Quart. Colo. School of Mines* **70**, 105–134.

Jeffrey, L. M., and Hood, D. W. (1958). *J. Mar. Res.* **17**, 247–271.

Junk, G. A., Svec, H. J., Vink, R. D., and Avery, M. J. (1974a). *Environ. Sci. Technol.* **8**, 1100–1106.

Junk, G. A., Richard, J. J., Grieser, M. D., Witiak, D., Witiak, J. L., Arguello, M. D., Vick, R., Svec, H. J., Fritz, J. S., and Calder (1974b). *J. Chromatogr.* **99**, 745–762.

Junk, G. A., Richard, J. J., Fritz, J. S., and Svec, H. J. (1976). *In* "Identification and Analysis of Organic Pollutants in Water" (L. L. Keith, ed.), pp. 135–153. Ann Arbor Sci. Publ., Ann Arbor, Michigan.

Kahn, L., and Wayman, C. H. (1964). *Anal. Chem.* **36**, 1340–1343.

Kaiser, R. (1971). *J. Chromatogr. Sci.* **9**, 227–235.

Kammerer, P. A., Jr., and Lee, G. F. (1969). *Environ. Sci. Technol.* **3**, 276–278.

Karger, B. L. (1972). *In* "Adsorptive Bubble Separation Techniques" (R. Lemlich, ed.), pp. 145–156. Academic Press, New York.

Karger, B. L., Poncha, R. P., and Miller, M. M. (1966). *Anal. Chem.* **38**, 764–767.

Karger, B. L., Grieves, R. B., Lemlich, R., Rubin, A. J., and Sebba, F. (1967). *Sep. Sci.* **2**, 401–404.

Karger, B. L., Snyder, L. R., and Horvath, C. (1973). "An Introduction to Separation Science." Wiley, New York.

Katz, S., Pitt, W. W., Jr., Scott, C. D., and Rosen, A. A. (1972). *Water Res.* **6**, 1029–1037.

Keith, L. H., ed. (1976). "Identification and Analysis of Organic Pollutants in Water." Ann Arbor Sci. Publ., Ann Arbor, Michigan.

Keith, L. H., Garrison, A. W., Allen, F. R., Carter, M. H., Floyd, T. L., Pope, J. D., and Thruston, A. D., Jr. (1976). *In* "Identification and Analysis of Organics Pollutants in Water" (L. H. Keith, ed.), pp. 329–373. Ann Arbor Sci. Publ., Ann Arbor, Michigan.

Keith, L. H., Lee, K. W., Provost, L. P., and Present, D. L. (1979). *Meas. Org. Pollut. Water* (C. E. Van Hall, ed.) *1978*, p. 85–107. American Society for Testing and Materials, Philadelphia.

Kennedy, D. C., and Minor, G. (1972). *Pap., 2nd Annu. Environ. Eng. Sci. Conf. 1972*.

Kepner, R. E., Van Straten, S., and Weurman, C. (1969). *J. Agric. Food Chem.* **17**, 1123–1127.

Kerr, R. A., and Quinn, J. G. (1975). *Deep-Sea Res.* **22**, 107–116.

Khare, M., and Dondero, N. C. (1977). *Environ. Sci. Technol.* **11**, 814–819.

Khym, J. S. (1974). "Analytical Ion-Exchange Procedures in Chemistry and Biology," p. 128. Prentice-Hall, Englewood Cliffs, New Jersey.

Kim, B. R., Snoeyink, V. L., and Saunders, F. M. (1976). *J. Water Pollut. Control Fed.* **48**, 120–133.

Kipling, J. J. (1965). "Adsorption from Solution of Non-Electrolytes." Academic Press, New York.

Kissinger, L. D., and Fritz, J. S. (1976). *J. Am. Water Works Assoc.* **68**, 435–437.

Klein, E., Smith, J. K., Wendt, R. P., and Desai, S. V. (1972). *Sep. Sci.* **7**, 285–292.

Klein, E., Smith, J. K., Weaver, R. E. C., Wendt, R. P., and Desai, S. V. (1973). *Sep. Sci.* **8**, 585–592.

Klein, E., Eichelberger, J., Eyer, C., and Smith, J. K. (1975). *Water Res.* **9**, 807–811.

Klocking, R., and Mucke, D. (1969). *Z. Chem.* **9**, 453–545.

Kopfler, F. C., Melton, R. G., Lingg, R. D., and Coleman, W. E. (1976). *In* "Identification and Analysis of Organic Pollutants in Water" (L. H. Keith, ed.), pp. 87–104. Ann Arbor Sci. Publ., Ann Arbor, Michigan.

Korenman, Y. I., and Bortnikova, R. N. (1977). *Zh. Anal. Khim.* **32**, 443–445.

Kroeff, E. P., and Pietrzyk, D. J. (1978). *Anal. Chem.* **50**, 502–511.

Krutz, M. (1975). *Z. Anal. Chem.* **273**, 123–126.

Kummert, R., Molnar-Kubica, E., and Giger, W. (1978). *Anal. Chem.* **50**, 1637–1639.

Kun, K. A., and Kunin, R. (1964). *J. Polym. Sci., Polym. Chem. Ed.* **6**, 2689–2701.

Kunin, R. (1977). *Polym. Eng. Sci.* **17**, 58–62.

Kuo, P. P. K., Chian, E. S. K., and DeWalle, F. B. (1977). *Water Res.* **11**, 1005–1011.

Kurtz, D. A. (1977). *Bull. Environ. Contam. Toxicol.* **17**, 391–398.

LaRose, R. H. (1974). *J. Assoc. Off. Anal. Chem.* **57**, 1046–1049.

Ledet, E. J., and Laseter, J. L. (1974). *Anal. Lett.* **7**, 553–562.

Leenheer, J. A., and Farrier, D. S. (1979). Oil Shale Symposium: Sampling, Analysis, and Quality Assurance. (C. Gale, ed.). Denver, Colo. EPA Report 600/9-80-022, p. 273–285.

Leenheer, J. A., and Huffman, E. W. D., Jr. (1976). *J. Res. U.S. Geol. Surv.* **4**, 737–751.

Leenheer, J. A., and Huffman, E. W. D., Jr. (1979). *Water-Resour. Invest. (U.S. Geol. Surv.)* **79–4**, 1–16.

Leenheer, J. A., and Malcolm, R. L. (1973). *Geol. Surv. Water-Supply Pap. (U.S.)* **1817-E**, 1–14.

Leenheer, J. A., Malcolm, R. L., McKinley, P. W., and Eccles, L. A. (1974). *J. Res. U.S. Geol. Surv.*, **2**, 361–369.

Leenheer, J. A., Malcolm, R. L., and White, W. R. (1976). *Geol. Surv. Prof. Pap. (U.S.)* **987**, 12.

Lemlich, R., ed. (1972). "Adsorptive Bubble Separation Techniques." Academic Press, New York.

Leoni, V., Puccetti, G., Colombo, R. J., and D'Ovido, A. M. (1976). *J. Chromatogr.* **125**, 399–407.

Ligon, W. V., Jr., and Johnson, R. L., Jr. (1976). *Anal. Chem.* **48**, 481–484.

Little, J. N., and Fallick, G. J. (1975). *J. Chromatogr.* **112**, 389–397.

Lovell, V. M., and Sebba, F. (1966). *Anal. Chem.* **38**, 1926–1928.

McAuliffe, C. (1971). *Chem. Technol.* **2**, 46–51.

MacCarthy, P., and O'Cinneide, S. (1974). *Soil Sci.* **25**, 420–428.

McClure, V. E. (1972). *J. Chromatogr.* **70**, 168–170.

McKinney, R., Jr. (1972). *Sep. Purif. Methods* **1**, 31–115.

Malcolm, R. L. (1968). *Geol. Surv. Prof. Pap. (U.S.)* **600-C**, C211–C216.

Malcolm, R. L. (1976). *J. Res. U.S. Geol. Surv.* **4**, 37–40.

Malcolm, R. L., and Durum, W. H. (1976). *Geol. Surv. Water-Supply Pap. (U.S.)* **1817-F**, 1–21.

Malcolm, R. L., and Leenheer, J. A. (1973). *Proc. Annu. Tech. Meet.—Inst. Environ. Sci.* **19**, 336–340.

Manka, J., Rebhun, M., Mandelbaum, A., and Bortinger, A. (1974). *Environ. Sci. Technol.* **8**, 1017–1020.

Mantoura, R. F. C., and Riley, J. P. (1975). *Anal. Chim. Acta* **76**, 97–106.

Martin, F. D., and Pierce, R. H., Jr. (1971). *Environ. Lett.* **1**, 49–52.

Matysik, G., and Soczewinski, E. (1977). *Sep. Sci.* **12**, 657–665.

Michaels, A. S. (1968). *Prog. Sep. Purif.* **1**, 297–334.

Middleton, F. M., Pettit, H. H., and Rosen, A. A. (1962). *Eng. Ext. Ser.* (*Purdue Univ.*) **112**, 454–460.

Mieure, J. P. (1977). *J. Am. Water Works Assoc.* **69**, 60–62.

Mieure, J. P., and Dietrich, M. W. (1973). *J. Chromatogr. Sci.* **11**, 559–570.

Mieure, J. P., Mappes, G. W., Tucker, E. S., and Dietrich, M. W. (1976). *In* "Identification and Analysis of Organic Pollutants in Water" (L. H. Keith, ed.), pp. 113–133. Ann Arbor Sci. Publ., Ann Arbor, Michigan.

Miget, R., Kator, H., Oppenheimer, C., Laseter, J. L., and Ledet, E. J. (1974). *Anal. Chem.* **46**, 1154–1157.

Musty, P. R., and Nickless, G. (1974). *J. Chromatogr.* **89**, 185–190.

National Academy of Sciences (1977). "Drinking Water and Health," pp. 489–856. Natl. Acad. Sci., Washington, D.C.

Natusch, D. F. S., and Tomkins, B. A. (1978). *Anal. Chem.* **50**, 1429–1434.

Niederschulte, U., and Ballschmiter, K. (1974). *Z. Anal. Chem.* **269**, 360–363.

Novak, J., Zluticky, J., Kubelka, V., and Mostecky, J. (1973). *J. Chromatogr.* **76**, 45–50.

Osterroht, C. (1974). *J. Chromatogr.* **101**, 289–298.

Packham, R. F. (1964). *Proc. Soc. Water Treat. Exam.* **13**, 316–329.

Parr, J. F., Willis, G. H., McDowell, L. L., Murphree, C. E., and Smith, S. (1974). *J. Environ. Qual.* **3**, 292–294.

Peake, E., Baker, B. L., and Hodson, G. W. (1972). *Geochim. Cosmochim. Acta* **36**, 867–883.

Pietrzyk, D. J., and Chu, C. H. (1977). *Anal. Chem.* **49**, 757–764.

Pietrzyk, D. J., Kroeff, E. P., and Rotsch, T. D. (1978). *Anal. Chem.* **50**, 497–502.

Pitt, W. W., Jr., Jolley, R. L., and Scott, C. D. (1975). *Environ. Sci. Technol.* **9**, 1068–1073.

Poncha, R. P., and Karger, B. L. (1965). *Anal. Chem.* **37**, 422–424.

Poulson, R. E., chair. (1978). *Prepr. Pap.—Am. Chem. Soc., Div. Fuel Chem.* **23**(2).

Raymond, A., and Guiochon, G. (1975). *J. Chromatogr. Sci.* **13**, 173–177.

Rebhun, M., and Manka, J. (1971). *Environ. Sci. Technol.* **5**, 606–609.

Renberg, L. (1978). *Anal. Chem.* **50**, 1836–1838.

Richard, J. J., and Fritz, J. S. (1974). *Talanta* **21**, 91–93.

Richard, J. J., and Junk, G. S. (1977). *J. Am. Water Works Assoc.* **69**, 62–64.

Riley, J. P., and Taylor, D. (1969). *Anal. Chim. Acta* **46**, 307–309.

Rogers, I. H., and Keith, L. H. (1974). *Fish. Mar. Serv.* (*Can.*) **465**, 1–21.

Rook, J. J. (1972). *Water Treat. Exam.* **21**, 259–274.

Rook, J. J. (1974). *Water Treat. Exam.* **23**, 234–243.

Rosen, A. A., and Middleton, F. M. (1955). *Anal. Chem.* **27**, 790–794.

Ryan, J. P., and Fritz, J. S. (1978). *J. Chromatogr. Sci.* **16**, 488–492.

Saxena, J., Kozuchowski, J., and Basu, D. K. (1977). *Environ. Sci. Technol.* **11**, 682–685.

Schaumburg, F. D. (1974). "Freeze Concentration of Toxic Pollutants for Bioassay," NTIS Rep. No. PB-235-911. Natl. Tech. Inf. Serv., Springfield, Virginia.

Schnitzer, M., and Khan, S. U., eds. (1978). "Soil Organic Matter." Am. Elsevier, New York.

Sebba, F. (1962). "Ion Flotation." Elsevier, Amsterdam.

Shapiro, J. (1961). *Science* **133**, 2063–2064.

Shapiro, J. (1967). *Anal. Chem.* **39**, 280.

Sherma, J. (1976). "Manual of Analytical Quality Control for Pesticides and Related Compounds," EPA Rep. 600/1-76-017. Environ. Prot. Agency, Washington, D.C.

Shriner, R. L., and Fuson, R. C. (1948). "The Systematic Identification of Organic Compounds," 3rd ed. Wiley, New York.

Sieburth, J. M., and Jensen, A. (1968). *J. Exp. Mar. Biol. Ecol.* **2**, 174–189.

Siegel, A., and Degens, E. T. (1966). *Science* **151**, 1098–1101.

Simpson, R. M. (1972). "The Separation of Organic Chemicals from Water." Rohm & Haas, Philadelphia, Pennsylvania.

Sirotkina, I. S., Varshall, G. M., Lur'e, Y. Y., and Stepanova, N. P. (1974). *Zh. Anal. Khim.* **29**, 1626–1632.

Smith, D., and Eichelberger, J. (1965). *J. Water Pollut. Control Fed.* **37**, 77–85.

Smith, R. G., Jr. (1976). *Anal. Chem.* **48**, 74–76.

Snoeyink, V. L., and Weber, W. J., Jr. (1967). *Environ. Sci. Technol.* **1**, 228–234.

Sourirajan, S., ed. (1977). "Reverse Osmosis and Synthetic Membranes. Theory-Technology-Engineering." Natl. Res. Counc., Ottawa, Canada.

Sridharan, N., and Lee, G. F. (1972). *Environ. Sci. Technol.* **6**, 1031–1033.

Stalling, D. L., Smith, L. M., and Petty, J. D. (1979). *Symp. Meas. Org. Pollut. Water,* (C. E. Van Hall, ed.) *1978* pp. 302–323. American Society for Testing and Materials, Philadelphia.

Stepan, S. F., and Smith, J. F. (1977). *Water Res.* **11**, 339–342.

Stepanenko, V. E. (1977). *Zh. Anal. Khim.* **32**, 788–796.

Stepanenko, V. E., and Golovina, Z. M. (1975). *Zh. Anal. Khim.* **30**, 890–894.

Stuber, H. A., and Leenheer, J. A. (1979). Establishment of Water Quality Monitoring Programme, 1978 (Everett, L. G., and Schmidt, K. D., eds.), pp. 266–272. American Water Resources Association, Minneapolis, Minnesota.

Suffet, I. H. (1973a). *J. Agric. Food Chem.* **21**, 288–294.

Suffet, I. H. (1973b). *J. Agric. Food Chem.* **21**, 591–598.

Suffet, I. H., and Faust, S. D. (1972). *J. Agric. Food Chem.* **20**, 52–56.

Suffet, I. H., Brenner, L., Coyle, J. T., and Cairo, P. R. (1978). *Environ. Sci. Technol.* **12**, 1315–1322.

Swift, R. S., and Posner, A. M. (1971). *Soil Sci.* **22**, 237–249.

Swinnerton, J. W., and Linnenbom, V. J. (1967). *J. Gas Chromatogr.* **5**, 570–573.

Symons, J. M., Bellar, T. A., Carswell, J. K., Demarco, J., Kropp, K. L., Robeck, G. G., Seeger, D. R., Slocum, C. J., Smith, B. L., and Stevens, A. A. (1975). *J. Am. Water Works Assoc.* **67**, 634–647.

Tanford, C. (1973). "The Hydrophobic Effect; Formation of Micelles and Biological Membranes." Wiley (Interscience), New York.

Taylor, C. G., Waters, J., and Williams, P. V. (1974). *Anal. Chem.* **69**, 373–387.

Thruston, A. D., Jr. (1978). *J. Chromatogr. Sci.* **16**, 254–259.

Thurman, E. M., Aiken, G. R., and Malcolm, R. L. (1978a). *Am. Chem. Soc., Div. Environ. Chem., Proc. Jt. Conf. Sens. Environ. Pollut., 4th, 1978* pp. 630–634.

Thurman, E. M., Malcolm, R. L., and Aiken, G. R. (1978b). *Anal. Chem.* **50**, 775–779.

U.S. Environmental Protection Agency (1977). "Sampling and Analysis Procedures for Screening of Industrial Effluents for Priority Pollutants," Available from EPA Environmental Monitoring and Support Laboratory, Cincinnati, Ohio.

Uthe, J. F., Reinke, J., and Gesser, H. (1972). *Environ. Lett.* **3**, 117–135.

Uthe, J. F., Reinke, J., and O'Bradovich, H. (1974). *Environ. Lett.* **6**, 103–115.

Van Rossum, P., and Webb, R. G. (1978). *J. Chromatogr.* **150**, 381–392.

Voznakova, Z., Popl, M., and Berka, M. (1978). *J. Chromatogr. Sci.* **16**, 123–127.

Walton, H. F. (1975). *Sep. Purif. Methods* **4**, 189–214.

Wang, L. K. (1975). *J. Am. Water Works Assoc.* **67**, 19–21.

Warren, C. B., and Malec, E. J. (1972). *J. Chromatogr.* **64**, 219–237.

Webb, R. G. (1975). "Isolating Organic Pollutants: XAD Resins, Urethane Foams, Solvent Extraction," EPA Rep. 660/4-75-003. Environ. Prot. Agency, Washington, D.C.

Weber, W. J., Jr., and Morris, J. C. (1963). *J. Sanit. Eng. Div., Am. Soc. Civ. Eng.* **89,** 31–59.

Welch, P. S. (1948). "Limnological Methods." McGraw-Hill (Blakiston), New York.

Wershaw, R. L., and Pinckney, D. J. (1973). *J. Res. U.S. Geol. Surv.* **1,** 701–707.

Westover, L. B., Tou, J. C., and Mark, J. H. (1974). *Anal. Chem.* **46,** 568–571.

Wetzel, R. G. (1975). "Limnology," pp. 538–621. Saunders, Philadelphia, Pennsylvania.

White, W. R., and Leenheer, J. A. (1975). *J. Chromatogr. Sci.* **13,** 386–389.

Wickbold, R. (1972). *Tenside Deterg.* **9,** 137–177.

Williams, P. M. (1971). *In* "Organic Compounds in Aquatic Environments" (S. D. Faust and J. V. Hunter, ed.), pp. 145–163. Dekker, New York.

Williams, P. M., and Zirino, A. (1964). *Nature (London)* **204,** 462–464.

Wu, C., and Suffet, I. H. (1977). *Anal. Chem.* **49,** 231–237.

Ziechmann, W. (1976). *Amazoniana* **6,** 135–144.

Zlatkis, A., Lichtenstein, H. A., and Tishbee, A. (1973a). *Chromatographia* **6,** 67–70.

Zlatkis, A., Bertschi, W., Lichtenstein, H. A., Tishbee, A., Shunbo, F., Liebich, H. M., Coscia, A. M., and Fleischer, N. (1973b). *Anal. Chem.* **45,** 763–767.

4   GAS CHROMATOGRAPHY

R. P. Schwarzenbach and W. Giger

Swiss Federal Institute for Water Resources
and Water Pollution Control (EAWAG)
Dübendorf, Switzerland

and

K. Grob, Jr.

Kantonales Laboratorium
Zürich, Switzerland

I. INTRODUCTION

Gas chromatography (GC) is presently the most important analytical technique for the qualitative and quantitative determination of individual organic compounds in the aquatic environment. Although the method is confined to those compounds that can be volatilized either directly or after derivatization [and therefore only to a small portion of the organic material (5–10%) typically present in natural and treated waters], it allows the analysis of many important classes of organic pollutants and natural products (e.g., hydrocarbons, halogenated hydrocarbons, pesticides, phenols, and fatty acids).

The gas chromatographic separation of a mixture of organic compounds is performed by partitioning the compounds between a gaseous mobile phase and a stationary phase that is confined to a long tube, the column. The stationary phase consists, in most cases, of a thin liquid film that is either distributed over an inert granular support (packed column) or supported as a thin coating on the inner surface of the column: the wall-coated open tubular column (WCOT), the porous-layer open tubular column (PLOT), or the support-coated open tubular column (SCOT).

In a gas chromatograph, the column is inserted between an injector and a detector and is continuously swept with the mobile phase, the carrier gas (Fig. 1). An analysis is performed by introducing the mixture of compounds into the column by means of an injecting system. Each compound is then swept toward the detector whenever it is present in the moving stream of the carrier gas. The compounds that are more soluble in or exhibit stronger affinities for the stationary phase require more time to reach the detector than those compounds that are less strongly retained by the stationary phase. Thus, compounds with different retention behavior may be separated. The detector, which is placed at the column exit, measures changes in the composition of the effluent and provides a signal for a strip chart recorder where the chromatogram is registered as a series of peaks. The detector signal may also be sent to an electronic integrator. During the analysis the temperature of the column is either kept constant

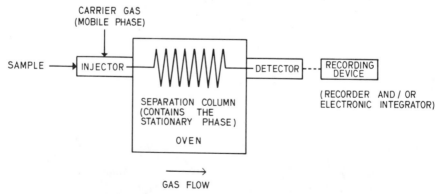

Fig. 1. Schematic diagram of a gas chromatograph.

(isothermal) or is continuously increased to a final value (temperature programmed).

When applying GC to water analysis, two things must be kept in mind:

1. The great complexity of the mixture of compounds present in the aquatic environment (e.g., Blumer, 1975), and
2. The low levels at which individual compounds occur (nmol/L and less).

Therefore, columns of very high separating power and very sensitive detector systems are prerequisites for proper analyses. Only high-resolution capillary GC can satisfactorily fulfill such requirements. Fortunately, the equipment and the technical knowledge that are necessary to perform capillary GC have become generally available (K. Grob, 1975; Novotny, 1978; Jennings, 1980; K. Grob, 1979; K. Grob and Grob, 1979).

Although capillary GC allows separation of very complex mixtures of compounds at very low concentration levels, preconcentration and fractionation steps must frequently be performed before the final gas chromatographic analysis (see also Chapter 3, this volume). Figure 2 shows schematically the steps involved in the determination of individual organic compounds in the aquatic environment by GC. It is clear that the type of compounds to be analyzed and the presence of interfering compounds (from the type of water sample and the selected preconcentration

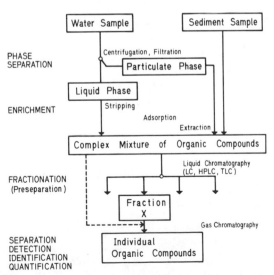

Fig. 2. Analytical scheme for the determination of individual organic compounds in the aquatic environment by gas chromatography.

and/or fractionation techniques) greatly influence the selection of the appropriate gas chromatographic equipment and the conditions of analysis (e.g., choice of injection technique, column, detector, and temperature). The compounds of interest may, for example, cover a wide range of volatility; they may be members of various compound classes with extreme differences in polarity or they may have very similar structures (e.g., isomers). Also, they may be major constituents of the mixture or be present only at trace levels. Consequently, often there is no general recipe for the optimum solution of a specific problem. A large amount of work, however, has been conducted in the field of gas chromatographic analysis of environmental samples. The results of these numerous studies provide sufficient background information to allow one to find an optimum gas chromatographic method for solving a given problem.

The aim of this chapter is to acquaint the reader with the most important practical aspects of the application of contemporary gas chromatography to water analysis. Because we believe that in most laboratories packed-column GC will be replaced by capillary GC in the near future and because various books dealing with the application of packed-column GC to water analyses already exist, we have confined this chapter mostly to capillary GC. Furthermore, rather than giving a general introduction, we have attempted to provide a critical evaluation of the state of the art of gas chromatographic techniques. The discussions and selection of examples are, of course, influenced by our personal opinions and experiences.

For a more detailed introduction to gas chromatography, the reader is referred to the existing literature (e.g., Tranchant, 1969; Littlewood, 1970; Jennings, 1980; Lee *et al.*, 1984; K. Grob, Jr., 1984a). The most important periodical journals containing theoretical and practical aspects as well as a wide range of applications of gas chromatography are the *Journal of Chromatography, Journal of High Resolution Chromatography & Chromatography Communications (HRC & CC), Chromatographia, Journal of Chromatographic Science,* and *Analytical Chemistry.*

II. GAS CHROMATOGRAPHIC TECHNIQUES AND EQUIPMENT

A. General Remarks

When applying gas chromatography to complex systems, the analyst is continuously confronted with various practical questions and problems to which appropriate answers are often difficult to find in textbooks. This is especially true for high-resolution capillary GC, which, for obtaining optimum results, requires experience and the knowledge of various methodological details. In the following discussion, major emphasis is therefore placed on those aspects of capillary GC that we feel are most critical for the proper application of this technique to water analysis. When possible, the reader is referred to easily accessible literature, whereas detailed information is given on those subjects (e.g., injection techniques) for which no appropriate references exist or have been found.

B. Definition of Gas Chromatographic Parameters

The theory of chromatographic processes has been the subject of many publications. An excellent general introduction to "separation science" has been published by Karger *et al.* (1973), and for a brief introduction to gas chromatographic theory, the textbooks by Tranchant (1969), Littlewood (1970), or Jennings (1980) are recommended. The definitions of some frequently used GC parameters are summarized in Table I and Fig. 3.

C. Capillary versus Packed-Column Gas Chromatography

A great number of comparisons between packed versus capillary GC have been made, and almost all cite only the much higher separation efficiency of capillary WCOT columns. Considering the incomplete separation usually obtained when analyzing an environmental sample on even the best capillary column, it appears to be a waste of time to work with a tool as "crude" as a packed column. However, for a fair comparison of the two methods, other technical aspects should also be considered. Some of them are still in favor of packed-column GC; working with capillaries requires more experience and more sophisticated techniques for obtaining optimum results. This may be one of the reasons why many laboratories (especially in the United States) use or have used SCOT columns, which yield better separation than packed columns and are easier to manipulate than capillary columns. However, even good SCOT columns exhibit less separation efficiency than do average WCOT columns. Thus, in our opinion this "compromise" (i.e., the use of SCOT columns) does not represent a favorable solution, and consequently, for complex mixtures such as usually encountered in water analysis, capillary GC should be employed. In the following section a few additional practical aspects concerning the comparison of packed versus glass capillary columns are discussed.

TABLE I

Definitions of Some Important Gas Chromatographic Parameters[a,b]

Retention time	t_R
Retention time of nonsorbed compound	t_M
Adjusted retention time	$t_R' = t_R - t_M$
Idealized peak width at baseline	w
Peak width at half height	$w_{0.5}$
Partition ratio (capacity factor)	$k = t_R'/t_M$
Phase ratio	$\beta = \dfrac{V_{stationary\ phase}}{V_{mobile\ phase}}$
Measures for the separation efficiency of a gas chromatographic system:	
1. Number of effective theoretical plates (N is a measure for the broadening of the peaks resulting from the passage through the separation system; valid only for isothermal chromatography)	$N_{eff} = 16(t_R'/w)^2 = 5.54(t_R'/w_{0.5})^2$
2. A more practical number is the Trennzahl (TZ)[c] (valid for isothermal and temperature-programmed chromatography)	$TZ = \dfrac{t_{R(B)} - t_{R(A)}}{w_{0.5(A)} + w_{0.5(B)}} - 1$
Separation factor of two compounds ($\alpha > 1$ by convention)	$\alpha_{C/B} = t_{R(C)}'/t_{R(B)}'$
Degree of separation: resolution	$R_s = \dfrac{2(t_{R(C)} - t_{R(B)})}{w_{(B)} + w_{(C)}}$
For $w_{(B)} \approx w_{(C)}$	$R_s = \dfrac{\sqrt{N_{eff}}}{4}\left(\dfrac{\alpha_{C/B}-1}{\alpha_{C/B}}\right)$

[a] For illustration, see Fig. 3.
[b] Terminology according to Jennings (1980).
[c] Kaiser (1976, 1977).

1. Injection Techniques

The importance of the injection, especially with regard to the resulting precision and accuracy of a gas chromatographic analysis, is often underestimated. In packed-column GC, the injection technique commonly used is, at least at first glance, extremely simple, and there is practically only one injection mode. When using capillary columns, the sample injection process becomes much more critical. Many variations of conventional vaporizing injectors have been designed for capillary columns. The precision and accuracy obtained with such injection systems is frequently inferior to that achieved with packed-column GC, because there are various additional sources of error that are difficult to control. Some other injection techniques have been successfully introduced, but none of them

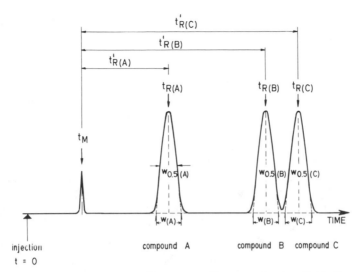

Fig. 3. Definition of gas chromatographic parameters (see Table I).

is universally applicable. An important development is the cold on-column injection technique, which eliminates a number of factors that cause lower precision and accuracy in packed as well as in capillary GC. It seems that a proper cold on-column injection can only be realized with capillary columns. If used correctly, this is probably the first injection technique that permits chromatographic separations with full "linearity," that is, without discrimination of parts of the sample (usually the high-boiling compounds). However, the fact remains that the injection technique in capillary GC requires more know-how than in packed-column GC, because there are factors such as the volatility of the solvent and its "compatibility" with the sample that which may strongly influence an analysis.

2. Columns

For the preparation of capillary columns, there is a much smaller choice of materials that are suitable for supporting the stationary phase than for packing classical columns, because in capillary columns the wall of the tubing itself must be coated with the liquid film. If one excludes metals as a material for capillary columns (for reasons of thermostability and catalytic activity of the columns), practically only glasses remain: soft alkali, borosilicate such as Pyrex and Duran, and various forms of pure silica (quartz) or fused silica glasses. Thus, the desired surface characteristics of the capillary tubing cannot be chosen from a wide variety of materials as is possible for packed columns. Furthermore, modifications of the surface are relatively difficult to achieve since they must be performed inside of the tubing, which means laborious and time-consuming procedures.

Capillary columns coated with nonpolar stationary phases are much more susceptible to polar contaminants than are comparable packed columns. Furthermore, capillaries usually show enhanced adsorption activity. Alcohols, for example, elute satisfactorily from most packed columns but may be troublesome on average nonpolar capillary columns. Free acids and bases are easier to chromatograph on packed columns. With regard to catalytic activity, both types of columns are similar; for example, DDT is eluted without degradation whereas phenylurea herbicides are labile on both types of columns. Because of the lower elution temperatures from the capillary columns, capillaries are preferable for these herbicides. Adding further examples probably would lead to a weak preference of the packed columns if adsorption properties were compared but to a preference of the thin-film capillaries if thermolabile solutes need to be analyzed.

3. Selectivity

So far, only a limited number of the many available stationary phases that can be used in packed columns have been successfully applied to capillary GC. This drawback is, however, not too critical because in capillary GC, selectivity is much less important than in packed-column GC. The high efficiency of the capillary column commonly allows the separation of compounds even on nonideal columns. This has the advantage that different types of samples may be analyzed on one column, which means (except for very difficult separations) less column changing and a simplification with regard to the selection of the stationary phase of optimum selectivity for a given problem.

4. Sensitivity

On capillary columns, the peaks obtained from the eluting compounds are much narrower than on packed columns. This leads to a higher peak-height to peak-area ratio and thus (assuming comparable noise) to a several-fold increase in sensitivity compared to packed columns. Therefore, much smaller amounts or more diluted solutions can be analyzed by capillary GC, which is of paramount importance for water analyses.

5. Conclusions

In summary, the operation and maintenance of a capillary gas chromatographic system requires more time and attention than does a packed-column system. In our opinion, it is not a technique that can be simply bought and applied immediately. Only with the more or less continuous involvement of a somewhat specialized operator can the full potential of the method be exploited. However, if the technique is properly used, there is no substitute for the unique performance of capillary GC.

D. Injection Techniques

1. Vaporizing versus Cold Injection

Various injector systems are used in capillary chromatography. They can be divided into two groups according to the mode of the sample transfer to the column: (1) conventional vaporizing injectors in which the sample transfer occurs (at least ideally) via the vapor phase, and (2) cold injectors, whereby the sample is introduced mechanically as a liquid, thus avoiding the evaporation step.

A third and recently developed type of injector combines some of the advantages of both the hot and the cold injectors. The sample is introduced into a cold injector by a syringe. The injector is then heated ballistically to transfer the sample into the column (programmed temperature vaporizer, or PTV). The idea was first developed by Vogt et al. (1979a,b), primarily for the injection of large sample sizes. The advantages of the method were later recognized by Poy et al. (1981) and Schomburg (1981). The technique was further developed (Vogt et al., 1981; Poy, 1982; Schomburg, 1983) but has not been investigated to a point that allows a comprehensive discussion on its routine use.

The design of conventional vaporizing injectors for packed columns is very simple. The injector consists basically of a small chamber with a septum cap, an inlet for the carrier gas, and a connection for the column. With the introduction of capillary GC, the split injector was developed to allow the transfer of very small amounts of sample to the capillary column. In this injector the sample vapors are split into a main stream that is vented and a small stream that enters the column. To obtain "linear splitting" (Ettre and Averill, 1961), this technique is assumed to require an instantaneous and complete evaporation of the compounds. This is achieved by heating the injector to a relatively high temperature (mainly because the heat transfer from the injector walls to the sample to be evaporated is relatively poor). After being transferred to the column, most of the sample is recondensed at the column entrance. Thus, the evaporation step only serves to effect the sample transfer. When working with thermally labile compounds, a high injector temperature is obviously disadvantageous. In many cases, however, this drawback is compensated by the simplicity and the practical usefulness of this type of injector. Also, there is no alternative in sight for splitting small sample volumes. Before going into details on the realization and the limitations of the splitting injector, we continue the discussion of the basic concepts of sample transfer.

In the early 1970s, trace analyses became more and more important. This required splitless injections for capillary columns to avoid sample losses from splitting (K. Grob and Grob, 1969). The splitless injector was constructed using the same hot injector cavity as for split injections. The great advantage of this kind of injector is its versatility. Depending on the flow through the split valve, it can either be operated in the splitless mode or at any desired split ratio.

Although this technique has been proven to be very useful and has been widely applied, the basic concept of the splitless injection is not satisfying. The sample transfer through the vapor phase has few advantages but leads to many problems including adsorption, losses through incomplete transfer, and thermal destruction of sample components.

One of the major problems of the classical vaporizing injectors (in capillary and in packed-column GC) is the injection by a syringe, because under normal conditions the evaporation of the sample occurs to a high degree inside the syringe needle (see Section II,D,2,a). The material that elutes from the needle is poorly representative of the sample because higher boiling materials remain in the dead volume of the syringe needle. In addition, the evaporation from a metal surface is not at all desirable because of possible catalytic degradation. Some types of vaporizing injectors, such as the "moving needle" or "solid" injectors, avoid introduction of the sample by syringe. In the case of the "moving needle," the sample is deposited at the tip of a glass needle that is kept at ambient temperature. The solvent is then blown off in the carrier-gas stream before the needle is moved into the hot injector. With this technique the sample transfer also occurs by evaporation, but the dead volume of the syringe is avoided (although this has never been used as an argument in favor of this injector). However, the moving-needle injection can only be used for relatively high-boiling compounds since a considerable portion of volatile compounds is lost when the solvent is evaporated. In a similar, usually automatic, "solid" injector, small pieces of capillary tubing that are filled with sample at ambient temperature are used. After the solvent has evaporated, the tubing is inserted into a cartridge and dropped into the hot injector cavity. The restrictions for this technique are similar to those for the moving-needle technique. Buser and Widmer (1979) suggested the use of noble-metal capsules that are filled with sample and sealed. The capsule is inserted into a cartridge, moved into the hot zone of the injector, and punctured by a sharp point. The sample is then "rinsed out" by the carrier gas.

Although there have been great efforts to develop precise and accurate vaporizing injectors, evaporation as the key step for the sample transfer still seems questionable. The concept of transferring the sample directly as a liquid into the column (i.e., by mechanical motion) is much more attractive because discrimination and chemical reactions resulting from the sampling can be avoided. Evaporation of the compounds occurs at the lowest possible temperature—the temperature that is required for the chromatographic process. It is interesting that this basically old concept (Zlatkis and Walker, 1963; Desty, 1965) has been implemented for general use only since the late 1970s (K. Grob and K. Grob, Jr., 1978a).

2. Injection Methods

a. The Problem with the Syringe Needle

The problems connected with the use of a syringe needle for the injection of samples into a vaporizing injector have been mentioned by a number of workers (Kruppa and Henley, 1971; Brötell et al., 1973; LeBel and Williams, 1979), but the impact of these problems on quantitative analysis has not been sufficiently recognized until the late 1970s. How is the liquid sample really introduced into the injector? If a liquid is "injected" into open air, one can easily observe how it leaves the needle. Quite often a drop remains hanging at the needle tip. Under the conditions of a hot injector, the liquid then probably evaporates, but this evaporation is incomplete if the needle is withdrawn too soon.

A second, much more severe problem, is caused by the fact that most samples contain high proportions of volatile solvents or sample components that evaporate inside the needle before leaving it. It is well known that the total volume of the injected liquid is determined by the volume indicated on the scale of the syringe and a part of the internal needle volume. It certainly must not be assumed that the part leaving the needle is always representative of the sample. Because the temperature of the needle is often below the temperature of the injector, one obtains the unfortunate situation that the place that needs to be hottest is actually the coldest spot in the system (K. Grob, Jr. and Neukom, 1980b). The needle temperature should be especially high because the needle represents a dead volume, in contrast to the rest of the injector.

As already mentioned, when using a syringe in conjunction with a vaporizing injector, the two major problems are (1) discrimination of high-boiling compounds and (2) catalytic degradation of thermolabile compounds.

It is well known that the syringe must be cleaned after each injection. The impurities to be rinsed out consist mainly of sample material left from the previous injection, usually representing high-boiling compounds that were missing in the previous chromatogram. The magnitude of such losses has been quantified by K. Grob, Jr. and Neukom (1979a). They found that the "missing peak area" for higher alkanes (up to C_{44}) corresponded nicely to the amount of material left in the syringe needle (which was determined by a second injection). They also found that the way the syringe is handled during the injection has a strong influence on the discrimination pattern. The strongest discrimination was obtained when the plunger was moved slowly [although this technique has advantages for other reasons (Yang *et al.*, 1978)]. If handled in this way, the sample slowly reaches the upper part of the syringe needle, which is only warmed by the septum area of the injector (often at 130–160°C when the injector is set to 350°C). These conditions favor a selective evaporation of the volatiles, corresponding to a distillation. The high-boiling material may be "piled up" almost completely in this zone of the needle so it never reaches the column.

The other injection methods tested can be judged by the same principles. The smallest losses occur if a maximum proportion of the sample is expelled from the needle as a liquid (droplets or fog).

The best results are shown by the "hot-needle" technique, where the sample is drawn back into the barrel of the syringe. The empty needle is then inserted and allowed to warm up for 3–5 s before the plunger is pushed down as rapidly as possible. The major part of the sample is expelled as a liquid into the injector by the explosion-like evaporation of volatiles along the hot needle wall. It should be noted, however, that considerable losses occur even with the hot-needle technique (Fig. 4).

Another aspect to consider is the precision of the injection, which is at least as important as the accuracy. Many authors mentioned that the quantitative data obtained by different operators seldom agree. This is probably largely the result of different losses in the needle resulting from the different injection speeds, that is, from injections from needles of

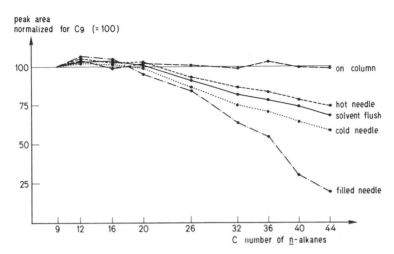

peak area
normalized for C9 (= 100)

Fig. 4. Discrimination of *n*-alkanes (C$_9$–C$_{44}$) by different injection techniques. Peak areas are normalized to the *n*-nonane peak. Injection: 1 μL, corresponding to the needle volume, 1 : 15 split, injector temp 350°C; for needle handling see text. (From K. Grob, Jr. and Neukom, 1979a.)

different temperatures. Hot-needle injections hardly suffer this problem, and therefore give by far the most reproducible results when using manual injection (K. Grob, Jr. and Neukom, 1979a).

An alternative to the hot-needle method is the "solvent flush" technique (Kruppa and Henley, 1971). At least one needle volume of pure solvent is drawn into the syringe prior to the sample. This solvent is supposed to rinse the high-boiling material from the needle into the injector. However, the efficiency of this process is poor because a cushion of evaporated solvent probably prevents the contact of the liquid with the needle walls. For a test sample containing *n*-alkanes covering a wide range of boiling points, the hot-needle technique was found to be clearly superior to the solvent flush method (K. Grob, Jr. and Rennhard, 1980).

Many years ago, most manufacturers of gas chromatographic equipment stressed the importance of having "all-glass" systems. Indeed, the necessity of avoiding contact of the sample with metal surfaces cannot be overemphasized. When injecting with a syringe, a metal surface is introduced at one of the most sensitive parts of the system. During the evaporation from the needle wall, degradations and alterations of labile components may occur. Thus, for an optimum injection, often an unsatisfactory compromise must be found between a high needle temperature (favoring degradation) and a low needle temperature (favoring discrimination).

b. Split Injection

Sampling with carrier-gas splitting was the first injection technique widely used in capillary GC and has been investigated by many workers (e.g., Ettre and Purcell, 1976; Jennings, 1975; Gaspar *et al.,* 1977; Schomburg *et al.,* 1981; K. Grob, Jr., 1981a). A high portion of the carrier gas bypasses the capillary-column inlet and is vented through the split valve. The preset split ratio is defined by the ratio of the amount of carrier gas entering the column and the amount vented through the split valve. This ratio can be adjusted by the split valve.

It is assumed that the sample is split by the same ratio, thus allowing the injection of a very small known quantity into the column. However, the true split ratio for the sample (the fraction that really reaches the column) is often very different from what has been preset for splitting the carrier gas. Because the vapors generated by the injection build up a zone of increased pressure, the flow rates through the column, the split and septum flush valve, and the carrier-gas line change, and it is unlikely that split and column flow are altered by the same proportion (K. Grob, Jr. and Neukom, 1979b). Furthermore, recondensation of the sample in the column inlet may drastically increase the quantity of sample reaching the column (K. Grob, Jr. and Neukom, 1982). Finally, incompletely evaporated sample parts may be split by a different ratio than the carrier gas and volatile components. This aspect may be substantially improved by packing the insert of the injector (e.g., with glass wool) in order to achieve full evaporation or at least the formation of an aerosol that is split reproducibly.

The split ratio may change during injection by more than a factor of 2, sometimes even up to a factor of 20. Bruderreck *et al.* (1967) and Jennings (1980) suggested that a buffer volume in the split line (serving as an expansion volume) minimizes the pressure increase during sampling. However, such devices help to keep the inlet pressure constant rather than the flow rates that actually determine the split ratio. The recondensation of sample in the column inlet may be reduced by mixing the sample vapor with carrier gas, that is, by using a rather large insert and a large distance between the needle exit and the column inlet.

Another drawback of the split injection is its "nonlinearity" (Ettre and Averill, 1961). The split ratio for different compounds in a mixture is often not the same, which means that the sample fraction entering the column is not representative of the total sample. Among the numerous papers written on this subject between 1960 and 1970, the one by Bruderreck *et al.* (1967) is probably the most comprehensive. If one assumes full evaporation of the sample as it passes the split point (the zone of the column head), the major problem is to create a sufficient sample homogeneity over the cross section of the injector liner. While this is achieved for the more volatile constituents of the sample, the heavier compounds are enriched in the center of the jet of sample created by the injection. A similar problem is caused by the different diffusion speeds of molecules of different molecular weight: The low molecular weight compounds are more easily deviated when the linear velocity of the sample vapor entering the column is not the same as the one directed to the split valve.

German and Horning (1972) pointed out that the major cause of nonlinear splitting is an insufficient evaporation of the sample. Under most circumstances, the sample probably passes the split point as a fog or an aerosol rather than a true vapor. Thus, a considerable part of the sample consists of small droplets that are not split by the same ratio as the gases and vapors. If the sample composition in the droplets is different from the composition in the vapor phase, a nonlinearity of the splitting is produced, leading to a loss in accuracy and precision. A second source of nonlinearity is related to the split ratio which fluctuates during the splitting process. If the sample composition varies from the front to the rear of the vapor cloud in the injector, the changing split ratio splits different sample parts by different split ratios (K. Grob, Jr., 1981a).

Several methods have been described for obtaining a more complete evaporation of the sample before it passes the split point. German and Horning (1972) proposed the use of a packed precolumn, whereas Jennings (1975) suggested a device that forces the vapor to change direction twice before entering the column. A much cheaper method is to fill an ordinary glass insert at least partly with glass wool. Schomburg *et al.* (1976) reported that better reproducibility and linear splitting was obtained when using glass wool. However, this method appears to improve quantitative analysis only in some cases; in others it even has a negative effect (K. Grob, Jr. *et al.*, 1981). A better linear splitting was achieved by using a high-boiling solvent (Schomburg *et al.*, 1977). We do not think that the problem of linear splitting has been solved. Filling the glass insert of the injector (with glass wool or a column packing material) or using the device designed by Jennings (1975) has the disadvantage of being discriminative as a result of adsorption of certain compounds, at least as soon as these surfaces become dirty, which may happen after a few injections of "real" samples. For the analysis of triglycerides, glass wool has been shown to be disasterous (K. Grob, Jr., 1979); the same holds for the analysis of polynuclear aromatic hydrocarbons (H. P. Maître and K. Grob, Jr., unpublished results, 1980).

However, despite all these problems, split injections are still widely used because they have four great advantages:

1. By splitting, only a small fraction of the sample reaches the column. This can be very useful because the maximum amount of a sample compound that may be chromatographed on a capillary column (yielding an approximately symmetric peak) is of the order of 10–100 ng. However, the minimum amount of sample that can be handled reasonably is of the order of 1 mg. Thus, where necessary, the split injection allows one to work with little or even no solvent.

2. Since the sample enters the column only during a very short period of time, the compounds to be chromatographed start as a narrow band. This means that there is little peak broadening caused by the injection. This is not realized in such a simple way by any other injection system used for capillary GC.

3. Unless small split ratios are used, the amount of sample reaching the column is small enough to avoid solvent effects. This is important if solvent effects cause peak distortion rather than a reconcentration of the bands (partial solvent trapping, K. Grob, Jr., 1982b).

4. The split injection allows analyses at a constant temperature. This is rarely the case in splitless or cold on-column injection. Fully isothermal chromatography gives a stable base line and readily reproducible absolute retention times, and is therefore a very convenient technique.

In summary, split injection is a useful technique that is rapid and easy for qualitative analysis and provides fairly accurate quantitative results as long as similar compounds of a limited volatility range are analyzed. For samples containing a broad variety of different constituents spanning a wide volatility range, the results are not accurate. Because discrimination effects tend to be poorly reproducible, many internal standards must be used for quantification, and this does not facilitate analyses. Although well-optimized conditions and careful calibration may substantially improve the results, it is probably more advisable to use less discriminative sampling techniques, particularly the cold on-column injection (see Section II,D,2,d).

c. Splitless Injection

Although there are many other injection techniques that do not split the sample, the term *splitless injection* was introduced by K. Grob and Grob (1969) for splitless injection using a vaporizing injector that is also suited for split injections. Further papers on this subject were published by K. Grob and Grob, Jr. (1974, 1978b) and Yang *et al.* (1978). This type of injection is carried out in three steps. First, the split valve is closed, then the sample is injected and finally after 30–90 s (the "splitless" period) the split valve is reopened in order to flush the injector. During the splitless period, at least 95% of the sample should be transferred from the injector into the capillary.

The sample transfer from the vaporizing chamber to the column requires quite a long time because it is impossible to prevent the sample vapor from mixing with carrier gas when injected. This vapor cloud may easily reach a volume of 0.5 to 1 mL. Furthermore, the transfer is coupled with an additional dilution process with carrier gas. Thus, at typical carrier-gas flow rates of 2 to 4 mL/min, a splitless period of 30 to 90 s is required depending on the geometry of the injector (K. Grob, Jr. and Romann, 1981). The sample transfer is never 100% complete because of the dilution process in the injector. If the split valve is not reopened (corresponding to the conditions of the direct injection used for packed columns) the solvent peak becomes extremely broad because traces of solvent enter the column for a long period of time. Sample components also tail, but because the peaks are much smaller than the solvent peak, the tailing is less obvious. This problem can be avoided when the last 1–5% of the sample vapor is vented through the split valve. Thus, the split exit is not used for sample splitting but for cleaning purposes.

Splitless injection is an important sampling technique for trace analysis (e.g., water extracts) because very diluted samples can be analyzed. A component present in the sample at 1 ppm (1 ng/μL) yields a fairly large flame-ionization detector signal. Furthermore, it seems that most discrimination processes are eliminated by splitless injection except for those caused by the syringe needle. In fact, it has been found that splitless sampling of triglycerides yielded less discrimination and a better reproducibility than a split injection (K. Grob, Jr., 1979).

One disadvantage of the long thermal stress in the injector has been demonstrated by Schomburg et al. (1976), who reported an increased rearrangement of divinylcyclobutane when using splitless injection. Many of the thermally induced alterations of molecules apparently happen on the hot metal surface of the syringe needle. In these cases, the same effects are encountered with split and splitless injection.

Finally, it should be noted that a dirty glass liner in the injector has a much greater impact on the sample transfer when using splitless rather than split injection (K. Grob and Grob, 1979).

When using splitless injection, the volume of the vaporizing chamber (glass liner) has to be sufficiently large to contain the whole sample vapor. In case of an overflow, sample vapors may enter the cold carrier-gas lines, thus causing memory effects in subsequent chromatograms. If the septum flush (see below) is kept open during the splitless period, the overflow is vented through this valve, which means that sample is lost. In any case, overloading the injector should be avoided. However, use of an excessively large injector volume is also not desirable because the sample transfer of a large volume of vapor is not feasible (K. Grob, Jr. and Romann, 1981).

The vaporizing chamber should be designed in a way that minimum dilution of the sample vapor with the carrier gas occurs. Thus, it should be long and narrow so it can be gradually filled with sample vapor from the bottom to the top by displacing the carrier gas backwards. An injector volume of ~1 mL allows one to inject 2–2.5 μL (including the amount eluted from the syringe needle) of the usual kinds of samples. A glass liner of 80-mm length and 3.6 mm id has been found to be a good compromise for routine applications.

The tip of the capillary column should be placed close to the bottom of the glass liner. Sample molecules that diffuse below the column entrance have little chance to reach the column and are therefore lost. The optimum situation seems to be when the capillary reaches ~5 mm into the glass liner. For splitless injections, the syringe needle has to be unusually long because its tip has to reach within 1–2 cm of the column inlet in order to fill the vaporizing chamber from the bottom to the top. If a short needle is used, a plug of more or less pure carrier gas hinders the sample vapors from entering the column. Thus, the splitless period is unnecessarily prolonged. For samples with less volatile solvents, it is advisable to surround the tip of the column with glass wool, filling the glass liner as far as the column sticks into the injector. This glass wool serves as a barrier for small sample droplets that are expelled from the syringe and otherwise may bypass the column inlet.

Vaporizing injectors, particularly if used for splitless injections, require a septum flush as described by K. Grob and Grob (1972). A small portion of the carrier gas is directed towards the septum and is vented through a needle valve (see Fig. 5). One of the purposes of the septum flush is to prevent the septum bleed from reaching the column. Its major purpose, however, is to solve another problem of the splitless injection: during the splitless period, a small amount of sample vapor may diffuse out of the vaporizing chamber and be adsorbed by the septum. When the vaporizing chamber is flushed by reopening the split valve after the splitless period, this material is not purged. It then diffuses slowly back into the carrier-gas stream and causes a broad tailing solvent peak. This can be prevented by flushing the septum with a few milliliters per minute of carrier gas.

Depending on the type of septum and the solvent used, a considerable amount of septum material is transferred into the vaporizing chamber when the syringe needle is inserted. This material produces false peaks ("ghost peaks") and therefore disrupts the analysis. The amount of

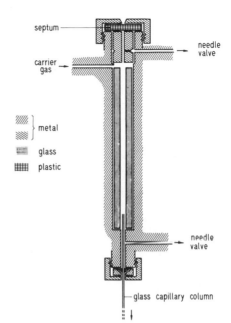

Fig. 5. Schematic diagram of an injection port applicable for all vaporizing injection techniques. Flow: upper needle valve, 5 mL/min. (From K. Grob and K. Grob, Jr., 1978b.)

"ghost material" entering the column depends on the split ratio. If the split valve is already closed when the needle is inserted, a maximum of ghost peaks is obtained. It is therefore advisable to open the split valve widely when the needle is introduced into the injector and to close it ~3 s afterwards (which corresponds to the warming period for a hot-needle injection) prior to the injection. This obviously requires that the sample be pulled back into the barrel of the syringe before the needle is inserted.

A major problem of the splitless-injection technique is the long time required for the sample transfer. This has a great impact on the initial peak width of a component. The peaks are 40–90 s wide at the beginning of the chromatographic process, which means that the high efficiency of a column may be wasted. An acceptable peak width at the column inlet may be roughly estimated as follows: a narrow sharp peak early in the chromatogram may have a half width of ~1 s. If one accepts that 10% of the peak broadening is caused by a poor injection technique, the initial peak width should be of the order of 0.3 s, which is about 100 times shorter than is obtained by the splitless injection. Thus, a reconcentration technique is required for producing narrow initial peak widths. This is achieved by techniques known as "cold trapping" and "solvent effect."

The cold trapping effect is obtained if the injection is carried out at a column temperature that is at least 60°C below the temperature of an isothermal analysis or the initial temperature of the temperature program. Under these conditions, the compounds of interest recondense at the capillary inlet, and because of their low volatility at this temperature, do not migrate further into the column.

The solvent effect also reduces the speed of migration of the components to almost zero. To produce a solvent effect, the solvent must recondense at the first part of the capillary column forming a thick layer (3–10 μm and greater) that functions like a thick film of stationary phase. The retention of the sample components in such a great amount of temporary stationary phase is extremely high. Thus, the compounds only start to migrate when the solvent has evaporated from the first coils of the column. For obtaining optimum results with the solvent effect, a sufficient (but not excessive) condensation of a suitable solvent must be achieved. For further details of this technique, the reader is referred to K. Grob and K. Grob, Jr. (1974, 1978b), Miller and Jennings (1979), K. Grob, Jr. (1982b, 1983b), and K. Grob, Jr. and Schilling (1983a,b) and, for a theoretical treatment, to Jennings *et al.* (1978) and Pretorius *et al.* (1983).

d. Cold On-Column Injection

The purpose of using cold on-column injection is to avoid the evaporation step. This is achieved by introducing the liquid sample mechanically into the capillary, a technique that was used as long ago as 1963 by Zlatkis and Walker (1963) for wide-bore columns. The idea was brought up again by Desty (1965) and Schomburg *et al.* (1977). The latter described the application of cold on-column injection to ordinary-sized capillary columns. Two different versions of injectors, a macro version using a crucible and a micro version using a micropipette for the sample transfer have been suggested. K. Grob and K. Grob, Jr. (1978a) developed an

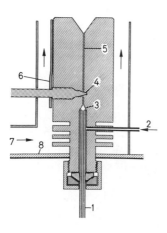

Fig. 6. Schematic diagram of an on-column injector. (1) Glass capillary column; (2) carrier-gas entrance; (3) centering cone; (4) stopcock valve; (5) channel (replacing the septum); (6) spring; (7) cooling-air entrance; (8) heat insulation. Stippling indicates plastic portion; larger shaded area is glass. (From K. Grob, 1978.)

injector based on a syringe needle with a very thin wall (0.23 mm od if made of steel and 0.17 mm if fused silica). This needle can be introduced into capillaries with inner diameters of >0.28 mm (see Fig. 6).

When using cold on-column injection, the following points should be considered. The sample must be deposited into the part of the capillary that is exposed to the oven temperature and not at the column entry which is located in the cold injector. Otherwise, high-boiling components are retained in the cold part of the column. This means that the sampling device (e.g., the micropipette or syringe needle) must penetrate sufficiently far into the capillary. Above the injection point, however, the capillary and the injector itself need to be cold enough to prevent any evaporation of the sample during the introduction of the sampling device (e.g., the syringe needle). This can be achieved by a continuous cooling of the injector block and by using an additional cooling device for the capillary above the injection point (Fig. 7).

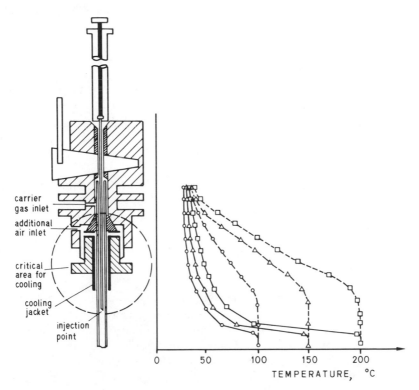

Fig. 7. Temperature profiles along the axes of the capillary column with secondary cooling system switched on (continuous curves) and off (dashed curves). Oven temperatures: ○, 100°C; △, 150°C; □, 200°. (From Galli *et al.,* 1979.)

During the injection, it is very important to obtain a full mechanical transfer of the liquid sample from the syringe to the column wall. If the transfer is incomplete, the remaining liquid slowly evaporates. The more volatile components are then preferentially introduced into the column, whereas the high-boiling material remains on the surface of the needle and is taken out of the column when the syringe is withdrawn. Furthermore, if liquid sample remains on the syringe needle, the cool part of the column entry may become contaminated. Whereas the high-boiling materials stay there, compounds of medium volatility slowly bleed off causing strong tailing of peaks (K. Grob, Jr. and Neukom, 1980b; K. Grob, Jr., 1984b).

A satisfactory sample transfer can only be achieved by a rapid injection. However, if ordinary sample volumes are injected rapidly, a volume of vapor may be generated that cannot be removed fast enough by the carrier gas. As a consequence, the pressure at the injection point increases and part of the vapor is pushed backward out of the column. Since such losses are not necessarily equal for the various sample components, erroneous results are obtained. To prevent such backflows, K. Grob (1978) suggested injecting as slowly as necessary to guarantee a sure removal of the vapors generated. However, this technique is not optimum because partial evaporation may occur in the syringe needle. There is only one better alternative, which is to inject rapidly at a column temperature at which the evaporation of the sample is slow enough to create vapor only as fast as it is removed through the column. It has been found that the maximum allowable column temperature during the injection is just about equal to the boiling point of the solvent (K. Grob, Jr. and Neukom, 1980a), or slightly above it if a secondary cooling or equivalent device is used (Trestianu and Galli, 1981).

The liquid introduced by the on-column injection technique forms a plug at the column inlet that is pushed further into the column by the carrier gas. It coats the wall of the column with a thick film. Usually, the flow of the liquid continues within the liquid layer forming waves that are pushed forward by the rapid flow of the carrier gas. This latter flow is also observed in splitless sampling if the solvent recondenses at the column inlet (solvent effect).

The flow of the solvent in the liquid phase spreads all sample components over a length of 15 to 30 cm, sometimes up to several meters per microliter of sample. This phenomenon is called "band broadening in space" (K. Grob, Jr., 1981b). It causes broadening and distortion, often even splitting of the signals of those solutes that elute more than a few tens of degrees above the initial column temperature. It is particularly pronounced if large sample volumes are injected and short columns are used. Two methods were proposed to avoid peak distortion by band broadening in space. (1) One can use a retention gap, that is, a column inlet not containing any stationary phase for a length that at least corresponds to that of the flooded zone (K. Grob, Jr., 1982a); the migration of the solute material within the retention-gap zone is accelerated and the solutes are reconcentrated at the beginning of the coated part of the column (K. Grob, Jr., 1983a). (2) Sisti *et al.* (1982) found that there is far less band broadening in space if, during the injection, the column temperature is held ~10°C above the boiling point of the sample solvent. This method is restricted to on-column sampling and requires a secondary cooling device (Galli *et al.*, 1979) or some equivalent. The method restricts the choice of the analysis temperature to temperatures above the boiling point of the solvent and may cause problems with the solvent effects on the solutes eluting early in the chromatogram. However, it is a useful technique whenever there are problems with the retention gap.

Ideally, the retention gap should be introduced into the column during the column preparation procedure. It is also possible to attach uncoated but deactivated precolumns to the column inlet, for example, by means of butt connectors. The retention gap may be excessively long (e.g., 3–5 m) without a risk of noticeable peak broadening under normal conditions (K. Grob, Jr., 1983a). This allows the cutoff of dirty sections several times.

Retention gaps are also useful to reduce the side effects of dirty samples. If the low-boiling by-products of the sample are deposited on top of a layer of stationary phase, they cause more adverse effects than on a bare surface, and rather often they even degrade the stationary phase, causing column bleeding (K. Grob, Jr., 1984c).

Presently, the on-column injection is by far the most accurate and precise sampling method. Discrimination caused by volatility can be practically eliminated (see Fig. 4). Thus, for quantitative analyses it should be used instead of the split or splitless injection wherever possible (K. Grob, Jr., 1979; Badings *et al.*, 1981; Schomburg *et al.*, 1981; Hiltunen *et al.*, 1982; Proske *et al.*, 1982).

There are only a few applications where it is not feasible to use on-column injection. If the sample is extremely contaminated with high-boiling or nonvolatile material, this dirt is deposited by the on-column injection exactly at the spot that is most sensitive, the column inlet. As mentioned, a retention gap should be used. However, if the deposited dirt exhibits a greater "retention power" than the stationary phase, peaks are nevertheless broadened and distorted. Peak distortion usually occurs if 1–30 μg of dirt have accumulated in the column inlet. It is essential to realize that this dirt cannot migrate through the gas phase. Thus, it remains within the flooded inlet section of the column. If this section is removed, the column performance is restored. This is particularly easy if precolumns of fused silica are used.

Another disadvantage of the cold on-column injection is that relatively large amounts of solvent enter the column, which also occurs with the splitless-injection technique. This forces the operator to make a careful selection of the solvent, mainly considering its volatility at the column temperature during the injection. For many applications a quick cooling of the column below the boiling point of the solvent is required. Furthermore, undesirable solvent effects, primarily partial solvent trapping, are observed if the solvent is not compatible with the stationary phase and/or the sample constituents.

E. Capillary Columns

1. Some Basics of Column Preparation

There are various ways to successfully prepare columns that are suitable for a wide range of applications, as can be seen in numerous publications. Differences of column quality usually become apparent only when dealing with difficult samples, for example, those containing very polar, extremely high-boiling, or labile compounds. Even for such applications, it is difficult to classify the differently prepared columns because each of them may perform outstandingly for a specific problem. In the following discussions, some of the most important aspects of column preparation and column characteristics are discussed. For a detailed description of specific column preparation methods, the reader is referred to the literature cited in this section and the books by Lee *et al.* (1984) and Schulte (1983).

a. Film Building

The tendency of a stationary phase to form a homogeneous film over the surface of a capillary tubing depends on the chemical "wettability" (the contact angle for that phase) and the geometry of the surface. It was found experimentally that nonpolar phases build homogeneous films on smooth surfaces, such as those usually obtained from glass. Polar phases, however, often do not spread out under these conditions; instead they form droplets that lead to columns of very low separation efficiencies. To obtain equivalent results with polar phases, the surface of the capillary tubing must be roughened. Some practical experiences on this topic have been summarized by K. Grob (1979). The separation efficiency of a column is highly dependent on the homogeneity of the film formed by the stationary phase.

Some of the problems related to the stability of the film of the stationary phase were overcome by the introduction of the "bonded" or immobilized coatings. Rigaud *et al.* (1976) studied the *in situ* synthesis of cross-linked and possibly surface-bonded silicon phases. R. Jenkins (unpublished results, 1980) achieved the first cross-linking of an ordinary stationary phase SE 54 in capillary columns by thermal treatment. K. Grob *et al.* (1981b,c) introduced the immobilization of primarily apolar silicone phases by the use of peroxides. Since then, techniques have been developed to immobilize most silicon stationary phases.

Films of immobilized stationary phases cannot contract to form droplets, thus rendering the columns much more stable (which is especially important for medium to polar stationary phases that tend to form labile films). Furthermore, immobilized films can be washed and resilylated (K. Grob and Grob, 1982). On the other hand, immobilized films cannot be extracted again, which creates problems for the preparation of retention gaps (K. Grob, Jr., 1982a) and exit sections free of stationary phases.

b. Chemical Characteristics

The problems encountered when using capillary columns usually arise from the chemical characteristics of the surface underneath the stationary phase rather than from the stationary phase itself. These surface characteristics are responsible for the adsorption of polar, acidic, basic, or (after special treatment) nonpolar compounds. In addition, they determine the catalytic activity that influences the thermostability of the stationary phase and the thermostability and chromatographic behavior of labile compounds. Thus, when selecting a column for the analysis of a specific group of compounds (e.g., acids) one should not only look for a suitable stationary phase, one should also be very concerned about the support surface. For example, when using a free fatty acid phase (FFAP) on a surface with basic adsorption sites, poor results are obtained for the determination of free acids. Unfortunately, for most commercially available columns, only the stationary phase is indicated.

There is still considerable discussion about what chemical properties of the surface create which chromatographic phenomena. It is generally thought that certain metal ions may function as adsorptive sites (e.g., as Lewis acids) thus causing tailing of polar compounds (Filbert and Hair, 1968). The silica framework of the glass itself also shows adsorptive behavior: the silanol groups affect the peak shapes of almost all types of compounds (except alkanes), whereas the siloxanes are somewhat less adsorptive. It is still very difficult or impossible, however, to find the exact cause(s) responsible for an observed adsorption.

Many procedures, usually very empirical, have been developed for surface deactivation. The aim of the classical deactivation procedures is to cover active sites (prior to the coating of the column) with nonvolatile organic compounds exhibiting similar adsorption behavior as the compounds to be chromatographed. Thus, the adsorption sites are blocked by these immobile compounds. In this case, it is not necessary to know anything about the structure of these sites and the mechanisms of adsorption. A similar procedure (which is also applied to packed columns) is sometimes used for acid/base deactivation. Columns used for acidic compounds are pretreated with acids, and analogously columns used for basic compounds are pretreated with bases. However, these procedures are only partially successful.

The other approach is to specifically eliminate active sites from the glass surface (e.g., metal ions can be washed out or complexed, or silanol groups can be eliminated or transformed by fluorination, complexation, or silylation). This method of surface modification is certainly more systematic but also more difficult.

The hardest problem to solve is the durability of the surface modification. Deactivation materials need to be thermostable. It should also be kept in mind that glass is a viscous liquid, able to create new active sites such as silanol groups or to allow metal ions to diffuse from the bulk of the glass to the surface. The importance of a thermostable deactivation for the thermostability of a column may be illustrated by the following two examples. (1) A column coated with Carbowax 20 M on a surface containing many silanol groups can only be used up to a temperature of 180°C, whereas corresponding "$BaCO_3$" (K. Grob and Grob, 1976) or "NaCl" columns (Alexander and Rutten, 1973) stand 250–260°C. (2) A classical nonpolar silicon column can be used up to 270°C (with basic support, only up to 240°C). On a persilylated support, the same stationary phase withstands heating to 370°C.

The most important methods used for surface modifications were reviewed by Verzele (1979). Two types of columns appear to be most interesting: (1) leached columns of ordinary glass where ions are extracted with hydrochloric acid producing a deep layer that is extremely rich in silanol groups, which are then persilylated (K. Grob et al., 1979, 1982), and (2) fused-silica columns (Dandeneau and Zerenner, 1979), which contain almost no metal ions or silanol groups. Their surface is, however, still active, probably from the siloxanes. Thus, they also require an appropriate deactivation.

The introduction of fused-silica columns has often been wrongly considered as the invention of a new technique. All of a sudden, as if the fused-silica columns created a new situation, capillary GC became commonly acceptable, especially in the United States. However, good fused-silica columns are of very similar quality to good glass columns. For the great majority of the applications, the two types of columns are equal in performance; for a few applications, one type has preferable characteristics, for others, the other type. There are very few chromatographic arguments to use fused-silica (or glass capillary) columns. Fused silica is more flexible than glass and eliminates the necessity to straighten end sections (which may be carried out with automatic devices). On the other hand, dirty spots may often be detected by visual inspection in glass but not in fused-silica columns. Furthermore, manipulations such as washing, recoating, and especially the preparation of new columns, is far easier with glass capillary columns.

2.   Column Qualities and Column Tests

It is a pity that many commercially available columns and the majority of published column-preparation methods are advertised like cars; their real qualities as well as their insufficiencies become obvious only when they are in use. Often the only quality parameter indicated is the separation efficiency (i.e., plate number or "Trennzahl") (see Table I), which are just the parameters that differ the least when comparing columns of a given geometry (i.e., column length, id). In the following subsections, we discuss some of the most important parameters that must be considered when describing the quality of a column and the testing procedures that may be applied for evaluating these parameters. Presently, column testing procedures are still under development. More should be known about the general mechanisms of adsorption, degradation, and rearrangements of solutes in order to find specific test compounds and subsequently to design better column-preparation procedures.

a.   Separation Efficiency

The traditional way of indicating the separation efficiency of a column is the theoretical plate number per unit column length (N_{eff}/L) (see Table I). However, this parameter has the disadvantages that it is not a direct measure for resolution, which is the parameter of primary interest for the separation of compounds (K. Grob, Jr. and Grob, 1981) and that it is defined only for isothermal conditions. Thus, a more practical number, the Trennzahl (TZ) suggested by Kaiser (1976, 1977) is preferred. The TZ describes the separation efficiency in terms of how well two peaks are separated under experimentally defined conditions (Table I).

The separation efficiency of a column depends mainly on the homogeneity of the liquid film. The most homogeneous films are obtained when the stationary phase is coated on a smooth surface. This is a major reason why PLOT or SCOT columns are much less efficient than wall-coated capillary columns (WCOT). Another factor that influences the efficiency of a column is its dimensions: The plate number increases (at least theoretically) linearly with the column length. This is not the case with the resolution. Theoretically, the resolution increases with the square root of the plate number (in practice it is even less). In addition, it has been found that the resolution scarcely improves when using columns longer than 100 m. Often the resolution even *decreases* in columns that are longer than 150–200 m. The inner diameter of the column also has a strong influence on the separation efficiency (Schutjes *et al.*, 1981). Very narrow columns are extremely efficient but difficult to use in practice. Finally, the film thickness has a relatively weak effect; thick films slow down the exchange of a solute molecule between the liquid phase and the gas phase and therefore decrease efficiency (Sandra *et al.*, 1983).

b. Adsorption

During the 1960s and early 1970s, adsorption was the major problem encountered when using capillary columns. The mechanisms of adsorption of polar compounds are still not exactly known. Based on our experience, all compounds with higher polarity than alkanes may be adsorbed. These compounds range from aromatic compounds to ketones to secondary and primary alcohols. The degree of adsorption seems to be relatively constant for a given class of compounds. Thus, the adsorptive properties of a column may be checked with one single sensitive compound, commonly a primary alcohol such as 1-octanol. For polyglycol phases (e.g., Carbowaxes, Ucons, and Pluronics) it may be better to use a diol (e.g., 2,3-butanediol) to obtain adequate information, because on these columns 1-octanol is usually perfectly eluted (K. Grob, Jr. *et al.*, 1978).

c. Acid–Base Effects

When dealing with compounds that exhibit acidic or basic properties, the acidic and basic sites of the support surface become very important, especially because of irreversible adsorption processes. Also, these sites are of concern when analyzing substances that readily undergo hydrolysis. The acid–base activity of glass capillary columns varies with different preparation procedures, but it seems that this cannot be avoided. Thus, usually only one type of compound (strongly acidic or strongly basic) is chromatographed properly, and in some cases neither class behaves properly. It is fairly common that capillary columns contain acidic as well as basic sites. Such columns are, of course, limited in their applicability to the analysis of environmental samples.

For the detection of acidic and basic sites, the so-called AP test with 2,6-dimethylphenol (DMP) and 2,6-dimethylaniline (DMA) (K. Grob and Grob, 1971) is frequently used. However, this test has been shown to be not entirely satisfactory since it is very difficult to determine absolute peak areas to check for irreversible adsorption. Comparing just the two test compounds may be misleading, since both peak areas may be reduced (K. Grob, Jr. *et al.*, 1978).

A complicating factor is that acidic or basic induced adsorption is very often irreversible, which leads to peaks with perfect shape but deficient areas. Tailing cannot be used to detect such adsorption. K. Grob, Jr. *et al.* (1978) suggested the use of a comprehensive test mixture containing DMP and DMA. By comparing peak heights in a temperature-programmed run, the deficiency of the original AP test (K. Grob and Grob, 1971) can be overcome. Furthermore, information on the presence of only weakly acidic or basic sites is obtained by using additional compounds that are stronger bases and acids (e.g., dicyclohexylamine and 2-ethylhexanoic acid).

d. Catalytic Degradation

It would be very useful to have a simple test for determining the catalytic activity of a support surface. Donike (1973) suggested chromatographing a series of silylated fatty acids in a temperature-programmed run. Losses from hydrolysis can be detected by comparison with alkanes that are included in the test mixture and the influence of temperature can be studied. The latter is of interest since the stability of labile compounds very strongly depends on the elution temperature [also shown for phenylureas (K. Grob, Jr. *et al.*, 1980)]. When composing test mixtures for the detection of catalytic activity, the major problem is to find compounds that do not decompose in the hot injector if a vaporizing injection technique is used. Also, compounds should be found for which different catalytic degradation reactions can be distinguished. Catalytic versus adsorptive effects were studied by K. Grob (1980) and Ahnoff and Johansson (1981).

e. *General Remarks on Column Testing*

The results of any kind of test depend strongly on the chromatographic parameters used. Adsorption, for example, generally decreases with increased amount of solute reduced retention and increased column temperature. Therefore, test results are only comparable if they have been obtained under comparable conditions. When testing a column, the conditions should be selected according to the task for which the column is to be used. Only then can it be decided whether a column is suitable for the quantitative analysis of a given sample. K. Grob, Jr. *et al.* (1978, K. Grob *et al.*, 1981c) developed a comprehensive, standardized quality test for glass capillary columns that yields comparable information for many stationary phases (Fig. 8). Information is obtained by one chromatographic run on separation efficiency (TZ), adsorption of polar, acidic, and basic compounds, and on film thickness, or more precisely the phase ratio β (see Table I).

3. Stationary Phases

It is not the purpose of this section to discuss the abundant literature on the numerous existing stationary phases and the parameters used to characterize them [e.g., McReynolds constants (McReynolds, 1970)]. We feel

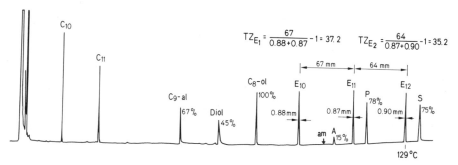

Fig. 8. Test chromatogram of a column (20 m × 0.30 mm) that was a failure: $BaCO_3$ on HCl-leached Pyrex converted into $BaSO_4$. Liquid phase: Pluronic L-64, a polyethylenepropylene glycol copolymer. Test after conditioning at 220°C for 1 night. Test mixture: C_{10}, Decane; C_{11}, undecane; $C_9\text{-}al$, nonanal; Diol, 2,3-butanediol; $C_8\text{-}ol$, 1-octanol; E_{10}, C_{10}-acid methylester; *am*, dicyclohexylamine; A, 2,6-dimethylaniline; E_{11}, C_{11}-acid methylester; P, 2,6-dimethylphenol; E_{12}, C_{12}-acid methylester; S, 2-ethylhexanoic acid. The AP test indicates "acidic", but both A and P are adsorbed. This shows that there are some basic active sites left (besides more acidic sites). No *am* elutes; S elutes to only 75% because the basic sites are also active to P. Adsorption by hydrogen bonding is acceptable (not good). The *al* seems reasonable at first glance, but one third of the material is adsorbed. The separation efficiency is relatively high. The film thickness is 0.18 μm as E_{12} elutes at 129°C. (From K. Grob, Jr. *et al.*, 1978.)

that in practical capillary GC (in contrast to packed-column GC), the selection of the stationary phase is often of only limited importance. When dealing with columns of similar polarities, the characteristics of the support surface and parameters such as the film thickness are usually much more critical than the exact choice of the stationary phase. Furthermore, it should be remembered that the "polarity" of a stationary phase is not constant. It depends on the support surface and column temperature (e.g., K. Grob and Grob, 1983a). In many laboratories using capillary columns, a major part of all GC analyses is usually performed on columns with the same stationary phase but with different supports and various film thicknesses. For a general overview of the most common stationary phases used in gas chromatography, we refer the reader to the literature (e.g., Lee *et al.*, 1984).

Unfortunately, it is rather common to transfer experiences with packed columns to capillary GC. However, the best stationary phases for capillary GC are not necessarily good phases for packed-column GC, and vice versa. Therefore, the usefulness of a stationary phase for capillary GC must be evaluated independently. The argument that the same stationary phases should be used for capillary and packed columns in order to allow direct transfer of retention data is often overstressed. In fact, chromatograms obtained from a packed column and from a capillary column with the same stationary phase may be quite different. Because of the usually big differences in elution temperatures, the sequence of the peaks may be very different for complex mixtures, which will require that peak identifications be redone.

When using capillary columns, it is generally worthwhile to start out with those stationary phases with which the best columns can be produced. Only when these columns fail (e.g., because of selectivity reasons or sample losses), is it advisable to switch to other stationary phases. In the following subsections, only the two most widely used types of stationary phases for capillary columns are discussed.

a. Silicone Gum Phases

Silicone phases exist over a wide range of viscosity, for example, OV-1 is a gum that can be cut into pieces, whereas OV-101 is a viscous liquid. Examples for gum phases are OV-1 and SE-30 (containing methyl substituents), and SE-52, SE-54, and OV-73 (which are slightly more polar because they contain ~5% phenyl and vinyl groups). Their most important characteristics are

1. Nonpolar silicone phases have the widest useful temperature range of all stationary phases. Below 20–30°C, the separation efficiency may

decrease slightly because of their high viscosity. The upper temperature limit depends strongly on the catalytic activity of the support surface, ranging from 250°C for alkaline surfaces to 380°C for persilylated glasses and ~330°C for fused silica.

2. Capillary columns coated with silicone oils lose some of their efficiency when heated above ~230°C. Since the gum phases or immobilized films do not show this behavior, they are preferable for reasons of efficiency. The loss of efficiency observed for oils is combined with a decrease in retention, suggesting breakup of the film and the formation of small droplets.

3. The gum phases are best suited for producing columns of a wide range of film thicknesses. The practical limits are ~0.03 and 3 μm (K. Grob, Jr. and Grob, 1977a; K. Grob and Grob, 1983b). The oil phases, however, do not provide homogeneous films if they are coated thinner than ~0.07 μm, and they tend to form droplets at >0.6 μm.

4. Silicone phases have the advantage of being relatively insensitive to oxygen. It is usually not necessary to use traps for removing oxygen that enters the system mainly through plastic membranes of pressure regulators or plastic tubing.

5. The bleeding of the nonpolar silicone phases (especially the gums) is usually low and produces only a few peaks in mass spectra, thus causing minimum disturbances when used in GC/MS analyses.

6. Nonpolar columns are generally relatively adsorptive, although this aspect has been significantly improved by the persilylation of the support. Therefore, they should not be continuously subjected to polar compounds (e.g., alcohols, amines, and acids). On the other hand, they are less catalytically active than polyglycols, for example, partly because of lower elution temperatures.

7. Their major drawback is their sensitivity toward polar dirt. Polar constituents of samples seem to cause disruption of the film of the stationary phase if the film has not been immobilized. The consequence is a decrease in retention, separation efficiency, and in most cases, an increase in adsorption. Once a nonpolar silicone column has been destroyed during use, it cannot be recoated successfully. This causes a fairly high consumption of such columns if many dirty samples are analyzed.

b. Nonionic Detergents

Nonionic detergents are a completely different type of stationary phase with some excellent characteristics that are quite complementary to those of the silicone phases. The most widely used nonionic detergents are the

Carbowaxes (polyethylenglycols). They are also the most polar ones. Less polar polyglycols, which contain an increasing portion of polypropyleneglycol, are the Ucons and the Pluronics. Finally, there are some phases in which the polyethyleneglycol chain is linked to an alkyl or aryl group at one end: Emulphor (ether bond), Marlazins (ammonium bond), and the Tritons. The polyglycol columns may be characterized as follows.

1. Detergents such as the polyglycols deactivate the support surface as far as polar adsorption is concerned. Therefore, they are well suited for the analysis of polar solutes such as alcohols, acids, or amines.

2. They form films that are hardly disturbed by dirt. Extremely dirty samples may cause coloring of the column (brown color), but they seldomly affect the chromatographic performance.

3. If a detergent column has deteriorated seriously, it may be washed and recoated, which is in contrast to the silicone phases. This requires about one-half day for a 20-m column. The recoating may be repeated many times, and usually the quality of the column improves. Because of their resistance to dirt and the ease of recoating them, the detergent columns may serve as work horses for the analysis of all sorts of unknown and crude samples.

4. Despite the fact that the polar Carbowaxes are the best known polyglycol phases, the less polar detergents are preferable for many applications. The retention of polar solutes is substantially reduced on less polar columns, thus allowing lower elution temperatures or a wider range of compounds to be analyzed. The adsorptive characteristics of the less polar phases are comparable to those of the Carbowaxes, and their separation efficiencies are usually somewhat higher.

5. The major drawback of the polyglycol columns is their strong bleeding. Since these polymers are mostly crude technical products, they contain a large portion of low molecular weight material, and the molecular weight range of the detergent itself is usually very broad. Furthermore, they are chemically not very stable and may catalytically decompose, especially if the support surface contains many silanol groups. The most frequently encountered cause of "column death" is apparently, however, a poorly understood depolymerization process that is initiated by traces of oxygen. The deterioration starts even at 40°C. The sensitivity to oxygen increases with increasing proportion of polypropyleneglycol.

6. The efficiencies of detergent columns are not as good as those obtained with silicone gum phases, because a roughened surface is required to ensure sufficient wettability. Resolution in terms of TZ may be ~10–20% lower.

F. Detector Systems

1. General Remarks and Definitions

Detectors play an important role in chromatographic analyses of environmental samples. Whereas the choice of good detector systems (especially of universal character) is still very limited in liquid chromatography, a variety of very sensitive and reliable detectors have been developed for gas chromatography. Most of them are commercially available and have been successfully adapted to capillary GC.

The most universal and most common detector used for trace analyses of organic water constituents is the flame ionization detector (FID). However, since many of the compounds of concern in the environment contain heteroatoms such as halogens, nitrogen, sulfur, or phosphorus, detector systems exhibiting selectively high sensitivities for compounds containing such elements are becoming increasingly important. With such selective detectors, two goals can be achieved simultaneously: (1) Information can be obtained about the presence or absence of certain compounds in a complex mixture, and (2) interferences of coeluting compounds of dissimilar chemical nature can be eliminated. Furthermore, in some cases, a much lower detection limit can be achieved for a specific group of compounds. Very often only the use of one or several selective detector(s) leads to satisfactory results because the mixture of compounds obtained by a particular sampling procedure is often still too complex to be completely resolved even by the best capillary column.

When comparing detector systems, characteristics such as sensitivity, selectivity, linear dynamic range, response time, and dead volume of the detector must be addressed. The *sensitivity* is a measure of the ability of the detector to produce an electrical signal from the presence of a compound in the effluent of the column, that is, the change in response per unit weight of material. In practice, a quantitative measure for the sensitivity of a detector is usually given by its (lower) detection limit. A common definition of the *detection limit* is the amount per time or the concentration of a compound producing a peak the height that is twice the noise level. It should be noted that the way detectors respond can be different. Some are mass sensitive, that is, they respond to the amount of material passing the detector per unit time, whereas others are concentration sensitive. The detection limit should therefore be expressed in the appropriate units (i.e., g/s and g/mL, respectively). Furthermore, some detectors show great differences in their sensitivity to different compounds. Thus, when reporting detection limits, the compounds with which they were established should be indicated.

If a detector exhibits a much greater sensitivity for members of a specific group of compounds (compared to all other chemicals) it is called a *selective detector* for this particular group. In practice, a detector can be considered to be selective for a certain group of compounds if the ratio of its response to these compounds compared to other substances exceeds $10^4:1$.

Another important property of a detector is its *linear dynamic range*. This parameter indicates the concentration range (or amount/time range) over which a constant sensitivity can be assumed. A common measure for the linear dynamic range of a detector is the ratio of the highest concentration (or amount/time) for which <3–5% deviation from linear behavior and the detection limit is observed. The linear dynamic range of a detector system is of great importance for quantitative determinations. Reasonable linear dynamic ranges ($>10^4$) for various detectors have been achieved only recently.

In capillary GC the *response time* of a detector is especially critical since the compounds are eluted in very short times (a few seconds). The time constant of the detection system must be small enough so that the separated compounds are detected individually without loss of resolution. The time constant is determined mainly by the detector volume, the dissipation constant of the sensing element, and the recording equipment. Because the effective geometric volume of a detector must be smaller than the volume of the carrier gas entering the detector during the half width time of the narrowest peak with the shortest retention time (Jentzsch and Otte, 1970), the detector volume can be a very critical point in capillary GC. This problem is, however, not encountered with flame-ionization type detectors (i.e., flame-ionization detector, thermionic detector, and flame-photometric detector). The problems with detectors exhibiting a large dead volume (e.g., electron capture detector) can be overcome by adding a make-up gas at the column exit.

There are, of course, several other parameters that influence the performance of a gas chromatographic detector (e.g., stability and reactivity). For more details, especially with regard to electronic aspects, the reader is referred to the literature (e.g., Jentzsch and Otte, 1970; David, 1974). A compilation of some of the most common practical troubles encountered when using gas chromatographic detectors has been published by McCown and Earnest (1978).

2. Flame Ionization Detector

The flame ionization detector (FID) can be considered to be the universal standard detector for gas chromatographic analyses. It is a mass-sensitive detector and responds to virtually all organic compounds with roughly the same sensitivity (same order of magnitude). Because of its high linear dynamic range (10^6), its moderately high sensitivity (detection limit $\sim 10^{-12}$ g/s), and its reasonably good stability, the FID is a reliable detector for quantitative work.

In capillary GC, the FID is always operated relatively close to its detection limit because of the restricted amount of sample that can be analyzed on capillaries. Thus, only a small span of its full linear dynamic range is used. The detector should therefore be optimized for this small range. It should be noted that column bleed and dirt may significantly reduce the detection limit of the FID and that its stability is slightly affected by changes in the carrier-gas flow rate (K. Grob, Jr., 1980).

The general operation of the FID is as follows (see Fig. 9A). The effluent of the column is mixed with hydrogen and burned at the tip of a jet in an excess of air. When organic matter enters the detector it is combusted in the flame. During this process, ions are formed that are collected by an electrode situated above the jet. This ion current is then amplified and recorded. The theory of flame ionization, which is not fully understood, will not be discussed here.

The response of the FID (peak area) is more or less proportional to the total number of "effective carbon atoms" of a compound i passing through the detector:

$$\text{Peak area } F_i = f_i \times \left(\Sigma\, C_{\text{effective}} \right)_i \times \text{mol}_i$$

Table II shows the contribution of various atoms to this effective carbon number. Thus, in principle it is possible to estimate the relative response of the FID for a specific compound. For an accurate quantification, however, individual response factors should be determined from a reference mixture.

Finally, it should be noted that the FID, like all flame detectors, is a destructive detector. This has to be kept in mind when using it in series with other detectors.

3. Electron Capture Detector

The most widely used selective detector in environmental analyses is the electron capture detector (ECD). It responds primarily to molecules exhibiting functional groups that have a tendency to "capture" free electrons, that is, groups containing atoms with high electron affinity. Thus, the ECD is very sensitive to most halogenated compounds, but it also

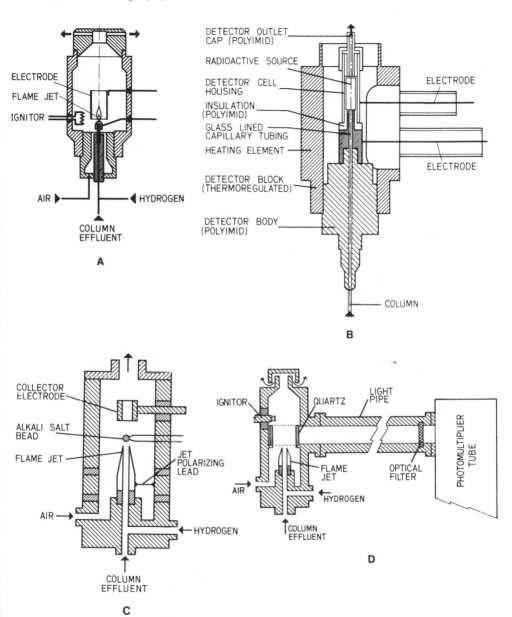

Fig. 9. Schematic diagrams of some common types of gas chromatographic detectors. (A) Flame-ionization detector (FID); (B) electron-capture detector (ECD) (from Brechbühler *et al.*, 1977); (C) alkali flame-ionization detector (AFID). (Adapted from Ettre, 1978.) (D) flame photometric detector (FPD). (Adapted from Ettre, 1978.)

TABLE II

Contribution of Various Atoms to the "Effective Carbon Number"[a]

Atom	Type	Effective carbon number contribution ($C_{effective}$)
C	Aliphatic	1.0
C	Aromatic	1.0
C	Olefinic	0.95
C	Acetylenic	1.30
C	Carbonyl	0.0
C	Nitrile	0.3
O	Ether	−1.0
O	Primary alcohol	−0.6
O	Secondary alcohol	−0.75
O	Tertiary alcohol, esters	−0.25
Cl	Two or more on single aliphatic C	−0.12 each
Cl	On olefinic C	−0.05
N	In amines	Similar to O in corresponding alcohols

[a] From David (1974).

shows enhanced response for nitro compounds, highly oxygenated compounds, and some aromatic compounds. It measures concentration and has a very low detection limit for many compounds [e.g., $\sim 10^{-13}$ g/mL for lindane (Pigliucci *et al.*, 1975)].

Two basic problems with the ECD are that it is very difficult to predict its response to a specific compound and its sensitivity for different compounds may vary by several orders of magnitude even within a group of structurally closely related substances. This is illustrated in Table III, which contains the relative ECD responses for a selected group of compounds.

Another drawback of the ECD is its relatively limited linear dynamic range. Only in the late 1970s were linear dynamic ranges of $\sim 10^4$ reported (Patterson, 1977; Brechbühler *et al.*, 1977; Ettre, 1978). In practice, however, the actual linear dynamic range is usually much smaller. For quantitative work, it is therefore absolutely necessary to calibrate the system for each individual compound.

The basic operation of the ECD is as follows (see also Fig. 9B). The effluent of the column is introduced into a cell containing a β-radiation source (e.g., Ti³H, Sc³H, or ⁶³Ni). If a compound with high electron affinity enters the cell, electrons are captured by the compound, thus forming negative ions. Because such negative ions recombine with the

TABLE III

Relative Response Factors of the ECD

Compound	Relative response[a]
Methylene chloride	0.02[b]
Chloroform	1[c]
Carbon tetrachloride	10[c]
Tetrachloroethene	3[c]
Bromoform	2[c]
1,1-Dichloroethene	0.05[c]
Chlorobenzene	0.001[d]
Bromobenzene	0.005[d]
Iodobenzene	0.5[d]
Nitrobenzene	0.5[d]
2,3-Butanedione	0.8[d]
Quinone	6[d]
Hydroquinone	0.0001[d]
Benzene	0.000001[d]

[a] Relative to chloroform (= 1).
[b] From Eklund et al. (1978).
[c] From Piet et al. (1978).
[d] From Ettre (1978).

positive ions present in the cell much faster than they do with electrons, a decrease in charged species occurs. This loss of ions by recombination can be observed as a reduction of the standing β current. Traditional electron capture detectors work with a constant pulser frequency (amount of electrons produced) and measure directly the decrease in standing β current. In modern detectors, this current is kept constant by adjusting the pulser frequency. This leads to a better linearity and is less critical with regard to the choice of the carrier gas (Patterson, 1977; Ettre, 1978). Detailed discussions of the theory of the ECD are given by Pellizzari (1974) and Lovelock (1974).

Although conventional electron capture detectors have relatively large dead volumes, they have been successfully used in combination with capillary columns (e.g., Eklund et al., 1978; Reinhard et al., 1979). The dead volume problem is overcome by adding an appropriate amount of make-up gas (e.g., nitrogen) at the column exit. It is important that this make-up gas be thermostated. Otherwise, during temperature-programmed runs, a baseline drift is caused by the nonconstant temperature difference between the make-up gas and the gas leaving the column oven. Since the ECD is a concentration-sensitive detector, the dilution of the carrier with make-up gas leads to a decrease in sensitivity. A slight improvement of this situation has been obtained with the introduction of micro electron-capture detectors (Fig. 9B) (Brechbühler *et al.*, 1977).

In practice, the ECD is rather delicate to operate since its stability is affected by various factors such as temperature changes, ECD active solvents, dirt, etc. Hence, careful operation and frequent recalibration are prerequisites for quantitative analyses.

4. Alkali Flame Ionization Detector

The alkali flame ionization detector (AFID), a thermionic detector, is a special type of FID in which the flame is "doped" by the slow vaporization of an alkali metal salt built into the detector (see Fig. 9C). The choice of the salt (e.g., Na_2SO_4, RbCl, KCl, and CsBr) and the design of the ionization chamber allow the selective enhancement of the response to organic molecules containing heteroatoms such as phosphorus, nitrogen, arsenic, or halogens. The thermionic detector is mostly used to determine nitrogen and/or phosphorus containing compounds (NPD; Maier-Bode and Riedmann, 1975; Kolb *et al.*, 1977; Burgett *et al.*, 1977; Ettre, 1978). The reactions occurring in the flame are not fully understood, but various possible mechanisms have been postulated (Maier-Bode and Riedmann, 1975; Kolb *et al.*, 1977). The NPD offers, compared to the conventional FIDs, remarkably improved sensitivity for nitrogen- and phosphorus-containing compounds. It is a mass-sensitive detector. Detection limits of approximately 10^{-14} g P/s and 10^{-14} g N/s can be achieved with a linear dynamic range of better than 10^4. Typical selectivities are $10^5 : 1$ for phosphorus (compared to carbon) and $10^4 : 1$ for nitrogen. Various versions of NP detectors have been designed and are commercially available. Like the FID, the NPD can be easily adapted to capillary GC. However, for instability reasons, hydrogen cannot be used as a carrier gas. Also, the NPD shows some sensitivity drifts with helium. An advantage of this detector is its insensitivity to common types of column bleed.

5. Flame Photometric Detector

The flame photometric detector (FPD) is an important detector for trace analysis of phosphorus- and sulfur-containing compounds. It operates on the principle that such compounds emit a characteristic radiation in a hydrogen-rich hydrogen–air flame. In the case of phosphorus containing compounds, green light is emitted from the excited-state HPO molecule (HPO*) formed, whereas blue light from S_2^* is observed for sulfur compounds.

Most present-day FPD systems are based on the work of Brody and Chaney (1966). The compounds present in the effluent of the column are burned in a hydrogen flame, and the emission from the flame is detected through appropriate filters by a photomultiplier. The characteristic wavelengths are 394 nm for sulfur and 526 nm for phosphorus. The schematic of a typical FPD is shown in Fig. 9D.

The advantages of a dual-flame photometric detector have been discussed (Patterson *et al.*, 1978; Patterson, 1978). In this version, a hydrogen-rich flame is used to decompose the compounds, and a second flame is used to form S_2^* and HPO* molecules. The authors claim a more uniform response of the detector, and state that with this configuration, the flame is not as readily extinguished by the solvent as happens frequently with single-flame detectors.

The FPD is a mass-sensitive detector, and its detection limit is approximately 10^{-10} g S/s and 10^{-12} g P/s. Whereas the response to phosphorus is directly proportional to the amount of phosphorus in the sample, a square law dependence is observed for sulfur, which is in agreement with theoretical expectations for S_2^* emissions (Patterson *et al.*, 1978). Some manufacturers try to overcome this difficulty by using so-called linearizing amplifiers that display the square root of the peak area. However, when using such linearizers with a fixed exponential factor of 2 for the analysis of sulfur compounds, errors may occur because the actual FPD response does not always obey the theoretical square law relationship (Farwell and Rasmussen, 1976; Burgett *et al.*, 1977).

The influence of the burner configuration on the response of the FPD to compounds containing heteroatoms has been evaluated by Kapila and Vogt (1979).

The linear dynamic range of the FPD is reasonable for phosphorus ($>10^3$) and poor for sulfur ($\sim 10^2$). The selectivity achieved with the S filter (394 nm) is (depending on the compound) between 10^3 and 10^6 S/C and between 10^2 and 10^3 S/P. With the P filter (526 nm), a selectivity of 10^4 to 10^5 P/C is obtained, but the detector shows almost no selectivity for P/S (only ~ 4).

The FPD is suited for use with capillary columns. For quantitative work, calibration curves must be established for each compound.

6. Miscellaneous Detectors

A few additional detector systems that have some application in the analysis of organic compounds in the aquatic environment are mentioned briefly.

The thermal conductivity detector (TCD) is a truly universal detector because it responds to almost any kind of material including the permanent gases. It is nondestructive and has therefore found a wide application in preparative gas chromatography. Because of its relatively low sensitivity ($\sim 10^{-9}$ g/mL), the TCD is of minor importance for trace analysis, except perhaps for the determination of certain gases. Its use in connection with capillary columns has been discussed by Proske *et al.* (1978).

Electrochemical detectors, such as the microcoulometric detector (MCD) and the electrolytic conductivity detector (ElCD) have been developed primarily to provide more specificity for halogenated pollutants (Ettre, 1978). The basic principle of these detectors is the electrochemical monitoring of specific inorganic species (i.e., HCl) derived from the combustion of the organic compounds present in the effluent. The combustion products are dissolved in a liquid and the ions formed are determined either by automatic titration (MCD) (Coulson and Cavanagh, 1960) or by measuring the electrical conductivity of the solution (ElCD) (Coulson, 1965).

The major advantage of electrochemical detectors is that extremely high selectivities can be achieved for specific groups of compounds by choosing appropriate combustion parameters [i.e., temperature, type of reaction gas (oxidative or reductive), reaction catalyst, and chemical abstractors] and the type of liquid for dissolving the products (Dolan and Hall, 1973; Hall, 1974; Pape *et al.*, 1977).

A new type of a photoionization detector (PID) has been described by Driscoll (1977). The ionization of the compounds is induced by a very stable UV lamp. The response of this detector is comparable to the FID, but it shows an enhanced sensitivity for aromatic compounds (Driscoll *et al.*, 1978). The PID has great potential as an important detector for trace organic analyses.

Another interesting type of detector system is the microwave plasma emission detector (MED) (e.g., Quimby *et al.*, 1978, and literature cited therein). With this detector, basically all elements present in organic molecules (including carbon and hydrogen) can be selectively determined. Its operation is as follows. The compounds eluting from the column are directed into a microwave discharge that is sustained in either argon or helium. The observation of the optical emission spectrum resulting from the fragmentation and excitation of compounds entering the plasma affords sensitive element selective detection. However, the MED is an expensive detector and reliable commercially available systems are still lacking.

The most powerful and important, but also most expensive and complex, detector for the gas chromatographic analysis of environmental samples is the mass spectrometer (MS). This "detector" is discussed extensively in Chapter 5 of this volume.

G.   Additional Technical Remarks

1.   Carrier Gas

Hydrogen is the ideal carrier gas for capillary GC except for those cases where it is excluded by the detection system (e.g., AFID). The most important advantage is the speed of analysis that the use of hydrogen permits. In order to obtain the same results on a given column, the iso-thermal retention time must be doubled if helium is used and even tripled if nitrogen is used instead of hydrogen. When using hydrogen, the time of analysis is reduced significantly, and thus prolonged heating of the column at high temperatures is avoided. This increases the lifetime of a column and decreases the danger of degradation of thermolabile com-pounds.

Another important advantage of hydrogen has been found experimen-tally. When helium instead of hydrogen was used as a carrier gas, the column showed a significantly reduced lifetime. The symptoms observed were increased bleeding followed by a slow deterioration. These are the symptoms observed when oxygen is present in the carrier gas. Oxygen may already be present in the gas cylinder, or as mentioned earlier, enter the system through plastic tubing or plastic membranes by diffusion pro-cesses. The increased column lifetimes observed when using hydrogen as the carrier gas may be caused by its reducing properties.

The chromatographic process itself is also influenced by the nature of the carrier gas. However, under optimized conditions, the separation efficiencies differ only little for different carrier gases, being slightly bet-ter for hydrogen than for helium or nitrogen (K. Grob and Grob, 1979).

2.   End-Piece Treatment

The end pieces of a capillary column should be more or less free of stationary phase. Therefore, when installing a new column, the stationary phase must be washed out of the end sections by an appropriate solvent for the following reasons:

1. End pieces fitted into the (vaporizing) injector and the detector are usually heated above the column temperature, which quite often means above the temperature limit of the stationary phase. Thus, the film may be disrupted (forming droplets) or decompose (leading to bleeding).

2. The capillary inlet is the part of the column that is most exposed to dirt. Dirt is less harmful when it is deposited onto a bare inlet instead of a film of stationary phase. For on-column injection and for splitless injec-tion with solvent effect, a retention gap is needed (see Section II,D).

3. Because the detector block of many gas chromatographs is not sufficiently heated over its full length, high-boiling material is retained in the cooler zones of the capillary. This effect is greater if stationary phase is present: the thickness of the "film" in these zones grows and the tendency to retain more material increases. This vicious circle ends with the formation of a droplet.

If the end piece of a column is dirty or breaks, a new one can be obtained by straightening the end of the column, again using a small flame. However, one has to be very careful not to carbonize the stationary phase inside the tubing. This can be avoided by introducing a stream of air from the other end of the column. The stationary phase is then burned. Since an extremely active surface remains, redeactivation is necessary. This is easily achieved by washing this section of the column with a dilute solution of a polyglycol stationary phase (e.g., 0.1% of Carbowax 1000 in methylene chloride). The column is then ready to be reused after only 10 min of conditioning. For high-temperature persilylated columns, electrical straightening under an inert gas (e.g., nitrogen) is preferable (K. Grob *et al.*, 1981a). This procedure does not affect the activity of the support or the stationary phase because of the lower temperature required as compared to straightening by flame.

The washing and deactivation of end pieces can be performed by using a 50- or 100-μL syringe attached to the capillary by a short piece of plastic tubing. As a safety measure, ~0.3 atm of pressure (air) should be applied to the other column end in order to prevent plugs of the solvent from being sucked further into the column (i.e., by a sudden drop of temperature).

3. Symptoms and Diagnosis of Defects

Retention is the most sensitive parameter for judging how seriously a gas chromatographic system has been damaged. If the retention in a temperature-programmed run drops by more than 2°C (for silicone phases) or 4°C (for other phases), the whole column is probably damaged: either a large amount of stationary phase has been lost by bleeding or (more often) the film of the phase has been disrupted. In such cases the column must be recoated (if possible) or replaced.

If a column produces bad results (e.g., loss of efficiency, tailing) but does not show reduced retention, there is considerable hope that the defect is repairable, because stable retention means that the major portion of the film is still intact. Sometimes a drastic decrease of the separation efficiency indicates a local disruption of the film or a deposit of dirt (including material from the stationary phase), usually in the first two or three coils of the column. If this is the case, the problem can be solved by breaking off or washing the first three coils of the column. It should be noted that the same effect can be caused by a droplet located inside of the detector end section of the capillary or in the detector itself (especially in connection pieces, also called "interface"). In most cases, such accumulations of material not only reduce the efficiency but also increase adsorption.

If only increased adsorption is observed, the diagnosis is more difficult. It is often possible to find out whether the defect is located in or outside the column part that is thermostated by the oven by the following procedure. A test mixture is analyzed twice under drastically different conditions but with comparable retention of the compounds. At oven high temperature and with low carrier-gas flow rate, a local defect outside the oven-thermostated column part shows a stronger effect than at low temperature with high-flow conditions. This is not observed if the general adsorptive activity of the column has been increased. In this case the column should be recoated, resilylated (K. Grob and Grob, 1982), or replaced. Local defects of the system (outside the oven-thermostated column) are usually a dirty injector insert, a contaminated column end, or an adsorptive spot in the detector.

In any case, it is very important to apply a sufficiently comprehensive testing procedure (see Section II,E,2). A new column should be tested before use to provide the necessary data concerning retention, efficiency, and adsorption behavior to be used for comparison after a possible alteration of the column.

H. Qualitative and Quantitative Analysis

1. Qualitative Analysis

Retention is the most common parameter used for the qualitative gas chromatographic analysis of organic compounds. However, in capillary GC it may be difficult to obtain sufficient reproducibility of absolute retention to allow the use of this parameter for a proper identification of a compound, because various factors that influence the absolute retention are very difficult to control (e.g., injection, initial temperature). It is therefore often preferable to employ relative retention data based on appropriate internal standard reference compounds. It is advantageous to choose internal standards that elute close to the compounds of interest.

When dealing with complex mixtures, it is often not sufficient to rely solely on relative retention data obtained from a single column. The use of a second column with a stationary phase of different polarity can often help to identify a compound properly. The most reliable method is the coinjection of the compound of interest with the sample. This requires, of course, that the compound is commercially available or that it can be relatively easily synthesized.

In cases where the compounds of interest can be selectively detected, it is very useful to employ two or more detectors coupled either in parallel or in series. When using such multidetection systems, several points must be kept in mind: The chromatographic efficiency should not be decreased by introducing additional dead volumes, and interface lines should be kept to a minimum to avoid the introduction of adsorptive sites. Since many detectors are destructive (see Section II,F), multidetection is commonly performed by splitting the effluent of the column, thus using the detectors in parallel. Various splitting devices for capillary GC have been developed (Hrivnac *et al.*, 1976; Bächmann *et al.*, 1977; Andersson and Bertsch, 1978). Hrivnac *et al.* (1976) described a quite reliable splitting device that allows the use of as many as four detectors in parallel. Recently, a dual-detector system with an ECD and a FID connected in series has been described (Poy, 1979). This detector system has the big advantage that no splitting of the column effluent is necessary.

As mentioned previously in this chapter, the most powerful method for the separation, identification, and structural elucidation of organic compounds in water is the combination of capillary GC with mass spectrometry (GC/MS). This method is addressed in Chapter 5 of this volume.

For use in routine analyses and for the exchange of retention data between laboratories, great efforts have been undertaken to develop simple and reproducible retention parameters. Some of these systems are reviewed in the literature (Ettre, 1973; Haken, 1976). The retention index introduced by Kovats (1958), in which retention is determined relative to a series of homologous *n*-alkanes, has found a wide application especially in packed-column GC. However, these systems yield satisfactory results only with well-defined chromatographic columns, isothermal conditions, and operational procedures (i.e., injection technique). In capillary GC, the degree of reproducibility of relative retention data is dependent on many factors, some of which cannot be easily controlled by the analyst (Lee *et al.*, 1979). The reproducibility increases with increasing similarity of the sample to chosen standard reference compounds. The use of chemically similar internal standards becomes even more important under temperature programming conditions in which the retention data are not only dependent on temperature but also on the rate of temperature programming (Van den Dool and Kratz, 1963; Majlat *et al.*, 1974). Furthermore, in temperature-programmed GC, the simple expression of elution behavior as retention relative to an arbitrary standard is not applicable, and the logarithmic relationship that exists under isothermal operation between

n-alkane carbon number and retention in the Kovats Index (Kovats, 1958) is replaced by the approximately linear relationship:

$$I = 100 \frac{t_{R(\text{substance})} - t_{R(C_z)}}{t_{R(C_{z+1})} - t_{R(C_z)}} + 100z, \tag{1}$$

where $t_{R(\text{substance})}$ is the retention time of the substance for which the retention index is to be determined, $t_{R(C_z)}$ and $t_{R(C_{z+1})}$ are retention times for the n-alkane standards that bracket the substance of interest, and z is the number of carbon atoms in the n-alkane standard that elutes just prior to the substance of interest (Lee et al., 1979).

On the basis of Eq. (1), Lee et al. (1979) proposed a new retention index system for polycyclic aromatic hydrocarbons (PAH) in which naphthalene, phenanthrene, chrysene, and picene are used as internal standards. In this system, z is the number of rings in the PAH standard that elutes just prior to the substance of interest. They have determined the PAH retention indices for over 200 PAH and have shown that, as a result of the chemical similarity of the chosen internal standards, a high degree of reproducibility is maintained under varying gas chromatographic conditions. It is hoped that similar retention index systems will be developed in the future for other compound classes in order to obtain standardized retention data that can be used by all laboratories.

2. Quantitative Analysis

When quantitatively determining specific organic compounds in water by GC, one has to be aware that the quality of the results (accuracy, precision) is not only dependent on the gas chromatographic step, but frequently even to a greater extent on the accuracy and precision of the sampling and pretreatment procedures. In the following, however, we focus on the gas chromatographic part of the analysis.

The quantitative determination of a compound by GC should, whenever possible, be based on internal standard reference compounds, preferably of similar chemical nature, by comparing the compound peak area or peak height with that of the standard. For example, when using vaporizing injection, the absolute amount of sample reaching the column depends greatly on factors such as the exact amount of solution in the syringe, the poor reproducibility of the amount of material leaving the syringe, the nonreproducible (unknown) split ratio, and the poorly reproducible rate of sample transfer in the case of splitless injection (see also Section II,D). Only with the cold on-column injection are reasonable results obtained with external standards. The use of appropriate internal standards has the additional advantage that losses and errors from the sampling procedure can be compensated.

For accurate quantification, it is usually necessary to use a "correction factor" for an individual compound. The correction factor for a compound, determined relative to a chosen internal standard, is composed of two parts. One part, the "detector response factor," takes into account the fact that detectors usually respond differently to different compounds (see Section II,F). The other part is a poorly defined factor incorporating discrimination caused by injection, losses in the column, and sensitivity changes during a chromatographic run.

Furthermore, the dependence of the correction factor on the concentration of the sample should be checked. Calibration curves must be established if the correction factor changes with concentration. This is especially important when using detectors with a poor linear dynamic range (e.g., ECD and FPD) (see Section II,F).

Correction factors as well as calibration curves need to be reestablished whenever important instrumental parameters are altered. Such alterations include column changes, changes in the injection procedure (split ratio, injector temperature, injector geometry, syringe needle length), and changes in the carrier-gas flow rate, solvent, injection mode, or carrier gas.

If the peak areas in a complex chromatogram cannot be accurately determined (see Section II,H,3), peak heights can be employed for quantification provided that internal standards are used that elute close to the compounds of interest.

3. Electronic Integrators

Because of the complexity of the chromatograms usually obtained when applying capillary GC to environmental samples, it is almost indispensable to use electronic integrators for the acquisition and the handling of the chromatographic data. Contemporary integrators store retention times, peak areas, and peak heights, and provide simple programs for calculating relative retention times and amounts or concentrations of individual components.

Unfortunately, commercially available electronic integrators have some significant drawbacks. Because only relatively simple peak-detection algorithms are applied, the areas of only partly resolved peaks (i.e., shoulders) are often incorrectly determined by the integrator. In such cases, it is recommended that peak heights be used for quantification. This requires, however, that the ratio of the peak height to the peak area of the compound of interest is about equal to that of the internal standard.

Another problem is that the automatic detection of internal standards by the integrator using preset "time windows" frequently fails because of the relatively poor reproducibility of the retention times in capillary GC. This is especially true for the compounds eluting soon after the solvent. Thus, a fully automatic computation based on the raw data is often not possible.

Finally, it should be pointed out that (at the end of 1980) most integrators are only able to store a very limited number of peaks (e.g., 100), which is quite often insufficient for capillary GC.

4. Automation and Computer Processing of Gas Chromatographic Data

In the field of water analyses, fully automated gas chromatographic systems have found only very limited application. This is mainly because of the relatively complicated steps involved in an analysis, (such as sampling, injection, and enrichment) and the complexity of the chromatograms usually obtained [difficult automatic data handling (cf. Section II,H,3)]. Thus, reasonable solutions have been found only for very simple routine applications (Dowty et al., 1976a,b; Lawler et al., 1977).

The storage and manipulation of chromatographic data by a computer offers certain advantages if many chromatograms must be interpreted. This can be the case in routine survey studies or in investigations where a large number of samples are analyzed. Programs that allow automatic reconstruction and display of chromatograms or parts of a chromatogram (Overton et al., 1978), or programs that create gas chromatographic profiles for a rapid qualitative screening of the chromatographic data (Suffet and Glaser, 1978) may then be of great help. However, computerized processing and interpretation of gas chromatographic data that goes beyond the calculation of retention and concentration data has so far not been a field of great activity, except for applications involving GC/MS (see Chapter 5, this volume).

I. Derivatization

When dealing with compounds of restricted volatility or compounds of extreme polarity that may interact with the chromatographic system, it is often possible to convert these compounds to more volatile substances by derivatization, thus enabling or facilitating their analysis by GC.

Derivatization is most frequently performed with compounds containing functional groups with active hydrogen atoms (amino, hydroxyl, or carboxyl groups) because these compounds may interact with other similar molecules or with the support underneath the liquid phase. The most common derivatization methods are silylation, acylation, and esterification. Comprehensive reviews on this topic have been published (Blau and King, 1977; Nicholson 1978a,b).

When using derivatization, one should be aware that the sample undergoes a chemical reaction. Many by-products that may complicate the analysis may be formed. Furthermore, reliable quantification may become more difficult because of an incomplete and poorly reproducible reaction or because of side reactions of the compounds of interest. Therefore, great care must be taken when carrying out a derivatization reaction.

Derivatization can also be used to permit or improve the analysis of a particular compound by a selective detector. For an overview of such applications of derivatization, the reader is referred to the literature (e.g., Ettre, 1978, and references cited therein).

III. APPLICATIONS

A. General Remarks

In this section, some applications of GC to the analysis of organic chemicals in the aquatic environment are discussed. Rather than cover a broad spectrum of substances, a few important compound classes have been selected to be discussed in some detail. Emphasis is placed on the most notorious pollutants including many chemicals that are included in the "Priority Pollutant" list issued by the U.S. Environmental Protection Agency. In addition, some examples of the application of recently developed novel gas chromatographic techniques are presented. These examples should stimulate future GC applications. An overview of various applications of capillary GC to the analysis of organic compounds occurring in different types of waters (groundwater, wastewater, surface water, drinking water) has been published by Lin (1981). Proceedings of international meetings contain many papers in which capillary GC methods are applied to determine organic pollutants in water samples (Keith, 1976a, 1981; Bjørseth and Angeletti, 1982; Angeletti and Bjørseth, 1984).

B. Volatile Halogenated Hydrocarbons

Volatile halogenated hydrocarbons such as halomethanes, halogenated ethanes, ethenes, and benzenes (Table IV) are introduced continuously into the environment by man (Pearson and McConell, 1975). They have been found to be ubiquitous contaminants in raw and drinking waters

TABLE IV

Some Volatile Halogenated Hydrocarbons Commonly Found in Raw and Drinking Water

Name	Formula	Boiling point (°C)	Forma-tion[a]	Use
Methyl chloride	CH_3Cl	−24.2	I, B	Chemical intermediate
Methyl iodide	CH_3I	42.4	I, B	Chemical intermediate, pharmaceutical
Methylene chloride	CH_2Cl_2	40.1	I	Solvent, coolant, chemical intermediate
Chloroform	$CHCl_3$	61.7	C, I	Solvent, pharmaceutical
Other trihalomethanes	$CHBrCl_2$	90	C	
	$CHBr_2Cl$	120		
	$CHBr_3$	149.5		
	$CHICl_2$	132		
Freon-11	CCl_3F	23.8	I	Propellants, refrigerants
Freon-12	CCl_2F_2	−29.8	I	
Carbon tetrachloride	CCl_4	76.5	I	Solvent, chemical intermediate
1,1,1-Trichloroethane	H_3C-CCl_3	74.1	I	Solvent
Vinyl chloride		−13.9	I	Chemical intermediate
cis-1,2-Dichloroethylene		60.3	I	Chemical intermediate, coolant
Trichloroethylene		87.0	I	Solvent, chemical intermediate
Tetrachloroethylene		121.2	I	Solvent (dry cleaning, metal degreasing)
Chlorobenzene		132.0	I	Chemical intermediate, solvent
1,4-Dichlorobenzene		174.0	I	Constituent of household products, deodorizer
1,2,4-Trichlorobenzene		213.5	I	Insulator

[a] (B) Biosynthesis; (I) industrial manufacturing; (C) water chlorination.

(Murray and Riley, 1973; Dowty *et al.*, 1975; Giger *et al.*, 1978). Since many of these chemicals are toxic and possibly carcinogenic (Dowty *et al.*, 1975; Simmon and Tardiff, 1978), simple and rapid methods for their routine individual determination in water at very low levels are necessary.

Although in a few cases volatile halogenated hydrocarbons have been analyzed by other methods [e.g., HPLC (Kummert *et al.*, 1978)], GC is the most common and obvious tool for their qualitative and quantitative determination. Many procedures involving GC have been reported (Table V). Great emphasis has been placed on the analysis of halomethanes since these compounds are also known to be formed during water treatment (e.g., Rook, 1977).

Common sampling procedures used for the determination of volatile halogenated hydrocarbons are direct aqueous injection, head-space analysis, extraction of the compounds with an organic solvent, and purge and trap techniques (Table V). Some of these sampling methods are compared by Varma *et al.* (1979) and Dressman *et al.* (1979). For details, the reader is referred to Chapter 3 of this volume.

Probably because of the relatively low retention of the haloforms on capillary columns and because of the availability of highly selective detector systems (ECD, MCD, and MED), many laboratories (especially in the United States) are still using packed columns with nonpolar to moderately polar stationary phases for the separation of volatile halogenated hydrocarbons. Such columns may yield satisfactory results in cases where only a few major compounds are of interest (e.g., chloroform after chlorination of water). Thus, for a rough screening for the presence of certain halomethanes in water, even the simplest method, the direct aqueous injection technique (DAI) (Nicholson *et al.*, 1977) using a packed column and electron-capture detection may be useful. It should be noted that DAI, in contrast with other sampling methods, usually indicates too high concentrations of certain halomethanes in raw and drinking water. A possible explanation for these findings is that halomethanes are formed in the injection port from precursors present in the water (Nicholson *et al.*, 1977; Pfaender *et al.*, 1978).

If a more detailed and precise analysis is required, other sampling techniques and gas chromatographic hardware must be employed. The advantages of capillary columns, even for the separation of the most volatile halomethanes, have been demonstrated by several laboratories (Piet *et al.*, 1978; Eklund *et al.*, 1978; Stock and Alberti, 1979). Figure 10 shows the separation of a standard mixture of some volatile halogenated hydrocarbons and the analysis of a pentane extract of an industrial wastewater on the same capillary column using electron-capture detection. It should be noted that the analysis was completed in <6 min. Better separa-

TABLE V

Summary of Methods Used for the Analysis of Volatile Halogenated Hydrocarbons

Reference	Sampling method[a]	Column[b]	Detector[c]	Detection limits (µg/L) for selected chlorinated hydrocarbons				Applications[d]
				CH_2Cl_2	$CHCl_3$	$CCl_2{=}CCl_2$	Cl—⬡—Cl	
Bellar and Lichtenberg (1974)	PT	p, Chromosorb 101	FID, MCD	—	0.5	—	—	DW, WW
Dietz and Traud (1973)	EX	p, various	ECD	—	0.1	0.01	—	SW, WW
Eklund et al. (1978)	EX, DAI	c, SE-52 (50 × 0.3) (see Fig. 10)	ECD	0.05	0.001	0.0002	—	DW, WW
K. Grob (1973)	PT	c, Pluronic 121 (50 × 0.27) (see Fig. 11)	FID, ECD	—	0.05	0.005	0.005	DW, GW, SW, WW
K. Grob and Habich (1983)	DAI	c, SE-54	ECD	2	0.2	0.1	—	DW, GW, SW
Henderson et al. (1976)	EX	p, 10% Squalane on Chromosorb W/AW	ECD	—	0.1	—	—	DW, GW

(continued)

TABLE V (Continued)

Reference	Sampling method[a]	Column[b]	Detector[c]	Detection limits (μg/L) for selected chlorinated hydrocarbons				Applications[d]
				CH_2Cl_2	$CHCl_3$	$CCl_2{=}CCl_2$	Cl—⟨C₆H₄⟩—Cl	
Kaiser and Oliver (1976)	HSA	p, 10% OV-1 on Chromosorb Q	ECD	—	0.1	—	—	SW
Mieure (1977)	EX	p, 3% OV-1 on Chromosorb W/HP	ECD	—	1	—	—	
Nicholson et al. (1977)	DAI, PT	p, Chromosorb 101	ECD	500 (DAI) 0.1 (PT)	1 (DAI) 0.1 (PT)	0.5 (DAI) 2 (PT)	500 (DAI)	DW
Piet et al. (1978)	HSA	c, OV-225	ECD	—	0.1	0.05	—	DW, GW, SW
Quimby et al. (1979)	PT	p, Tenax	MED	0.1	0.1	—	—	
Richard and Junk (1977)	EX	p, 4% SE-30/6% OV-210	ECD, FID	—	0.1	—	—	DW
Stock and Alberti (1979)	EX	c, Ucon LB-550 (100 × 0.27)	ECD	0.1	0.01	0.005	0.1	

[a] PT, Purge and trap; DAI, direct aqueous injection; EX, extraction; HSA, head-space analysis.

[b] p, Packed column; c, capillary column (m × mm).

[c] See Section II,F.

[d] DW, Drinking water; SW, surface water; GW, groundwater; WW, wastewater.

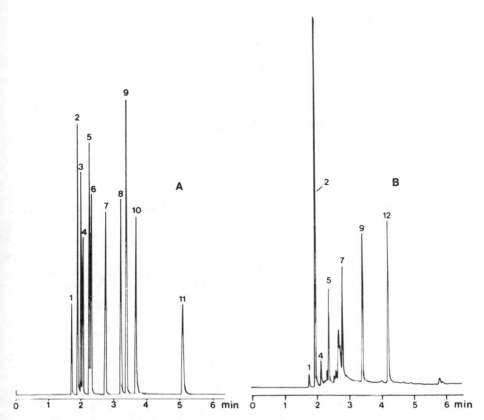

Fig. 10. Gas chromatograms of volatile halogenated hydrocarbons. (A) Mixture of reference compounds; (B) extract of an industrial wastewater. (1) Dichloromethane; (2) chloroform; (3) 1,1,1-trichloroethane; (4) carbon tetrachloride; (5) 1,1,2-trichloroethane; (6) bromodichloromethane; (7) bromotrichloromethane; (8) chlorodibromomethane; (9) tetrachloroethylene; (10) dichloroiodomethane; (11) bromoform; (12) unknown.

Column:	SE-52, 33 m × 0.3 mm id
Carrier gas:	Helium, 36 cm/s
Injection:	Split ratio 1 : 20, temp 200°C
Detector:	ECD, 250°C, scavenger gas flow 30 L/min
Oven:	Column temp 50°C, interface temp 250°C

(From Eklund *et al.*, 1978.)

tion could have been obtained by using a longer column (Stock and Alberti, 1979), or a column with a thicker film, as K. Grob and Grob (1979) demonstrated for the analysis of gasoline.

At present, the detector most widely used for the determination of volatile halogenated hydrocarbons is the ECD, despite its disadvantages (see Section II,F,3). Compounds with higher carbon to halogen ratios (e.g., halogenated ethanes, ethenes, benzenes) can be determined by a flame-ionization detector in combination with an efficient preconcentration method and a high-resolution capillary column. This is demonstrated in Fig. 11. The compounds were preconcentrated by the Grob stripping technique (K. Grob and Zürcher, 1976), the separation was performed on a Pluronic 121 glass capillary column (K. Grob Jr. and Grob, 1977b), and

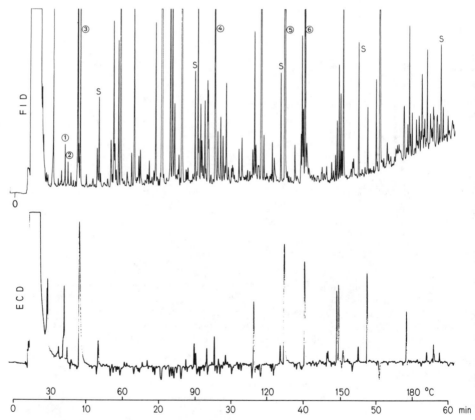

Fig. 11. Gas chromatogram of volatile organic compounds in a sample from Lake Zürich, Switzerland. (1) Trichloroethylene; (2) 1,2-dichloropropane; (3) tetrachloroethylene; (4) 1,4-dichlorobenzene; (5) 1,2,4-trichlorobenzene; (6) 1,2,3-trichlorobenzene; (S) 1-chloro-n-alkanes ($C_{6,8,10,12,14}$, 50 ng/L each), internal standards.

Column:	Pluronic 121, 50 m × 0.27 mm id
Carrier gas:	Hydrogen, 0.8 atm
Injection:	Splitless, temp 220°C
Detector:	FID and ECD after stream splitting, ECD with nitrogen as scavenger gas
Oven:	Ambient 5 min, then 4°/min to 180°C

the effluent of the column was split to allow a dual flame-ionization–electron-capture detection.

When using selective detection, volatile halogenated hydrocarbons can be identified from retention data by comparison with a chromatogram obtained from a standard mixture (Fig. 10A), preferably on two different columns with stationary phases of different polarity. If a flame-ionization detector is employed for quantification, it is very useful to have a selective detector in parallel (or in series, e.g., ECD–FID) in order to aid identification (see Fig. 11).

For quantification of the compounds, it is essential to calibrate the whole analytical procedure for each individual compound of interest, and it is necessary to use internal standards. It should be kept in mind that the response of selective detectors (especially the ECD) may vary by orders of magnitude for different compounds and that the linear dynamic range may be very small.

It is usually very easy to obtain good results for an analysis of relatively clean water (drinking water), but problems caused by the sample matrix may arise when dealing with heavily polluted waters (wastewater). Finally, it should be stressed that great care must be taken during handling and storage of the water samples. Because of the high air/water partition coefficients of the volatile halogenated hydrocarbons, significant losses may occur by evaporation.

In summary, the most efficient and reliable method for analyzing halomethanes in raw and drinking waters is probably liquid extraction with subsequent GC analysis using a capillary column and electron-capture detection. A powerful DAI method using a thick-film glass capillary column, cold on-column injection, and electron-capture detection has been published (K. Grob and Habich, 1983). This method is particularly well suited for the analysis of C_1- and C_2-polyhalogenated hydrocarbons. For the detection of other volatile halogenated hydrocarbons, the Grob-stripping method (K. Grob and Zürcher, 1976) employing the GC hardware (Fig. 11) should provide the best results. For real samples, an overall precision of better than ~10% relative standard deviation can be obtained for concentrations of greater than ~5–10 times the detection limit (e.g., Schwarzenbach et al., 1979).

C. Petroleum Hydrocarbons

Petroleum-derived products (mainly fossil fuels and lubricating oils) are of major concern as water pollutants. Acute contamination cases (oil spills) involve locally concentrated large amounts of petroleum compounds producing separated oil layers and/or saturated water solutions.

Chronic inputs, in contrast, cause widespread contamination at low or trace levels. The water analyst must be prepared to deal with both situations.

One of the first and often the most important goal of many investigations of oil pollution is the evaluation of the source of contamination. To successfully answer this question, the investigator must know something about the composition of various potential source materials. In addition, much attention must be given to the different processes that affect petroleum compounds once they are introduced into the environment (i.e., evaporation, dissolution, adsorption, desorption, and biodegradation).

Crude oil and its fractions are composed mainly of hydrocarbons with only minor amounts of oxygen, nitrogen, and sulfur compounds. Despite their great complexity, many petroleum-hydrocarbon mixtures contain a few prevailing components that provide easily recognizable gas chromatographic patterns ("fingerprints"). A homologous series of n-alkanes and some characteristic branched hydrocarbons dominate in many crude oils, No. 2 fuel oils, and kerosenes. Monoaromatic hydrocarbons are characteristic constituents of gasoline. In contrast, lubricating oils usually do not have any characteristic major components.

Various methods have been developed to determine petroleum hydrocarbons in the aquatic environment. Infrared spectrometry is often applied to measure the total concentrations of hydrocarbons (American Public Health Association, 1976) and, to a limited degree, to estimate the relative abundance of different types of hydrocarbons (e.g., Zürcher and Thüer, 1978). Ultraviolet absorption and fluorescence spectrometry, thin layer chromatography, and gravimetric methods are also valuable techniques for determining total hydrocarbon amounts or gross compositions (e.g., total aliphatic hydrocarbons). For a detailed characterization of petroleum hydrocarbons, however, gas chromatography is clearly the method of choice.

The upper traces of Figs. 12 and 13 show capillary gas chromatograms of gasoline and No. 2 fuel oil, respectively. These two petroleum fractions are of great importance as potential pollutants of natural waters.

It should be noted that in the No. 2 fuel-oil sample, the high separation power of capillary GC is needed to reliably separate two of the most prominent branched hydrocarbons, pristane (2,4,6,10-tetramethylpentadecane) and phytane (2,4,6,10-tetramethylhexadecane) from their adjacent n-alkanes, n-heptadecane and n-octadecane, respectively. The significance of these isoprenoid hydrocarbons is discussed later in this section. Similarly, the separation of m- and p-xylene in gasoline can only be easily achieved on capillary columns. These examples clearly show that capillary GC must be used if individual constituents of petroleum-

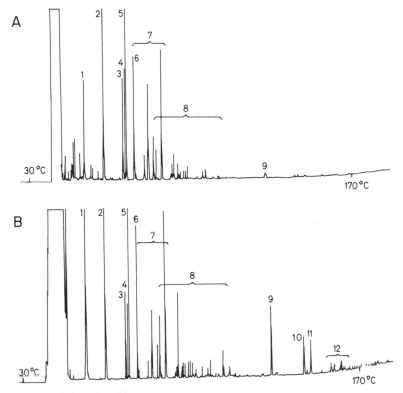

Fig. 12. Gas chromatograms of gasoline (A) and of the water-soluble fraction of No. 2 fuel oil (B). (1) Benzene; (2) toluene; (3) ethylbenzene; (4) *p*-xylene; (5) *m*-xylene; (6) *o*-xylene; (7) C$_3$-benzenes; (8) C$_4$-benzenes; (9) naphthalene; (10) 2-methylnaphthalene; (11) 1-methylnaphthalene; (12) C$_2$-naphthalenes.

Column:	Ucon HB, 56 m × 0.32 mm id
Carrier gas:	Hydrogen, 0.6 atm
Injection:	1.5 μL, splitless
Detector:	FID
Oven:	Ambient during injection and elution of the solvent, then 3°/min from 30 to 180°C

(From Zürcher and Thüer, 1978.)

derived hydrocarbons are to be determined. The lower separation efficiency of packed columns is sufficient, however, if only the boiling-point range of a particular petroleum fraction is of interest.

Petroleum hydrocarbons that have entered the aquatic environment are altered by microbiological degradation. It is well established that straight-chain hydrocarbons are more rapidly degraded than branched

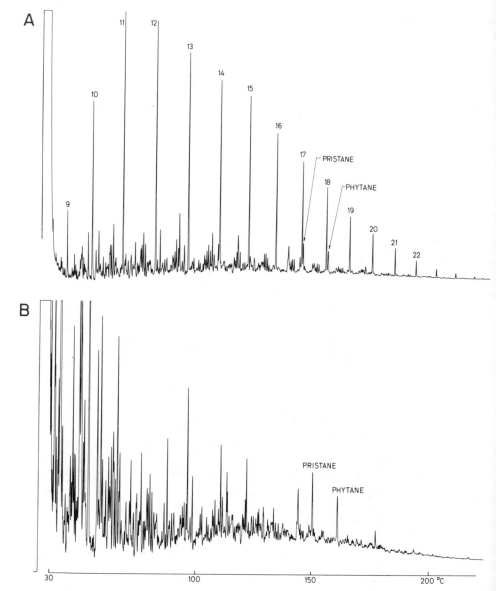

Fig. 13. Gas chromatograms of No. 2 fuel oil (A) and of a biodegraded oil layer (B). Numbers 9–22 refer to the chain lengths of the corresponding *n*-alkanes.

Column: OV-73, 17 m × 0.31 mm id
Carrier gas: Hydrogen, 0.5 atm
Injection: Splitless, temp 250°C
Detector: FID
Oven: Ambient during injection and elution of the solvent, then 2°/min from 30 to 260°C

and cyclic hydrocarbons. These differences in persistence can lead to drastic changes in the fingerprint of a hydrocarbon mixture. The lower trace in Fig. 13 is a capillary gas chromatogram of an oily layer that was sampled from a heavily contaminated groundwater. The n-alkanes, although still present, show much lower relative abundances than in a typical petroleum mixture (see upper trace in Fig. 13). The more resistant isoprenoid hydrocarbons are left as predominant compounds. The ratio of n-alkanes to isoprenoids (e.g., n-heptadecane to pristane) can be used to assess the degree of biodegradation of a petroleum mixture (Blumer and Sass, 1972).

Differential dissolution and adsorption also can cause strong changes in the composition of petroleum-hydrocarbon mixtures. Capillary GC is an excellent technique for detection and reliable interpretation of these alterations. Zürcher and Thüer (1978) gave some good examples of such alterations. Of particular significance is the fact that the water-soluble fraction of No. 2 fuel oil consists of alkylated benzenes and naphthalenes (Fig. 12) and is at first glance very similar to gasoline or its water-soluble part. This is caused by the fact that aromatic hydrocarbons are much more soluble in water than are n-alkanes. Therefore, benzenes and naphthalenes are enriched in the aqueous phase. The higher relative abundance of naphthalenes in the water-soluble part of No. 2 fuel oil still allows the differentiation between gasoline and No. 2 fuel oil as source materials (for details, see Zürcher and Thüer, 1978).

One feature frequently encountered when analyzing environmental hydrocarbon mixtures by GC is the so-called hump or unresolved complex mixture (UCM). This broad, envelope-type GC signal appears when the separation efficiency of the gas chromatographic method cannot cope with the compositional complexity of the investigated hydrocarbon mixture. Many minor and trace components give rise to overlapping peaks that produce this very broad detector signal, sometimes over a quite large boiling range. The literature contains many "humpy" gas chromatograms, particularly those occurring when packed columns were used and mixtures of biodegraded petroleum hydrocarbons were analyzed. Figure 14 demonstrates that petroleum-hydrocarbon mixtures can be so complex that even high-resolution capillary GC produces chromatograms with UCM signals. By using columns with even higher separation efficiencies, the hump could be flattened but could not be resolved completely into individual peaks. Blumer (1975) gave an excellent description of the basic facts leading to the UCM problem. A number of workers have tried to quantify the UCM signal by means of integration. Such results should be assessed very critically because the UCM signal obviously depends very much on the particular chromatographic performance.

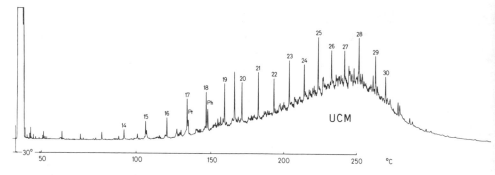

Fig. 14. Gas chromatograms of hydrocarbons associated with highway-runoff particulates. The numbers 14–30 refer to the chain lengths of the corresponding *n*-alkanes. Pr, pristane; Ph, phytane; UCM, unresolved complex mixture.

Column:	SE-52, 20 m × 0.32 mm id
Carrier gas:	Hydrogen, 0.4 atm
Detector:	FID
Oven:	30°C during injection and elution of the solvent, then 3°/min from 50 to 250°C

(From Zürcher *et al.*, 1980.)

Another problem is the differentiation between petroleum-derived and indigenous biogenic hydrocarbons. This question arises, for instance, when studying low levels of hydrocarbons in sediments and soils. Gas chromatography plays an important role in answering such questions because compositional differences between the two different hydrocarbon groups are difficult to detect by other methods. It is even feasible to separate diastereomeric hydrocarbons. Figure 15 shows such an example: This chromatogram of No. 2 fuel oil was obtained with a 62-m OV-1 capillary column. The signals of the isoprenoid hydrocarbons pristane and phytane are split into two peaks of equal height. In the case of pristane, it has been shown that the split is caused by an epimerization of the original $6(R),10(S)$-pristane to a mixture of $6(R),10(S)$- and $6(R),10(R)$- and/or $6(S),10(S)$-pristane (Patience *et al.*, 1978; Giger *et al.*, 1984a). A separation of the enantiomers is, in principle, not feasible if an optically inactive stationary phase is used. The same analysis of recently biosynthesized pristane shows only the $6(R),10(S)$ form (Patience *et al.*, 1978). Hence this high-resolution gas chromatographic analysis can distinguish between biogenically formed and geochemically altered pristane.

D. Polycyclic Aromatic Hydrocarbons

Polycyclic aromatic hydrocarbons (PAH) have long been recognized as hazardous environmental chemicals. Their occurrence in the aquatic en-

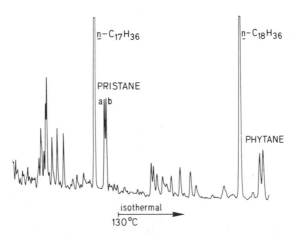

Fig. 15. Partial gas chromatogram of No. 2 fuel oil. (a) $6(R),10(R)$- and $6(S),10(S)$-pristane; (b) $6(R),10(S)$-pristane (see also Giger *et al.*, 1984a).

Column:	OV-1, 62 m × 0.27 mm id
Carrier gas:	Hydrogen, 1.0 atm
Injection:	Splitless, temp 270°C
Detector:	FID
Oven:	15 min isothermal at 125°C, then 0.3°/min to 130°C

vironment has been studied extensively (Neff, 1979). Although thin-layer chromatographic methods have been used in the majority of the studies, an increasing number of investigators are using GC to determine PAH in environmental samples (e.g., Giger and Schaffner, 1978; Grimmer *et al.*, 1978; Bjørseth and Eklund, 1979; Wakeham *et al.*, 1980).

Because of their low water solubility and therefore high tendency to adsorb on surfaces, PAH are predominantly associated with particulates (suspended matter and sediments). Figure 16 shows as an example the analysis of a recent lake sediment for PAH by capillary GC (for details, see Giger and Schaffner, 1978; Wakeham *et al.*, 1980). The gas chromatogram reveals the great compositional complexity of these compounds in environmental samples. The major peaks are unsubstituted PAH ranging from two-ring (naphthalene) to seven-ring (coronene) compounds. Two or more isomers with the same number of condensed aromatic rings occur in this PAH mixture. Phenanthrene and anthracene can be readily separated by the glass capillary column used. However, some other isomeric unsubstituted PAH, such as triphenylene and chrysene (peak No. 10) or the several benzofluoranthenes (peak No. 11), are not resolved. No significantly better separation of these isomeric PAH is achieved by using longer nonpolar columns (e.g., 50 m). These isomeric PAH can be better resolved on more polar columns with higher selectivities for these com-

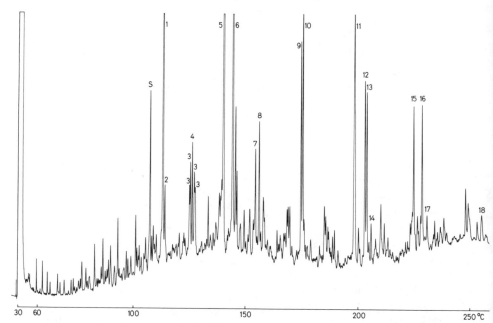

Fig. 16. Gas chromatogram of polycyclic aromatic hydrocarbons isolated from the surface sediments of a lake. (1) phenanthrene; (2) anthracene; (3) methylphenanthrenes; (4) 4,5-methylene-phenathrene; (5) fluoranthene; (6) pyrene; (7) benzo[*a*]fluorene; (8) benzo[*b*]fluorene; (9) benz[*a*]anthracene; (10) chrysene/triphenylene; (11) benzofluoranthenes; (12) benzo[*e*]pyrene; (13) benzo[*a*]pyrene; (14) perylene; (15) indeno[1,2,3-cd]pyrene; (16) benzo[*ghi*]perylene; (17) anthanthrene; (18) coronene; (S) 1-chlorotetradecane, internal standard.

Column:	SE-52, 20 m × 0.3 mm id, coated on barium carbonate interlayer
Carrier gas:	Hydrogen, 0.8 atm
Injection:	Splitless, temp 270°C
Detector:	FID
Oven:	Ambient during injection and elution of the solvent (methylene chloride), then 2.5°/min from 60 to 250°C

(From Giger and Schaffner, 1978.)

pounds (e.g., polyphenyl ether, Carbowax, or Silar phases). However, such columns have reduced practical temperature ranges and lower intrinsic separation efficiencies (see Section II,E,3). Thus, the choice of a 20-m long SE-52 column is a compromise between maximum resolution and practicality (analysis time and column temperature). Glass capillary columns coated after persilylation of the glass surface (K. Grob *et al.*, 1979) can also be used for the determination of PAH. They offer additional advantages because of their high temperature limits that allow shorter analysis times.

The gas chromatogram in Fig. 16 also shows a large number of poorly resolved minor peaks that are caused by substituted PAH. It has been shown by GC/MS techniques that these PAH mixtures contain great numbers of homologous and isomeric alkylated PAH (Giger and Schaffner, 1978). Methylphenanthrenes (peaks No. 3) and 4,5-methylenephenanthrene (peak No. 4) can be readily detected even with nonspecific flame ionization detection.

If the objective is to elucidate the structures of minor constituents present in the complex PAH assemblage, an additional prefractionation step should be used prior to the GC analysis. This can be accomplished by adsorption chromatography, preferably by high-performance liquid chromatography. Thus, PAH can be fractionated according to their aromatic ring systems. These PAH fractions can then be analyzed again by capillary GC using a nonpolar liquid phase (Wakeham et al., 1979; Giger et al., 1984a).

When considering the complexity of the PAH mixtures in environmental samples (e.g., Fig. 16), it seems obvious that for quantitative determinations of individual PAH, the most efficient separation technique available, namely capillary GC, is needed; packed-column GC or HPLC do not have sufficient separation power. GC/MS techniques have been used to compensate for the insufficient separation power of packed columns. By detecting the PAH by single-ion monitoring of their molecular ions, the interference of various compounds can be eliminated. However, this method does not work in the case of isomeric PAH having identical molecular ions (e.g., phenanthrene and anthracene, or benzofluoranthenes, benzo[a]pyrene, benzo[e]pyrene, and perylene).

The procedure of Giger and Schaffner (1978) allows the quantitative determination of individual PAH with relative standard deviations of better than 20% in the concentration range of 0.1–1 μg of individual component per gram of dry sediment. Their results were obtained with capillary columns that were coated with SE-52 according to the barium carbonate procedure of K. Grob et al. (1977) and by using the Grob-type splitless injector (see Section II,D,2,c). However, significant improvements, particularly for the heavier PAH, could be achieved by applying the cold on-column injection technique (see Section II,D,2,d).

For the qualitative determination of PAH, Lee et al. (1979) developed a new retention index system that is based on the series naphthalene, phenanthrene, chrysene, and picene (see Section II,H,1). This index system offers significant advantages over the conventional n-alkane system and should facilitate the qualitative determination of PAH. Hence, the exchange of PAH retention data should become more feasible.

Wakeham (1979) applied glass capillary GC to determine nitrogen-containing heterocyclic aromatic hydrocarbons (azaarenes) in environmental samples. The use of a nitrogen-selective detector (NPD) enabled the detection of trace amounts of azaarenes obtained after an extensive cleanup procedure. Figure 17 shows an example of an analysis of a lake sediment for azaarenes. The complexity of the azaarene mixture is even greater than that of the PAH because the aza nitrogens may occupy different ring positions. Wakeham (1979) presents evidence that only a small fraction of the total azaarenes could be analyzed by this gas chromatographic procedure because of insufficient sample volatility. Developments of more thermostable persilylated columns and the use of cold on-column injection should solve at least part of this problem.

E. Phenols

A review on the occurrence of phenols and phenolic compounds in aquatic ecosystems (Buikema *et al.*, 1979) shows that various phenolic

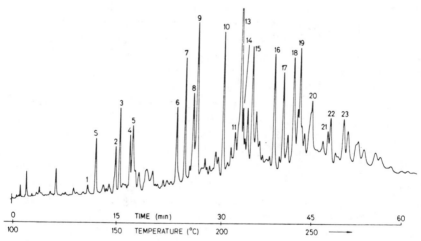

Fig. 17. Gas chromatogram of azaarenes isolated from surface sediments of a lake. (1) 4-azafluorene; (2) 7,8-benzoquinoline; (3) acridine; (4) phenanthridine; (5) 5,6-benzoquinoline; (6) 2-azafluoranthene; (7) 7-azafluoranthene; (8) 1-azapyrene; (9) $C_{15}H_9N$; (10) 3,4-benzacridine; (11) $C_{17}H_{11}N$; (12) $C_{17}H_{11}N$; (13) 1,2-benzacridine; (14) $C_{17}H_{11}N$; (15) $C_{17}H_{11}N$; (16,17,18,19,20) $C_{19}H_{11}N$; (21) $C_{21}H_{13}N$; (22) 1,2,3,4/1,2,5,6-dibenzacridine; (23) 1,2,7,8-dibenzacridine; (S) octadecanoic nitrile, internal standard.

Column: SP 2340, 20 m × 0.3 mm id
Carrier gas: Helium, 2.2 atm
Injection: Splitless
Oven: Ambient during injection and elution of the solvent (methylene chloride), then 4°/min from 100 to 250°C

(From Wakeham, 1979.)

chemicals have been found in many types of waters. Chlorinated phenols, for example, can be formed during water chlorination and pulp-bleaching operations. They are of much concern because of their low taste and odor thresholds and their high toxicity.

Many phenolic compounds are amenable to gas chromatographic analysis. In most studies, packed-column GC has been applied to extracts containing either the free phenols or phenol derivatives like ethers and esters. A review on the determination of phenolics in the aquatic environment has been published by Renberg (1982). The following selection of references should help the reader find more detailed information.

Dietz and Traud (1978), Keith (1976b), Lindström and Nordin (1976), Mousa and Whitlock (1979), and Shackelford and Webb (1979) analyzed different wastewaters for phenolic compounds. Lamparski and Nestrick (1978) proposed the use of heptafluorobutyl derivatives in order to improve the sensitivity of the electron capture detection of phenolic compounds. Baird et al. (1977) and Bartle et al. (1977) demonstrated the gas chromatographic analysis by direct aqueous injection of waters containing phenols in the parts per million range. Several of these publications address the problems of low and poorly reproducible extraction yields for certain phenolic compounds that are caused by their high water solubility. Difficulties related to derivatization processes are also discussed.

The complexity of mixtures of phenolic compounds occurring in certain wastewaters, such as pulp-mill effluents, is such that only capillary GC can cope with these samples (Keith, 1976a). There is no question that capillary GC work is needed if industrial effluents must be analyzed for individual phenolic compounds. Krijgsman and van de Kemp (1977) and Wegmann and Hofstee (1979) demonstrated the applicability of capillary GC to the determination of chlorophenols in surface waters of the Netherlands. The phenols were determined as phenolacetates on a SE-30, 35 m × 0.35 mm id glass capillary column with isothermal operation and electron capture detection. Renberg and Lindström (1981) used reversed-phase adsorption with octadecyl-modified silica to quantitatively recover polychlorinated phenols from water. After desorption with acetone, the chlorophenolic compounds were determined as their acetyl derivatives, using quartz capillary-column gas chromatography.

If the capillary column is properly selected (slightly acidic and/or very inert, persilylated), it is even feasible to determine phenols as free phenols. Figure 18 shows the gas chromatogram of phenols extracted from a biologically treated pulp-mill wastewater (Giger and Schaffner, 1981; Leuenberger et al., 1983). This example demonstrates that derivatization is not necessary for the determination of phenols by glass-capillary GC.

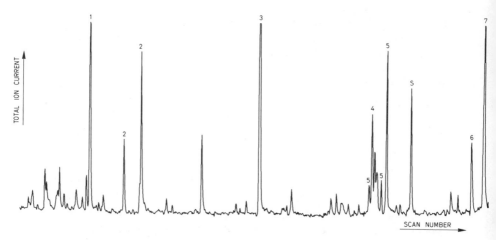

Fig. 18.   Total ion chromatogram of phenols extracted from a biologically treated waste-water of a sulphite pulp mill. (1) 2-methoxyphenol (guaiacol); (2) dichlorophenols; (3) 2,4,6-trichlorophenol; (4) 2,3,4,6-tetrachlorophenol; (5) trichloroguaiacols; (6) pentachlorophenol; (7) tetrachloroguaiacol; (S) 2,4,6-tribromophenol, internal standard.

Column:	OV-73, 26 m × 0.3 mm id
Carrier gas:	Helium, 0.7 atm
Injection:	Splitless, temp 275°C
Detector:	Total ion current of a Finnigan mass spectrometer coupled directly by means of a fused silica capillary
Oven:	Ambient during injection and elution of the solvent (methylene chloride), then 2.5°/min to 180°C

(From Giger and Schaffner, 1981.)

Hydroxy- and nitrophenols can be similarly gas chromatographed as free phenols.

F. Pesticides

Gas chromatography has played an important role in the discovery of the persistence of pesticides in the environment. GC is still the most common final separation and detection technique used for the determination of many pesticides. Although the vast majority of pesticide residue work deals with biological media (agricultural materials, food, organisms), a large amount of data on pesticide concentrations in water samples, suspended matter, and sediments has been published. Widespread pollution of the freshwater and marine environment was detected (e.g., reviews in *J. Water Pollut. Control Fed.*). In addition, some cases of acute pollution of waterways have been reported (e.g., Greve and Wit, 1971).

As a consequence of the environmental significance of the pesticides, standard methods for their determination have been issued by several government agencies. The details on these methods can be found in Goerlitz and Brown (1972), American Public Health Association (1976), U.S. Dept. Health, Education and Welfare (1976), U.S. Environmental Protection Agency (1978, 1979).

The standard methods are based exclusively on packed-column GC. The separations are performed on glass columns of 90- to 200-cm length packed with single and mixed organosilicone and polyester stationary phases ranging from low to high polarity, for example, SE-30, QF-1, DC-200, OV-210, OV-17, DEGS, Carbowax 20 M, GE-XE60, Versamid 900, and DC-200/QF-1, OV-17/QF-1, and SE-30/OV-210 mixtures. It is recommended that samples should be examined on two or even three columns of different polarity before results are considered conclusive. Glass columns are preferred because they minimize sample decomposition and because they are easier to pack. After packing, columns are conditioned at an elevated temperature to minimize liquid-phase bleeding and to assure reproducible results. Sometimes the injection of larger quantities of the pesticide being determined helps to improve the reliability of the analyses.

Columns coated with relatively low percentages of liquid phase generally give superior resolution and sensitivity but become contaminated more easily and are more prone to interactions between the sample and the solid support. Furthermore, for certain pesticides, such as DDT and endrin, problems caused by sample decomposition may occur.

Various selective detectors are used for the detection of pesticides. Organochlorine pesticides (chlorinated hydrocarbon insecticides, chlorinated phenoxy acids, and pentachlorophenol) are usually determined with a tritium or ^{63}Ni electron capture detector (ECD) or with a MCD. Although the ECD is less specific than the MCD, it has the advantage of being much more sensitive for many halogenated compounds. ^{63}Ni detectors are operable at higher temperatures ($>300°C$) than tritium detectors, thus problems caused by high-boiling materials condensing in the detector can be reduced. Tritium detectors are less expensive, and contaminated foils can be easily exchanged or removed for cleaning.

For the detection of organophosphorus pesticides, the alkali flame ionization detector (AFID or NPD) or the flame photometric detector (FPD) in the phosphorus mode are used. Sulfur-containing pesticides may be selectively detected by MCD or FPD detectors. Nitrogen-containing pesticides are usually detected selectively with an AFID or a Coulson electroconductivity detector (CECD) in the nitrogen mode. Labile, polar carbamate pesticides or their hydrolysis products are often derivatized with a halogen-containing reagent and the resulting derivative sensitively detected with the ECD.

Taylor and Thier (1979) published a status report on cleanup and determination procedures for pesticide-residue analysis. This review covers many new developments in gas chromatographic determinations of pesticides. The reader is also referred to three special issues of the Journal of Chromatographic Sciences (May, June, and July 1975), which gave an excellent overview on the state of the art of pesticide analysis at that time.

A number of difficulties have hampered the use of capillary GC for quantitative determinations of pesticides. Among them were problems associated with the quantitative injection (e.g., reproducibility of stream splitting, see Section II,D), irreversible adsorption on active surfaces in the capillary columns, and the lack of suitable specific detectors with sufficiently small dead volumes. The current literature, however, shows that most of these problems have been solved satisfactorily and that capillary GC is well suited to analyze all the different pesticide classes that are amenable to gas chromatography. Franken and Vader (1973) and Badings et al. (1975) demonstrated the use of glass-capillary GC for the determination of organochlorine pesticides. In both studies, an ECD was used, and the separations were performed under isothermal conditions. The stationary phases were SE-30, DEGS/phosphoric acid (3 : 1), and PMPE. Organophosphorus insecticides and their metabolites were determined by capillary GC by De Potter et al. (1978) and Hild et al. (1978). The complexity of the mixtures obtained, particularly when metabolites were analyzed, demonstrated the necessity of capillary GC for such applications.

Brechbühler et al. (1977) developed a micro ECD for capillary GC. This ECD has been successfully used for the determination of organochlorine pesticides, chlorinated benzenes, and chlorinated anilines. A mixture of 19 chlorinated pesticides was analyzed by packed-column and by glass-capillary GC. The advantages of high-resolution GC are clearly demonstrated.

The same micro ECD has been applied by Giger et al. (1976) for the determination of polychlorinated biphenyls (PCBs) and polycyclic aromatic hydrocarbons in rainwater. The detection of these two compound classes was performed simultaneously by FID and ECD using a stream-splitting device (see Section II,H,1). This dual-detection technique enabled the simultaneous determination of both compound classes. A similar system has been applied by Reinhard et al. (1979) for the determination of PCBs in various samples of an advanced waste treatment facility.

Brodtmann and Koffskey (1979) reported the use of a 60-m SE-30 glass capillary column for the determination of chlorinated hydrocarbon insecticides in river and drinking water. They compare capillary GC to conventional procedures using packed columns and conclude that the vastly superior resolution of capillary GC not only permits the analysis of sample extracts without prior column chromatography cleanup but also enables the user to resolve certain insecticides not resolvable on the most commonly used packed columns.

The methodology of Brodtmann and Koffskey, however, has some drawbacks. Their use of a split injection and of an isothermal oven temperature are not optimum. To prolong the lifetime of the capillary columns, we recommend injection of nonprecleaned samples only from rather pure water samples (like river or drinking water) but certainly not from any kind of wastewater.

Tuinstra and Traag (1979) applied capillary GC to determine PCBs and organochlorine pesticide residues in agricultural products. To protect the capillary column against less volatile substances, they used a precolumn of ~170-mm length and 3-mm id, filled with 2 cm of inert column-packing material (coated with 3% SE-30). The samples were injected by the splitless technique at 100°C, and the separation was performed under isothermal conditions at 220°C. Practical aspects concerning connection of the capillary, automatic, splitless injection, reproducibility of the retention time, and quality of the column were discussed with respect to separation, adsorption, and degradation of DDT. Figure 19 shows an example of a pesticide analysis of a soil extract by capillary GC as reported by Tuinstra and Traag (1979).

Polychlorinated biphenyls (PCBs) are very important environmental pollutants and belong to the same compound class as many chlorinated hydrocarbon insecticides. The PCBs interfere in the analysis for chlorinated hydrocarbon pesticides if no special cleanup is performed. PCBs are very complex technical-grade mixtures containing large numbers of isomers that cannot be separated into individual components by packed-column GC. Capillary GC is clearly much more suited for this purpose (see, e.g., Schulte and Acker, 1974; Tuinstra and Traag, 1979; Onuska and Comba, 1980).

Chlorinated dibenzo-p-dioxins have become environmental pollutants of particularly high significance because some of them are known to be extremely toxic and are suspected to be teratogenic and mutagenic. 2,3,7,8-Tetrachlorodibenzo-p-dioxin (2,3,7,8-TCDD) is one of the most

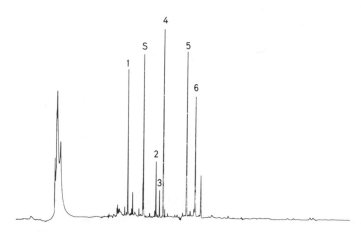

Fig. 19. Capillary gas chromatogram of chlorinated hydrocarbon pesticides extracted from a soil. (1) Hexachlorobenzene; (2) aldrin; (3) telodrin; (4) isodrin; (5) dieldrin; (6) endrin; (S) internal standard.

Column: cp-sil 7 Chrompack, 25 m × 0.25 ınm id
Carrier gas: Helium, 2 mL/min
Injection: 5 μL, splitless, temp 205°C
Detector: ^{63}Ni-ECD at 330°C
Oven: 100°C during injection, 4 min after injection up to 220°C at 40°C min, isothermal at 220°C

(From Tuinstra and Traag, 1979.)

toxic substances known. These chemicals are released into the environment either as contaminants in products like the herbicide 2,4,5-trichlorophenoxyacetic acid or by accidents during the manufacturing of chlorinated phenols.

Theoretically, there are 75 different polychlorinated debenzo-p-dioxins, including 22 TCDD isomers. Since different isomers of a given dioxin vary significantly in their toxicity, the determination of individual dioxins at ultratrace levels is necessary.

Buser (1977) developed an efficient method for the determination of 2,3,7,8-TCDD by capillary GC. Figure 20 shows gas chromatograms of a test mixture containing nine TCDD isomers. The chromatograms were obtained on capillaries coated with OV-17 and OV-101. Figure 20 shows that neither column is able to separate all isomers, but a combination of both allows a differentiation between the compounds. The compounds are selectively detected by single-ion monitoring using the molecular ion (m/e = 320). It should be noted that in this case the selective detection of individual components cannot be improved by single-ion monitoring be-

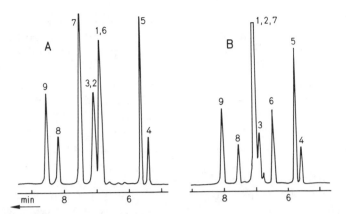

Fig. 20.   Partial mass fragmentograms (m/e 320) showing the separation of nine tetra-chlorodibenzo-*p*-dioxins (TCDD). (1) 2,3,7,8-TCDD; (2) 1,2,3,8-TCDD; (3) 1,2,3,4-TCDD; (4 and 5) 1,3,6,8- and 1,3,7,9-TCDD or vice versa; (6 and 7) 1,4,6,9- and 1,2,6,9-TCDD or vice versa; (8 and 9) 1,2,6,7- and 1,2,8,9-TCDD or vice versa.

Columns: (A) OV-17, 25 m × 0.37 mm id
              (B) OV-101, 27 m × 0.35 mm id
Carrier gas: Helium, 0.6 atm
Injection: 2 µL in *n*-tetradecane, splitless, temp 275°C
Detector: Mass spectrometer, single-ion monitoring, m/e 320
Oven: 207°C, isothermal

(From Buser, 1977.)

cause these isomers have very similar, if not identical, mass spectra. Therefore, the high separation power of capillary GC is indispensable.

Buser (1977) has taken advantage of the solvent effect (see Section II,D) by using a high-boiling solvent (*n*-tetradecane, bp 253°C), splitless injection, and isothermal operation at column temperatures of 207 to 220°C. Under the conditions chosen, the TCDDs were eluted within a 4–10 min range. Separation efficiencies and peak shapes were at least as good as when using a lower boiling solvent with injection at low temperature followed by temperature programming of the columns. No adverse effects have been observed that could have been caused by the high-boiling solvent.

G.   Miscellaneous

This final section contains a few examples of very recent applications of capillary GC. These examples have been selected to illustrate some future trends in this field. Great efforts are now being made to extend the scope of compounds amenable to GC by developing very inert high-temperature capillary columns. Thus, compounds with high molecular weights, such as wax esters (Fig. 21) or triacylglycerides (Fig. 22), can now be determined by capillary GC. Furthermore, with such columns compounds containing polar functional groups can be chromatographed without derivatization. Figure 23 shows two gas chromatograms of a mixture of free steroids on a SE-52 column using two different types of injectors. The two chromatograms also show the advantages of the on-column injection.

Low molecular weight acids (C_1–C_7) in water can now be determined by directly injecting the aqueous solution into a fused-silica capillary column (Fig. 24) (Dandenau *et al.*, 1979). The chromatogram shown in Fig. 24 was obtained on a column coated with Carbowax 20M without prior deactivation of the fused-silica surface. A detection limit of ~1 mg/L can be achieved.

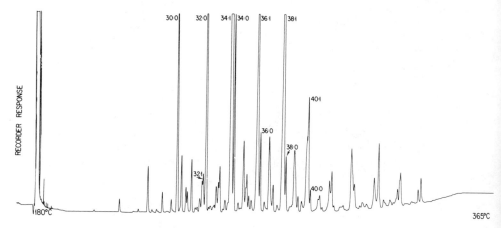

Fig. 21. Gas chromatogram of wax esters extracted from mixed zooplankton. Numbers refer to total-carbon numbers and number of double bonds, respectively.

Column: SE-52, persilylated, 25 m × 0.3 mm id
Carrier gas: Hydrogen, 1.5 atm
Injection: On-column, 100°C
Detector: FID
Oven: 2°/min., 180–365°C

(From Wakeham and Frew, 1982.)

Surfactants, which are important water pollutants, have so far not been amenable to direct GC analysis. Anionic surfactants (alkyl-benzenesulfonates) and their metabolites have been desulfonated and then determined by capillary GC (Leidner *et al.*, 1976).

Nonionic surfactants have been determined after ether cleavage by treatment with hydrogen bromide (Tobin *et al.*, 1976). Favretto *et al.* (1978) determined polyoxyethylene nonionic surfactants by high-temperature packed-column GC (up to 350°C oven temperature). The highly inert, high-temperature capillary columns can now be used to determine nonylphenoxyethoxylates with up to 12 ethoxylate groups (Geiser, 1980). Figure 25 shows a capillary gas chromatogram of nonylphenols and nonylphenoxyethoxylates that have been extracted from a secondary sewage effluent. These compounds are bioresistant metabolites that can pass

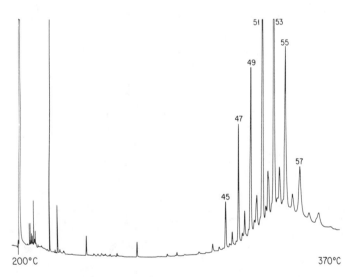

Fig. 22. Gas chromatogram of triacylglycerides extracted from mixed zooplankton. Numbers refer to total carbon numbers.

Column: SE-52, persilylated, 25 m × 0.3 mm id
Carrier gas: Hydrogen, 1.6 atm
Injection: On-column, 150°C
Detector: FID
Oven: 4°/min, 200–370°C

(From Wakeham and Frew, 1982.)

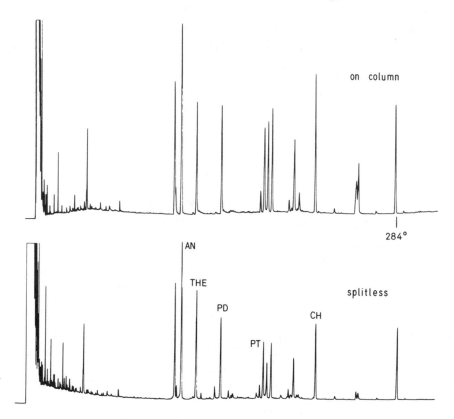

Fig. 23. Gas chromatograms of free steroids. AN, Androsterone; THE, tetrahydrocortisone; PD, pregranediol; PT, pregranetriol; CH, cholesterol.

Column: SE-52 (0.15 μm), persilylated Pyrex, 15 m × 0.32 mm id
Carrier gas: Hydrogen, 3.5 mL/min at 25°C
Injection: 2 μL containing ~20 μg per constituent, on-column, 120°C; splitless
Detector: FID
Oven: 90°C during injection and 40 s of splitless operation, then rapidly to 160°C, 4°/min, 160–300°C

(From K. Grob, 1980.)

through a biological sewage treatment plant (Stephanou and Giger, 1982; Giger *et al.*, 1984b). Nonylphenol has been found as a major component in digested sewage sludges (up to 2.5 g/kg dry sludge). These determinations were performed by glass-capillary gas chromatography (Giger *et al.*, 1984b; Schaffner *et al.*, 1984).

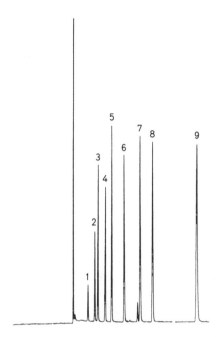

Fig. 24.   Gas chromatogram of volatile acids in aqueous solution. (1) acetic acid; (2) propionic acid; (3) isobutyric acid; (4) butyric acid; (5) isovaleric acid; (6) valeric acid; (7) isocaproic acid; (8) caproic acid; (9) heptanoic acid.

    Column: Carbowax 20M on undeactivated fused silica, 26 m × 0.2 mm id
    Injection: Aqueous solution
    Detector: FID
    Oven: 155°C

(From Dandenau *et al.*, 1979.)

The combined use of high performance liquid chromatography (HPLC) and high-resolution gas chromatography has been demonstrated to offer many advantages for the analyses of complex mixtures of organic pollutants (Giger *et al.*, 1984a). Of particular benefit is the complementary application of normal- and reversed-phase separation HPLC and the use of HPLC as a preparative separation before detailed analyses by capillary GC. Applications are presented for polycyclic aromatic hydrocarbons, oil hydrocarbons, alkylphenols, and alkylphenolethoxylates.

This limited number of final examples shows that continuous progress is made in capillary GC and that many new applications of this powerful technique can be expected.

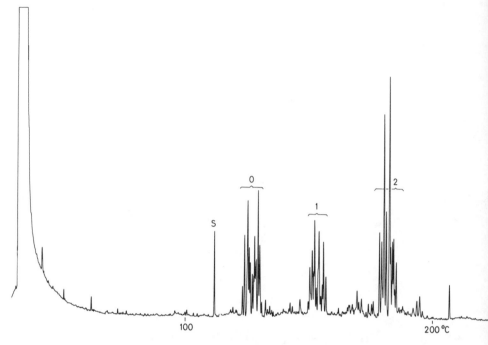

Fig. 25. Gas chromatogram of nonylphenol and nonylphenoxyethoxylates extracted from a secondary sewage effluent. (0) Nonylphenol; (1) nonylphenoxyethanol; (2) nonylphenoxyethoxyethanol; (S) 2,4,6-tribromophenol, internal standard.

Column:	OV-73, persilylated, 26 m × 0.31 mm id
Carrier gas:	Hydrogen, 0.5 atm
Injection:	Splitless, temp 300°C
Detector:	FID
Oven:	Ambient during injection and elution of the solvent (pentane), then 2°/min, 40–280°C

(From Stephanou and Giger, 1982.)

ACKNOWLEDGMENTS

We are indepted to R. H. Bromund for reviewing the manuscript. Very valuable advice and comments were given by K. Grob. This contribution was in part supported by the Swiss Department of Commerce (Project COST 64b). The manuscript was typed by B. Hauser and B. Schneider; the figures were drawn by H. Bolliger.

REFERENCES

Ahnoff, M., and Johansson, L. (1981). *Proc. Int. Symp. Capillary Chromatogr., 4th, 1981* pp. 487–504.
Alexander, G., and Rutten, G. A. F. M. (1973). *Chromatographia* **6**, 231–233.

American Public Health Association (1976). "U.S. Standard Methods for the Examination of Water and Wastewater," 14th ed., APHA, Washington, D.C.

Andersson, E. L., and Bertsch, W. (1978). *HRC CC, J. High Resolut. Chromatogr. Chromatogr. Commun.* **1**, 13–17.

Angeletti, G., and Bjørseth, A., eds. (1984). "Analysis of Organic Micropollutants in Water." Vol. 2. Reidel Publ., Dordrecht, Netherlands.

Bächmann, K., Emig, W., Rudolph, J., and Tsotsos, D. (1977). *Chromatographia* **10**, 684–686.

Badings, H. T., van der Pol, J. J. G., and Wassink, J. G. (1975). *Chromatographia* **8**, 440–448.

Badings, H. T., de Jong, C., and Wassink, J. G. (1981). *HRC CC, J. High Resolut. Chromatogr. Chromatogr. Commun.* **4**, 644–646.

Baird, R. B., Carmona, L. G., and Jenkins, R. L. (1977). *Bull. Environ. Contam. Toxicol.* **17**, 764–767.

Bartle, K. D., Elstub, J., Novotny, M., and Robinson, R. J. (1977). *J. Chromatogr.* **135**, 351–358.

Bellar, T. A., and Lichtenberg, J. J. (1974). *J. Am. Water Works Assoc.* **66**, 739–744.

Bjørseth, A., and Angeletti, G., eds. (1982). "Analysis of Organic Micropollutants in Water," Reidel Publ., Dordrecht, Netherlands.

Bjørseth, A., and Eklund, G. (1979). *HRC CC, J. High Resolut. Chromatogr. Chromatogr. Commun.* **1**, 22–26.

Blau, K., and King, G. S. (1977). "Handbook of Derivatives for Chromatography," Heyden, London.

Blumer, M. (1975). *Angew. Chem.* **14**, 507–514.

Blumer, M., and Sass, J. (1972). *Science* **176**, 1120–1122.

Brechbühler, B., Gay, L., and Jaeger, H. (1977). *Chromatographia* **10**, 478–486.

Brodtmann, N. V., Jr., and Koffskey, W. E. (1979). *J. Chromatogr. Sci.* **17**, 97–110.

Brody, S. S., and Chaney, J. E. (1966). *J. Gas Chromatogr.* **4**, 42–46.

Brötell, H., Ehrsson, H., and Gyllenhaal, O. (1973). *J. Chromatogr.* **78**, 293–301.

Bruderreck, H., Schneider, W., and Halasz, I. (1967). *J. Gas Chromatogr.* **5**, 217–225.

Buikema, A. L., McGinniss, M. J., and Cairns, J. (1979). *Mar. Environ. Res.* **2**, 87–181.

Burgett, C. A., Smith, D. M., and Bente, H. B. (1977). *J. Chromatogr.* **134**, 57–64.

Buser, H. R. (1977). *Anal. Chem.* **49**, 918–922.

Buser, H. R., and Widmer, H. M. (1979). *HRC CC, J. High Resolut. Chromatogr. Chromatogr. Commun.* **2**, 177–183.

Coulson, D. M. (1965). *J. Gas Chromatogr.* **3**, 134–137.

Coulson, D. M., and Cavanagh, L. A. (1960). *Anal. Chem.* **32**, 1245–1247.

Dandeneau, R. D., and Zerenner, E. H. (1979). *HRC CC, J. High Resolut. Chromatogr. Chromatogr. Commun.* **2**, 351–356.

Dandenau, R. D., Bente, P., Rooney, T., and Hishes, R. (1979). *Am. Lab.*

David, D. J. (1974). "Gas Chromatographic Detectors." Wiley, New York.

De Potter, M., Müller, R., and Willems, J. (1978). *Chromatographia* **11**, 220–222.

Desty, D. M. (1965). *Adv. Chromatogr.* **1**, 199–228.

Dietz, F., and Traud, J. (1973). *Vom Wasser* **41**, 137–155.

Dietz, F., and Traud, J. (1978). *Vom Wasser* **51**, 235–257.

Dolan, J. W., and Hall, R. C. (1973). *Anal. Chem.* **45**, 2198–2204.

Donike, M. (1973). *Chromatographia* **6**, 190–195.

Dowty, B. J., Carlisle, D., Laseter, J. L., and Steven, J. (1975). *Science* **187**, 75–77.

Dowty, B. J., Green, L. E., and Laseter, J. L. (1976a). *Anal. Chem.* **48**, 946–949.

Dowty, B. J., Green, L. E., and Laseter, J. L. (1976b). *J. Chromatogr. Sci.* **14**, 187–190.

Dressman, R. C., Stevens, A. A., Fair, J., and Smith, B. (1979). *J. Am. Water Works Assoc.* **71**, 392–396.

Driscoll, J. N. (1977). *J. Chromatogr.* **134**, 49–55.

Driscoll, J. N., Ford, J., Jaramillo, L. F., and Gruber, E. T. (1978). *J. Chromatogr.* **158**, 171–180.

Eklund, G., Joseffson, B., and Roos, C. (1978). *HRC CC, J. High Resolut. Chromatogr. Chromatogr. Commun.* **1**, 34–40.

Ettre, L. S. (1973). *Chromatographia* **6**, 489–495.

Ettre, L. S. (1978). *J. Chromatogr. Sci.* **16**, 396–417.

Ettre, L. S., and Averill, W. (1961). *Anal. Chem.* **33**, 680–684.

Ettre, L. S., and Purcell, J. E. (1976). *Prog. Anal. Chem.* **8**, 119–149.

Farwell, S. O., and Rasmussen, R. A. (1976). *J. Chromatogr. Sci.* **14**, 224–234.

Favretto, L., Stancher, B., and Tunis, F. (1978). *Analyst* **103**, 955–962.

Filbert, A. M., and Hair, M. L. (1968). *J. Gas Chromatogr.* **6**, 218–221.

Franken, J. J., and Vader, H. L. (1973). *Chromatographia* **6**, 22–27.

Galli, M., Trestianu, S., and Grob, K., Jr. (1979). *HRC CC, J. High Resolut. Chromatogr. Chromatogr. Commun.* **2**, 366–370.

Gaspar, G., Arpino, P., and Guiochon, G. (1977). *J. Chromatogr. Sci.* **15**, 256.

Geiser, R. (1980). Ph.D. Thesis, No. 6678. ETH Zürich.

German, A. L., and Horning, E. C. (1972). *Anal. Lett.* **5**, 619–628.

Giger, W., and Schaffner, C. (1978). *Anal. Chem.* **50**, 243–249.

Giger, W., and Schaffner, C. (1981). *In* "Advances in the Identification and Analysis of Organic Pollutants in Water" (L. H. Keith, ed.), pp. 141–154. Ann Arbor Sci. Publ., Ann Arbor, Michigan.

Giger, W., Reinhard, M., Schaffner, C., and Zürcher, F. (1976). *In* "Identification and Analysis of Organic Pollutants in Water" (L. H. Keith, ed.), pp. 433–452. Ann Arbor Sci. Publ., Ann Arbor, Michigan.

Giger, W., Molnar, E., and Wakeham, S. (1978). *In* "Aquatic Pollutants" (O. Hutzinger, L. H. van Lelyveld, and B. C. J. Zoeteman, eds.), pp. 101–123. Pergamon, Oxford.

Giger, W., Ahel, M., and Schaffner, C. (1984a). *In* "Analysis of Organic Micropollutants in Water" (G. Angeletti and A. Bjørseth, eds.), pp. 91–109. Reidel Publ., Dordrecht, Netherlands.

Giger, W., Brunner, P., and Giger, W. (1984b). *Science* (submitted for publication).

Goerlitz, D., and Brown, E. (1972). "Methods for Analysis of Organic Substances in Water," U.S. Geol. Survey Techniques of Water Resources Investigations, Book 5, Chapter A3.

Grant, D. W., and Clarke, A. (1974). *J. Chromatogr.* **97**, 115–129.

Greve, P. A., and Wit, S. L. (1971). *J. Water Pollut. Control Fed.* **43**, 2338–2348.

Grimmer, G., Böhnke, H., and Borwitzky, H. (1978). *Fresenius' Z., Anal. Chem.* **289**, 91–95.

Grob, K. (1973). *J. Chromatogr.* **84**, 255–273.

Grob, K. (1975). *Chromatographia* **8**(9), 423–433.

Grob, K. (1978). *HRC CC, J. High Resolut. Chromatogr. Chromatogr. Commun.* **1**, 263–267.

Grob, K. (1979). *HRC CC, J. High Resolut. Chromatogr. Chromatogr. Commun.* **2**, 599–604.

Grob, K. (1980). *HRC CC, J. High Resolut. Chromatogr. Chromatogr. Commun.* **3**, 585–586.

Grob, K., and Grob, G. (1969). *J. Chromatogr. Sci.* **7**, 584–585.

Grob, K., and Grob, G. (1971). *Chromatographia* **4**, 422–424.

Grob, K., and Grob, G. (1972). *Chromatographia* **5**, 3–12.
Grob, K., and Grob, G. (1976). *J. Chromatogr.* **125**, 471–485.
Grob, K., and Grob, G. (1979). *HRC CC, J. High Resolut. Chromatogr. Chromatogr. Commun.* **2**, 109–117.
Grob, K., and Grob, G. (1982). *HRC CC, J. High Resolut. Chromatogr. Chromatogr. Commun.* **5**, 349–354.
Grob, K., and Grob, G. (1983a). *Chromatographia* **17**, 481–485.
Grob, K., and Grob, G. (1983b). *HRC CC, J. High Resolut. Chromatogr. Chromatogr. Commun.* **6**, 133–139.
Grob, K., and Grob, K., Jr. (1974). *J. Chromatogr.* **94**, 53–64.
Grob, K., and Grob, K., Jr. (1978a). *J. Chromatogr.* **151**, 311–320.
Grob, K., and Grob, K., Jr. (1978b). *HRC CC, J. High Resolut. Chromatogr. Chromatogr. Commun.* **1**, 57–64.
Grob, K., and Habich, A. (1983). *HRC CC, J. High Resolut. Chromatogr. Chromatogr. Commun.* **6**, 11–15.
Grob, K., and Zürcher, F. (1976). *J. Chromatogr.* **117**, 285–294.
Grob, K., Grob, G., and Grob, K., Jr. (1977). *Chromatographia* **10**(4), 181–187.
Grob, K., Grob, G., and Grob, K., Jr. (1979). *HRC CC, J. High Resolut. Chromatogr. Chromatogr. Commun.* **2**, 31–35, 677–678.
Grob, K., Grob, G., Brechbühler, B., and Pichler, P. (1981a). *J. Chromatogr.* **205**, 1–11.
Grob, K., Grob, G., and Grob, K., Jr. (1981b). *J. Chromatogr.* **211**, 243–246.
Grob, K., Grob, G., and Grob, K., Jr. (1981c). *J. Chromatogr.* **219**, 13–20.
Grob, K., Grob, G., Blum, W., and Walther, W. (1982). *J. Chromatogr.* **244**, 197–208.
Grob, K., Jr. (1979). *J. Chromatogr.* **178**, 387–392.
Grob, K., Jr. (1980). *HRC CC, J. High Resolut. Chromatogr. Chromatogr. Commun.* **3**, 286–290.
Grob, K., Jr. (1981a). *Proc. Int. Symp. Capillary Chromatogr., 4th, 1981* pp. 185–199.
Grob, K., Jr. (1981b). *J. Chromatogr.* **213**, 3–14.
Grob, K., Jr. (1982a). *J. Chromatogr.* **237**, 15–23.
Grob, K., Jr. (1982b). *J. Chromatogr.* **251**, 235–248.
Grob, K., Jr. (1983a). *J. Chromatogr.* **270**, 17–22.
Grob, K., Jr. (1983b). *J. Chromatogr.* **279**, 225–232.
Grob, K., Jr. (1984a). "Injection Techniques in Capillary Gas Chromatography." Hüthig, Heidelberg (in preparation).
Grob, K., Jr. (1984b). *J. Chromatogr.* **283**, 21–25.
Grob, K., Jr. (1984c). *J. Chromatogr.* **287**, 1–14.
Grob, K., Jr., and Grob, K. (1977a). *Chromatographia* **10**, 250–255.
Grob, K., Jr., and Grob, K. (1977b). *J. Chromatogr.* **40**, 257–259.
Grob, K., Jr., and Grob, K. (1981). *J. Chromatogr.* **207**, 291–297.
Grob, K., Jr., and Neukom, H. P. (1979a). *HRC CC, J. High Resolut. Chromatogr. Chromatogr. Commun.* **2**, 15–21.
Grob, K., Jr., and Neukom, H. P. (1979b). *HRC CC, J. High Resolut. Chromatogr. Chromatogr. Commun.* **2**, 563–569.
Grob, K., Jr., and Neukom, H. P. (1980a). *J. Chromatogr.* **189**, 109–117.
Grob, K., Jr., and Neukom, H. P. (1980b). *J. Chromatogr.* **198**, 64–69.
Grob, K., Jr., and Neukom, H. P. (1982). *J. Chromatogr.* **236**, 297–306.
Grob, K., Jr., and Rennhard, S. (1980). *HRC CC, J. High Resolut. Chromatogr. Chromatogr. Commun.* **3**, 627–633.
Grob, K., Jr., and Romann, A. (1981). *J. Chromatogr.* **214**, 118–121.
Grob, K., Jr., and Schilling, B. (1983a). *J. Chromatogr.* **259**, 37–48.

Grob, K., Jr., and Schilling, B. (1983b). *Chromatographia* **17**, 357–368.
Grob, K., Jr., Grob, G., and Grob, K. (1978). *J. Chromatogr.* **156**, 1–20.
Grob, K., Jr., Neukom, H. P., and Hilling, P. (1981). *HRC & CC* **4**, 203–208.
Haken, J. K. (1976). *Adv. Chromatogr.* **14**, 367–407.
Hall, R. C. (1974). *J. Chromatogr. Sci.* **12**, 152–160.
Henderson, J. E., Peyton, G. R., and Glaze, W. H. (1976). *In* "Identification and Analysis of Organic Pollutants in Water" (L. H. Keith, ed.), pp. 105–111. Ann Arbor Sci. Publ., Ann Arbor, Michigan.
Hild, J., Schulte, E., and Thier, H. P. (1978). *Chromatographia* **11**, 397–399.
Hiltunen, R., Laakso, I., Hovinen, S., and Derome, J. (1982). *J. Chromatogr.* **237**, 41–48.
Hrivnac, M., Frischknecht, W., and Cechova, L. (1976). *Anal. Chem.* **48**, 937–940.
Jennings, W. G. (1975). *J. Chromatogr. Sci.* **13**, 185–187.
Jennings, W. G. (1980). "Gas Chromatography with Glass Capillary Columns," 2nd ed. Academic Press, New York.
Jennings, W. G., Freemann, R. R., and Rooney, T. A. (1978). *HRC CC, J. High Resolut. Chromatogr. Chromatogr. Commun.* **1**, 275–276.
Jentzsch, D., and Otte, E. (1970). "Detectors in Gas Chromatography" (in German). Akad. Verlagsges., Frankfurt am Main.
Kaiser, K. L. E., and Oliver, B. G. (1976). *Anal. Chem.* **48**, 2207–2209.
Kaiser, R. E. (1976). *Chromatographia* **9**, 337–352.
Kaiser, R. E. (1977). *Chromatographia* **10**, 455–465.
Kapila, S., and Vogt, C. R. (1979). *J. Chromatogr. Sci.* **17**, 327–332.
Karger, B. L., Snyder, L. R., and Horváth, C. (1973). "An Introduction to Separation Science." Wiley, New York.
Keith, L. H., ed. (1976a). "Identification and Analysis of Organic Pollutants in Water." Ann Arbor Sci. Publ., Ann Arbor, Michigan.
Keith, L. H. (1976b). *In* "Identification and Analysis of Organic Pollutants in Water" (L. H. Keith, ed.), pp. 671–707. Ann Arbor Sci. Publ., Ann Arbor, Michigan.
Keith, L. H., ed. (1981). "Advances in the Identification and Analysis of Organic Pollutants in Water," Vols. 1 and 2. Ann Arbor Sci. Publ., Ann Arbor, Michigan.
Kolb, B., Auer, M., and Pospisil, P. (1977). *J. Chromatogr. Sci.* **15**, 53–63.
Kovats, E. (1958). *Helv. Chim. Acta* **41**, 1915–1932.
Krijgsman, W., and van de Kemp, C. D. (1977). *J. Chromatogr.* **131**, 412–416.
Kruppa, R. F., and Henley, R. S. (1971). *Am. Lab.* 41–44.
Kummert, R., Molnar-Kubica, E., and Giger, W. (1978). *Anal. Chem.* **50**, 1663–1669.
Lamparski, L. L., and Nestrick, P. J. (1978). *J. Chromatogr.* **156**, 143–151.
Lawler, G. C., Loong, W.-A., Fiorito, B. J., and Laseter, J. L. (1977). *J. Chromatogr. Sci.* **15**, 532–536.
LeBel, G. L., and Williams, D. T. (1979). *J. Assoc. Off. Anal. Chem.* **62**, 1353–1355.
Lee, M. L., Vasslaros, D. L., White, C. M., and Novotny, M. (1979). *Anal. Chem.* **51**, 768–773.
Lee, M. L., Yang, F. J., and Bartle, K. D. (1984). "Open Tubular Column Gas Chromatography," Wiley, New York.
Leidner, H., Gloor, R., and Wuhrmann, K. (1976). *Tenside Deterg.* **13**, 122–130.
Leuenberger, C., Coney, R., Graydon, J. W., Molnar-Kubica, E., and Giger, W. (1983). *Chimia* **37**, 345–354.
Lin, D. C.K. (1981). *In* "Application of Capillary Chromatography" (W. Jennings, ed.), pp. 123–174. Dekker, New York.
Lindström, K., and Nordin, J. (1976). *J. Chromatogr.* **128**, 13–26.
Littlewood, A. B. (1970). "Gas Chromatography: Principles, Techniques, Applications," 2nd ed. Academic Press, New York.

Lovelock, J. E. (1974). *J. Chromatogr.* **99**, 3–12.

McCown, S. M., and Earnest, C. M. (1978). *Int. Lab.* pp. 37–47.

McReynolds, W. O. (1970). *J. Chromatogr. Sci.* **8**, 685–691.

Maier-Bode, H., and Riedmann, M. (1975). *Residue Rev.* **54**, 113–181.

Majlat, P., Erdos, Z., and Takacs, J. (1974). *J. Chromatogr.* **91**, 89–103.

"Manual of Analytical Quality Control for Pesticides and Related Compounds in Human and Environmental Samples." U.S. Environmental Protection Agency, Washington, D.C., 1979.

"Methods for Benzidine, Chlorinated Organic Compounds, Pentachlorphenol and Pesticides in Water and Wastewater." U.S. Environmental Protection Agency (Pending interim issuance, 1978).

Mieure, J. P. (1977). *J. Am. Water Works Assoc.* **69**, 60–62.

Miller, R. J., and Jennings, W. (1979). *HRC CC, J. High Resolut. Chromatogr. Chromatogr. Commun.* **2**, 72–73.

Mousa, J. J., and Whitlock, S. A. (1979). *ASTM Spec. Tech. Publ.* **STP 686**, 206–220.

Murray, A. J., and Riley, J. P. (1973). *Nature (London)* **242**, 37–38.

Neff, J. M. (1979). "Polycyclic Aromatic Hydrocarbons in the Aquatic Environment." Appl. Sci. Publ., Ltd., London.

Nicholson, A. A., Meresz, O., and Lemyk, B. (1977). *Anal. Chem.* **49**, 814–819.

Nicholson, J. D. (1978a). *Analyst* **103**, 1–28.

Nicholson, J. D. (1978b). *Analyst* **103**, 193–222.

Novotny, M. (1978). *Anal. Chem.* **50**, 16A–25A.

Onuska, F. I., and Comba, M. (1980). *In* "Hydrocarbons and Halogenated Hydrocarbons in the Aquatic Environment" (B. K. Afghan and D. Mackay, eds.), pp. 285–301. Plenum, New York.

Overton, E. B., Shele, C. F., and Laseter, J. L. (1978). *HRC CC, J. High Resolut. Chromatogr. Chromatogr. Commun.* **1**, 109–110.

Pape, B. E., Rodgers, D. M., and Flynn, T. C. (1977). *J. Chromatogr.* **134**, 1–24.

Patience, R. L., Rowland, S. J., and Maxwell, J. R. (1978). *Geochim. Cosmochim. Acta* **42**, 1871–1875.

Patterson, P. L. (1977). *J. Chromatogr.* **134**, 25–37.

Patterson, P. L. (1978). *Anal. Chem.* **50**, 345–348.

Patterson, P. L., Howe, R. L., and Abu-Sheemays, A. (1978). *Anal. Chem.* **50**, 339–344.

Pearson, C. R., and McConnell, G. (1975). *Proc. R. Soc. London, Ser. B* **198**, 305–332.

Pellizzari, E. D. (1974). *J. Chromatogr.* **98**, 323–361.

Pesticide Analytical Manual, Vol. I (Multiresidues) and Vol. II (Individual Residues). U.S. Dept. of Health, Education and Welfare; Food and Drug Administration.

Pfaender, F. K., Jonas, R. B., Stevens, A. A., Moore, L., and Hass, J. R. (1978). *Environ. Sci. Technol.* **12**(4), 438–441.

Piet, G. J., Slingerland, P., de Grunt, F. E., van den Heuvel, M. P. M., and Zoeteman, B. C. J. (1978). *Anal. Lett.* **5**, 437–448.

Pigliucci, R., Averill, W., Purcell, J. E., and Ettre, L. S. (1975). *Chromatographia* **8**(4), 165–175.

Poy, F. (1979). *HRC CC, J. High Resolut. Chromatogr. Chromatogr. Commun.* **2**, 243–245.

Poy, F. (1982). *Chromatographia* **16**, 345–348.

Poy, F., Visani, S., and Terrosi (1981). *J. Chromatogr.* **217**, 81–90.

Pretorius, V., Phillips, C. S. G., and Bertsch, W. (1983). *HRC CC, J. High Resolut. Chromatogr. Chromatogr. Commun.* **6**, 232–235, 273–274, 321–322.

Proske, M. G., Bender, M., Schirrmeister, H., and Böttcher, G. (1978). *Chromatographia* **11**, 715–719.

Proske, M. G., Bender, M., Schomburg, G., and Hübinger, E. (1982). *J. Chromatogr.* **240,** 95–106.

Quimby, B. D., Uden, P. C., and Barnes, R. M. (1978). *Anal. Chem.* **50,** 2112–2118.

Quimby, B. D., Delaney, M. F., Uden, P. C., and Barnes, R. M. (1979). *Anal. Chem.* **51,** 875–880.

Reinhard, M., Dolce, C. J., McCarty, P. L., and Argo, D. C. (1979). *J. Environ. Eng. Div. (Am. Soc. Civ. Eng.)* **105,** 675–693.

Renberg, L. (1982). *In* "Analysis of Organic Micropollutants in Water" (A. Bjørseth and G. Angeletti, eds.), pp. 286–297. Reidel Publ., Dordrecht, Netherlands.

Renberg, L., and Lindström, K. (1981). *J. Chromatogr.* **214,** 327–334.

Richard, J. J., and Junk, G. A. (1977). *J. Am. Water Works Assoc.* **69,** 62–64.

Rigaud, M., Chebroux, P., Durand, J., Maclouf, J., and Madani, C. (1976). *Tetrahedron Lett.* **44,** 3935.

Rook, J. J. (1977). *Environ. Sci. Technol.* **11,** 478–482.

Sandra, P., Temmermann, I., and Verstappe, M. (1983). *HRC CC, J. High Resolut. Chromatogr. Chromatogr. Commun.* **6,** 501–504.

Schaffner, C., Brunner, P., and Giger, W. (1984). *In* "Environmental Effects of Organic and Inorganic Contaminants in Sewage Sludge." Reidel Publ., Dordrecht, Netherlands (in press).

Schomburg, G. (1981). *Proc. Int. Symp. Capillary Chromatogr., 4th, 1981* p. 921A.

Schomburg, G. (1983). *J. Chromatogr. Sci.* **21,** 97–105.

Schomburg, G., Dielmann, R., Husmann, H., and Weeke, F. (1976). *J. Chromatogr.* **122,** 55–72.

Schomburg, G., Dielmann, R., Husmann, H., and Weeke, F. (1977). *J. Chromatogr.* **142,** 87–102.

Schomburg, G., Husmann, H., and Rittmann, R. (1981). *J. Chromatogr.* **204,** 85–96.

Schulte, E., and Acker, L. (1974). *Z. Anal. Chem.* **268,** 260–267.

Schulte, E. (1983). "Praxis der Kapillar-Gaschromatographie," Springer, Berlin.

Schutjes, C. P. M., Vermeer, E. A., Rijks, J. A., and Cramers, C. A. (1981). *Proc. Int. Symp. Capillary Chromatogr., 4th, 1981* pp. 687–702.

Schwarzenbach, R. P., Molnar-Kubica, E., Giger, W., and Wakeham, S. G. (1979). *Environ. Sci. Technol.* **13,** 1367–1373.

Shackelford, W. H., and Webb, R. G. (1979). *ASTM Spec. Tech. Publ.* **STP 686,** 191–205.

Simmon, V. F., Tardiff, R. G. (1978). *In* "Water Chlorination" (R. L. Jolley, ed.), Vol. II, pp. 417–431. Ann Arbor Sci. Publ., Ann Arbor, Michigan.

Sisti, G., Munari, F., and Trestianu, S. (1982). *Pittsburg Conf., 1982* Paper 633.

Stephanou, E., and Giger, W. (1982). *Environ. Sci. Technol.* **16,** 800–805.

Stock, W., and Alberti, J. (1979). *Vom Wasser* **52,** 75–86.

Suffet, I. H., and Glaser, E. R. (1978). *J. Chromatogr. Sci.* **16,** 12–18.

Taylor, I. S., and Thier, H. P. (1979). *Pure Appl. Chem.* **51,** 1603–1613.

Tobin, R. S., Onuska, F. I., Brownlee, B. G., Anthony, P. H. J., and Comba, M. E. (1976). *Water Res.* **10,** 529–535.

Tranchant, J. (1969). "Practical Manual of Gas Chromatography." Elsevier, Amsterdam.

Trestianu, S., and Galli, M. (1981). *J. Chromatogr.* **203,** 193–205.

Tuinstra, L. G. M. T., and Traag, W. A. (1979). *HRC CC, J. High Resolut. Chromatogr. Chromatogr. Commun.* **2,** 723–728.

"U.S. Standard Methods for the Examination of Water and Wastewater," 14th ed. American Public Health Association, Washington, D.C., 1976.

Van den Dool, H., and Kratz, P. D. (1963). *J. Chromatogr.* **11,** 463–471.

Varma, M. M., Siddique, M. R., Doly, K. T., and Machis, A. (1979). *J. Am. Water Works Assoc.* **71**, 389–392.

Verzele, M. (1979). *HRC CC, J. High Resolut. Chromatogr. Chromatogr. Commun.* **2**, 647–653.

Vogt, W., Jakob, K., and Obwexer, H. W. (1979a). *J. Chromatogr.* **174**, 437–439.

Vogt, W., Jakob, K., Ohnesorg, A. B., and Obwexer, H. W. (1979b). *J. Chromatogr.* **186**, 197–205.

Vogt, W., Jakob, K., Ohnesorg, A. B., and Schwertfeger, G. (1981). *J. Chromatogr.* **217**, 91–98.

Wakeham, S. G. (1979). *Environ. Sci. Technol.* **13**, 1118–1123.

Wakeham, S. G. and Frew, N. (1982). *Lipids* **17**, 831–843.

Wakeham, S. G., Schaffner, C., and Giger, W. (1980). *Geochim. Cosmochim. Acta* **44**, 403–413, 415–429.

Wakeham, S. G., Schaffner, C., and Giger, W. (1979). *Adv. Org. Geochem.*, A. G. Douglas and T. R. Maxwell, eds., Pergamon Press, Oxford, pp. 353–363.

Wegmann, R. C. C., and Hofstee, A. W. M. (1979). *Water Res.* **13**, 651–657.

Yang, F. J., Brown, A. C., III, and Cram, S. P. (1978). *J. Chromatogr.* **158**, 91–109.

Zlatkis, A., and Walker, J. Q. (1963). *J. Gas Chromatogr.* **1**, 9–11.

Zürcher, F., and Thüer, M. (1978). *Environ. Sci. Technol.* **12**, 838–842.

Zürcher, F., Thüer, M., and Davis, J. M. (1980). *In* "Hydrocarbons and Halogenated Hydrocarbons in the Aquatic Environment" (B. K. Afghan and D. Mackay, eds.), pp. 373–385. Plenum, New York.

5 ORGANIC MASS SPECTROMETRY

J. Ronald Hass

National Institute of Environmental Health Sciences
Research Triangle Park, North Carolina

and

Daniel L. Norwood

Department of Environmental Sciences and Engineering
School of Public Health
University of North Carolina at Chapel Hill
Chapel Hill, North Carolina

I. INTRODUCTION

Mass spectrometry is the field of science that deals with the formation of gaseous ions, their separation on the basis of mass-to-charge ratio and the subsequent detection of the mass-resolved ion beam. The study of gaseous ions began with the work of Sir J. J. Thomson, who by the early 1920s foresaw the potential of this as an analytical technique [1]. The widespread use in chemical analysis awaited the development of commercial instrumentation. By the 1940s, mass spectrometry was an accepted method for characterizing crude oils. The development of gas chromatography offered a technique that was less expensive and more reliable for these analyses, so that by the early 1950s it appeared mass spectrometry would revert to being a curiosity used mainly by physicists. However, advances in electronics led to more reliable and convenient instruments, and at the same time recognition of systematic explanations for fragmentation reactions [2] and the development of high-resolution instruments for elemental composition measurements from submilligram samples [3] resulted in the firm establishment of mass spectrometry in qualitative analysis laboratories by the early 1960s. The present eruption of interest in mass spectrometry can be traced to its successful combination with gas chromatography [4]. However, even this powerful combination would have seen rather limited application had it not been for the simultaneous development of relatively inexpensive computers. The 1970s saw the development of the gas chromatograph/mass spectrometer/data system (GC/MC/DS) into integrated systems, some of which have an high degree of automation [5] and can be considered a single instrument.

The GC/MS/DS instrument is the most powerful method available for the identification and/or determination of trace organic constituents in complex matrices. An experienced mass spectrometrist can tentatively identify most of the components in mixtures of more than a hundred components in as little as a few hours. Compared to older methods based on, for example, comparative chromatography using a number of columns even with selective detectors, the results have been spectacular. Many previously unidentified components were readily assigned structures and earlier mistaken identifications corrected. It is not surprising that GC/MS/DS has been embraced as a panacea. Indeed, the inclusion of both GC and MS data apparently meets the usual criterion that two different identification techniques give the same result before one can have confidence in an identification. In a strict sense this is not true, because GC is fundamentally a separation method. Valid identifications based on GC/MS/DS data can be achieved if and only if it can be established that all compounds that interfere with the mass spectral measurements have been resolved by the gas chromatograph.

It is widely recognized that any improved chromatographic resolution improves the quality of the final product. The resolution of the mass spectrometer is equally important and is traditionally defined in terms of peak sharpness, that is, the ability to separate ion beams of slightly differing masses. While this is a useful parameter for performance comparisons among various spectrometers, it is less useful for the selection of the most suitable technique or instrument for a particular analysis. Obviously, the higher the resolution of the mass spectrometer, the greater the confidence one can have that isobaric ions have been resolved, and the greater the likelihood that interferences caused by chromatographically unresolved components are eliminated. Less obvious is the importance of ionization technique selection. With the development of chemical ionization (CI), the mass spectrometrist has available a variety of ionization techniques that opens the possibility of making use of subtle chemical features of the sample matrix. The recent development of commercial instruments that permit selective study of the fragmentation reactions from mass-selected parent ions makes another tool available to the analytical mass spectrometrist.

It is the purpose of this chapter to review the basic principles of organic mass spectrometry, instrumentation, and data interpretation, and to describe recent developments in the field. The level of presentation is intended to be useful to the broad spectrum of scientists presently engaged in the analysis of water for trace organic chemicals. Thus, it is hoped that a scientist inexperienced in mass spectrometry is able to use this information for the evaluation of results from various mass spectral techniques and as a guide to the selection of the methods most appropriate for his applications. Although we hope to be fair and critical in our treatment of subject material, the authors make no apology for their biases which result from experience in this field.

II. PRINCIPLES OF ORGANIC MASS SPECTROMETRY

A. Introduction

In this section the chemical and physical principles underlying mass spectrometry are discussed. Most books on organic mass spectrometry discuss in greater or lesser detail the basics governing gas-phase ion chemistry. The authors acknowledge their debt to Williams and Howe [6] and recommend this book as a highly readable introduction for those interested in a more detailed treatment than is possible within the confines of this chapter. Physical chemists reading this section are referred to the recent review edited by Bowers [7] for a more comprehensive treatment of gaseous thermodynamics, unimolecular reaction rate theory, etc. For the purposes of this chapter, our attention is restricted to those areas the authors consider most important for the chemist primarily interested in spectral interpretation. As a matter of convenience, the discussion is organized along the same lines as the major functional units of a typical mass spectrometer, that is, ion formation, mass analysis, and ion detection.

B. Ion Formation

Ions may be formed by the direct ionization of an uncharged molecule, fragmentation of another ion, a gas-phase chemical reaction (e.g., proton transfer), or sputtering from a surface. With the exception of sputtering, these mechanisms involve excitation of a gaseous species followed by chemical reactions. Although the same general principles apply to all ionization methods, it is more convenient to discuss them individually.

1. Electron Impact

The first widely used method of sample ionization for organic compounds involved the excitation of gaseous neutral molecules by a beam of energetic electrons. Such an electron-impact (EI) ionization source is shown schematically in Fig. 1. Electrons are generated thermally by passing a current of a few amperes through the filament. By biasing the filament with a voltage V_{el} positive with respect to the ion volume (ion source block), a beam of electrons with an approximately known kinetic energy is formed. The electron-entrance aperture serves to define the diameter of the electron beam, which is collimated by the source magnet. By collecting the electrons that pass through the source, a means is available by which the electron current through the source might be regulated. This

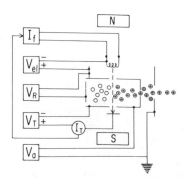

Fig. 1. Schematic diagram of an electron-impact source. The filament power supply provides a current I_f, which is regulated by the trap current I_t. After being accelerated by V_{el} and shaped by the electron-entrance hole, the electron beam is collimated by the source magnet. The sample is introduced into this beam of electrons and ionized. The positive ions are then pushed from the source by the repeller voltage V_r and accelerated by V_a.

current is used as a reference to the filament power supply. Thus, should the current be reduced, for example by increased losses resulting from the introduction of sample into the source, the filament current is increased by just the amount necessary to compensate for these losses. In this manner, the linear response of the ion source can be maintained over five or six orders of magnitude of sample concentration.

The interaction of the electron beam with the gaseous sample results in ion formation. For the moment, we restrict our attention to positive ions that are forced from the source by biasing the repeller positive with respect to the ion volume. The ions are then accelerated by maintaining the ion volume at a potential that is positive with respect to the mass analyzer. The magnitude of this acceleration potential V_a depends on the nature of the mass analyzer and may be from a few to a few thousand volts. Typically, a number of electrostatic lenses are included to shape the beam and/or position it properly prior to acceleration.

Let us return to the interaction of the electron beam with the sample. The ion source is maintained at a high vacuum (background pressure, typically on the order of 10^{-7} torr; sample pressure, typically $10^{-5}-10^{-4}$ torr) so that the mean free path of the sample ions greatly exceeds source dimensions. Thus, we are dealing with the interaction of electrons with otherwise isolated gaseous species. Because the interaction time between a 50-eV electron and a gaseous molecule has been estimated to be approximately 10^{-15} s [8] and the fastest carbon–hydrogen bond vibrational frequencies are on the order of 10^{-13} s, it seems safe to assume that the change in internuclear coordinates during the excitation process is negligible. This approximation is known as the Franck–Condon rule [9]. The consequences of this are best seen diagramatically (Fig. 2). Here we have plotted the potential energy diagram for a diatomic molecule and the

Fig. 2. Energy transfer during electron-impact ionization. The excitation of a neutral molecule leads to the formation of molecular ions in a variety of vibronic states. Some molecular ions are formed with too little energy to fragment; others may be formed with sufficient energy to fragment directly; still other ions may fragment after energy redistribution.

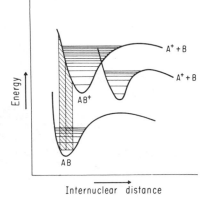

corresponding molecular ion. The same principles apply for polyatomic molecules except that graphic representation requires multidimensional surfaces. Because ionization normally reduces the electron density in bonding molecular orbitals, ions generally have greater equilibrium internuclear distances than do the corresponding neutral molecules. Thus, if we start with the molecule in its lowest vibrational energy state and add sufficient energy to cause ionization, according to the Franck–Condon approximation, the ion is formed initially with the same internuclear coordinates as the neutral molecule. This is possible only if the ion is formed in a vibrationally excited state. If we further complicate the situation by including the possibility of a second ionic state and assume that energy transfer between the states is rapid compared to any fragmentation reactions*, we have the basis of the quasi-equilibrium theory (QET). That is, energy transfer between the vibronic states of the excited molecular ion is sufficiently rapid that the energy distribution within this ion can be assumed to have reached equilibrium before any fragmentation reactions occur. This model was originally developed as the Rice–Ramsperger–Kassel–Marcus (RRKM) theory of unimolecular reaction kinetics [10,11]. A number of simplifications to this theory lead to Eq. (1) [6,12],

$$k(E) = \eta\left(\frac{E - E_0}{E}\right)^{n-1} \tag{1}$$

where $k(E)$ is the rate constant for the reaction in question, η the frequency factor, E the internal energy, E_0 the threshold energy for the reaction, and n the number of harmonic oscillators (i.e., bonds). The frequency factor is related to either the "stiffness" of the bond being broken for a simple cleavage reaction or the degree of order involved in the transition state of a rearrangement reaction [6,12].

Examination of Eq. (1) reveals many of the features of mass spectral fragmentation reactions. Simple cleavage reactions may occur from a wide variety of geometries so that η is relatively large. In comparison, rearrangement reactions normally require specific geometries and have smaller values of η. However, the rearrangement reaction also involves new bond formation so that this bond-formation energy is available to help offset the threshold energy for bond rupture [13]. Thus, rearrangement reactions typically have lower thresholds (E_0) than simple cleavage reactions and dominate the mass spectrum at low energy. At higher en-

* This is simply the Born–Oppenheimer approximation, which states that electronic motion is fast compared to nuclear motion, therefore, many transitions between vibronic states of equivalent energy occur in the minimum time necessary for a fragmentation reaction.

ergy, the $(E - E_0)/E$ term approaches unity so that η becomes the major influence on relative rates. Because η for simple cleavage reactions tends to be greater than η for rearrangements, these reactions are more important at higher energy (E). In summary, the rate constants (and hence the mass spectrum) depend on the degree of internal excitation, which is related to energy of the ionizing electrons.

Obviously, one wants to operate the ion source at a sufficiently high electron energy that small changes have a negligible effect on the spectrum. Empirically, one can simply measure the mass spectra of a few compounds to determine the proper V_{el}. In general, the spectrum is a strong function of V_{el} for $V_{el} \leq 30$ V and changes slowly at $V_{el} \geq 50$ V. By convention, mass spectra are typically measured at $V_{el} = 70$ V. It should be pointed out that the actual kinetic energy of the electrons is affected by the other potentials within the source. Also, the electron beam has an energy spread as a result of the voltage drop along the filament as well as the intrinsic Maxwell–Boltzman energy distribution resulting from the high temperature of the filament. At higher V_{el}, these factors have little effect on the measurement of spectra for analytical purposes. However, at lower V_{el} these factors severely limit the accuracy with which ionization or appearance energies can be measured with an EI source.

2. Chemical Ionization

Chemical ionization (CI) is distinguished from EI in that the gas conductance of the source is minimized so that a relatively high pressure can be maintained without causing undue pressure in the remainder of the instrument. It seems fitting that CI should come from the petroleum industry [14] as did the early analytical applications of EI mass spectrometry. From a chemical point of view, the distinguishing feature of CI is that sample ionization is the result of a chemical reaction. Some reagent gas is added to the ion source to give a pressure in the range of 0.1 to 1 torr and is ionized by EI. Since the mean free path of the ions at this higher pressure is much less than source dimensions, the result of the ion–molecule collisions is the formation of a standing ion current characteristic of reagent gas as illustrated for CH_4 [Eqs. (2–6)].

$$CH_4 + e \longrightarrow CH_4^+ + 2e \tag{2}$$

$$CH_4^+ \longrightarrow CH_3^+ + H\cdot \tag{3}$$

$$CH_4^+ + CH_4 \longrightarrow CH_5^+ + \cdot CH_3 \tag{4}$$

$$CH_3^+ + CH_4 \longrightarrow C_2H_5^+ + H_2 \tag{5}$$

$$C_2H_5^+ + CH_4 \longrightarrow C_3H_7^+ + H_2 \tag{6}$$

The reagent gas is ionized by EI and undergoes its usual fragmentation reactions. The products of these processes undergo collisions with reagent-gas neutral molecules to form a series of ion–molecule reaction products [Eqs. (4–6)], which create the ion current actually measured. Since a large number of collisions are involved, thermodynamic considerations are the controlling factors in contrast to kinetic arguments as in the case of EI. We now add a trace of sample (so as not to disturb the steady-state distribution of reagent-ion current) which should never exceed ~0.1% of the reagent gas. Since this concentration constitutes such a small fraction of the reagent gas, direct EI ionization of the sample is negligible and the primary mode of sample ionization is by ion–molecule reaction with reagent ions. For example, consider the following four sample ionization processes, where RH represents the reagent, S is the sample, F is a reagent-fragment ion, and X is any species transferred in an ion–molecule reaction:

$$RH_2^+ + S \longrightarrow SH^+ + RH \tag{7}$$

$$R_2H^+ + S \longrightarrow SRH^+ + R \tag{8}$$

$$FRH^+ + S \longrightarrow SFH^+ + R \tag{9}$$

$$RX^+ + S \longrightarrow SX^+ + R \tag{10}$$

As noted above, thermodynamics is the controlling factor in the observed chemistry. That is, SH^+ (protonated molecular ion) is observed for any sample molecules S that have a higher proton affinity than that of the reagent gas. CI is thus considered a milder ionization mode than EI and in many cases provides molecular weight information. In general, sample ionization may be effected by transfer of any charged entity [e.g., X^+ as in Eq. (10)]. This reaction then occurs any time the affinity of S for X^+ exceeds the affinity of R for X^+. Thus, in principle, a judicious selection of R and X^+ should permit the selective ionization of one sample component (e.g., S_1) in the presence of a second component (e.g., S_2) so that S_1X^+ is found but S_2X^+ is not. For this to be practical, extensive tables of X^+ affinities are necessary. At present, the only X^+ for which a substantial quantity of affinity data is available is the proton. Use of reagent gases with different proton affinities has been applied successfully for the selective ionization of compounds more basic than NH_3 in the presence of compounds more acidic than NH_3 [15]. Thus

$$NH_3^+ + NH_3 \longrightarrow NH_4^+ + NH_2 \tag{11}$$

$$NH_4^+ + S \longrightarrow SH^+ + NH_3 \tag{12}$$

The reaction shown in Eq. (12) is observed for alkyl amines but not for carboxylic acids.

More subtle differences in structure can be distinguished on the basis of ion–molecule reactivity. For example, if a bulky X^+ is chosen, it can be anticipated that steric hindrance at the reaction site might lead to reduced reactivity in comparison with reactivity at similar, but less hindered, sites. For example

$$(H_3CCOCOCH_3)^{\ddagger} + H_3CCOCOCH_3 \longrightarrow (H_3CCOCOCH_3)COCH_3^+ + H_3CCO \qquad (13)$$

$$(H_3CCOCOCH_3)COCH_3^+ + S \longrightarrow SCOCH_3^+ + H_3CCOCOCH_3 \qquad (14)$$

Acetyl-ion transfer has been studied for a number of systems [16]. As expected, the reaction in Eq. (14) is important for molecules with either π- or nonbonding electrons available and is not observed for saturated molecules. The reactivity of $(H_3CCOCOCH_3)COCH_3^+$ with a number of stereochemical isomers has been studied under medium-pressure CI conditions [17]. In each case, it was found that less hindered isomers gave a substantially higher ion current for the acetylation product than did the more hindered isomer.

Hunt *et al.* [18] reported that the use of D_2O as a reagent gas for CI resulted in the exchange of "active" hydrogens in a variety of molecules. Under typical CI conditions, hydrogens bound to the heteroatoms in alcohols, phenols, carboxylic acids, amines, amides, and mercaptans are reported to undergo essentially complete exchange for deuterium in the ion source. A minor complication arises from the incorporation of as much as 15% deuterium in ketones, aldehydes, and esters. For relatively simple systems, this presents no practical limitation. However, the presence of a large number of these groups could lead to an overestimation of the number of active hydrogens by one. Results from D_2O CI therefore should be interpreted as upper limit values for the number of active hydrogens.

Clearly, the potential exists for a systematic qualitative analysis scheme based on ion–molecule reactions that is analogous to the more traditional schemes for qualitative organic analysis. At present, the analyst is restricted by the limited knowledge of gas-phase ion chemistry. No one should be surprised that proton-transfer reactions are by far the most thoroughly studied examples since this gives a very convenient comparison for gas-phase acid–base properties with similar condensed-phase properties. Thus an avenue is available for the study of solvation effects, etc., on what should be a simple reaction. Nor should it be surprising that most analytical CI involves proton-transfer chemistry. Although the value of the resulting confirmation of molecular weight cannot be denied, it is also apparent that much more information about an unknown is accessible by the study of its reactivity with gaseous ions. Further analytical development awaits more results from basic research in gas-phase ion–molecule chemistry.

Thus far we have discussed only positive-ion formation. There is no intrinsic reason negative ions are not formed simultaneously with positive ions; indeed, they are. Although negative ions were observed early in the development of mass spectrometry [1], they received little attention until the development of CI sources. Following the pioneering work of Dougherty and Weisenburger [19], other scientists have shown negative ions to be useful for a variety of analytical problems [20,21]. There are good reasons for this course of events. These result from the fundamental nature of negative ion formation and the properties of the EI source. Recall that the EI source typically operates at 10^{-5}–10^{-7} torr pressure and 70 eV electron energy. The formation of a negative molecular ion under these conditions must involve resonance electron capture, since no third body is available as an energy sink:

$$AB + e \longrightarrow AB^- \qquad (15)$$

This process can occur if two conditions are met simultaneously. First, the product ion must be more stable than the separated neutral molecule and electron. That is, the sample must have a positive electron affinity, which is a requirement for any negative ion formation. Second, the electron must have the proper energy to fit into an unoccupied molecular orbital. The magnitude of this energy is a property of each chemical species and generally falls in the range of 0–2 eV. And herein lies the problem with negative molecular ion formation under EI conditions: The population of electrons with energies in the range 0–2 eV is very low, even when minimal V_{el} is applied. Recall that the electron beam is accelerated not only by V_{el} but also any other voltage that is applied to the source and/or penetrates from the outside. Thus, it is difficult to achieve electron-beam energies below 5–10 eV.

There are additional negative ion formation processes important under EI conditions. Dissociative electron capture results in the direct formation of a negative fragment ion [Eq. (16)].

$$AB + e \longrightarrow A^- + B\cdot \qquad (16)$$

The rupture of the bond requires the addition of the bond dissociation energy. The formation of two particles as products permits some excess energy to be dissipated as translational energy of the products. Unfortunately, the negative ion current is carried by the most stable products. Thus, one finds Cl^-, CN^-, etc. as the major products of dissociative electron capture. An even higher energy process is ion-pair production in which the incident electron acts simply as an energy source [Eq. (17)].

$$AB + e \longrightarrow A^+ + B^- + e \qquad (17)$$

Once again, the negative ions formed tend to be small stable species that offer little structural information. In summary, negative ion mass spectrometry under EI conditions suffers from poor sensitivity and a preponderance of ions that give little information about the sample.

The addition of CI reagent gas into the ion source greatly alters conditions in favor of negative ion formation. First, the reagent gas serves as an energy sink. Thus, electrons with high initial kinetic energies undergo a number of collisions, each one inelastic [Eq. (18)].

$$e + R \longrightarrow R^* + e \tag{18}$$

The average kinetic energy of the electron beam is reduced at the expense of a slight warming of the reagent gas. The ionization of the reagent gas [Eq. (19)] also leads to the formation of low kinetic energy electrons:

$$R^* \longrightarrow R^{\ddot{+}} + e \tag{19}$$

Therefore, the reagent gas serves to greatly increase the population of electrons in the critical 0- to 2-eV range necessary for resonance electron capture. The second important effect of the reagent gas is the stabilization of short-lived collision complexes between the sample and more energetic electrons [Eqs. (20) and (21)].

$$e + S \longrightarrow (S^{\overline{\cdot}})^* \tag{20}$$

$$(S^{\overline{\cdot}})^* + R \longrightarrow S^{\overline{\cdot}} + R^* \tag{21}$$

The collisional frequency under negative CI conditions is on the order of 10^9 to 10^8 s^{-1}. Thus a collision complex with a lifetime >1 ns has a good chance of undergoing one or more stabilizing collisions. The major effect of adding the CI reagent gas is to increase the probability of negative ion formation by resonance electron capture so that a major fraction of the negative ion current is now carried by species indicative of molecular weight.

If we define CI as a means of sample ionization by ion–molecule reactions then the process described obviously is not chemical ionization but rather ionization by electron capture. This problem has been recognized and more descriptive names have been proposed [22]. However, negative ion chemical ionization (NCI) seems to be firmly entrenched so that it is perhaps easier to change the definition to fit the usage. Thus, at least for the purposes of this chapter, NCI refers to negative ion formation under CI conditions and may or may not involve formation of an ion–molecule reaction product. In fact, the gaseous ion chemistry of negative ions is a rapidly developing area.

Not surprisingly, the observation of ion–molecule reactions under NCI conditions depends explicitly on the nature of the reagent gas. A comprehensive review of negative ion chemistry is beyond the scope of this chapter. The interested reader is referred to the proceedings of a recent NCI conference [23] and a review of the literature prior to 1979 [24]. However, it is instructive to consider an example. The EI mass spectra of chlorinated dibenzo-*p*-dioxins are similar [25], although minor differences are measurable [26]. Furthermore, the spectra are dominated by ions resulting from the loss of one or more Cl, perhaps in combination with the loss of CO; that is, the reactions are not isomer specific for the dioxins. When the NCI spectra are measured with $\sim 10\%$ O_2 present in the reagent gas mixture, two ion–molecule reactions are observed [Eqs. (22) and (23)] [20,25].

$$+ O_2 \rightarrow \qquad + ClO^{\cdot} \qquad (22)$$

$$+ O_2 \rightarrow \qquad (23)$$

The ether cleavage reaction [Eq. (23)] can be expected to be specific for dioxins as a compound class, and is certainly more useful than a molecular weight or even an elemental composition measurement for identification purposes. The Cl displacement reaction [Eq. (22)] was found to be more abundant for every case in which the molecular ion contained a peri chlorine; otherwise, the ether cleavage product was the more abundant ion [25].

Negative ion spectra also contain fragment ions. Since the factors affecting relative negative ion stabilities generally are different from those affecting positive ion stabilities, one would expect differences in the ion chemistry. For example, Hunt and co-workers [27] reported that for short-chain permethylated polypeptides, the methane positive ion spectrum consisted of the protonated molecular ions with fragment ions that resulted from cleavage of the amide linkages with charge retention by both the carbonyl and amine fragments. The negative-ion spectra exhibited fragments resulting from cleavage between the amide nitrogen and the alkyl carbon with charge retention by the amide group. The two spectra provided complementary data. By measuring both the positive and negative ion spectra simultaneously [27], it is possible to rapidly sequence short-chain polypeptides. The point of this discussion is that NCI is not to be viewed so much as a replacement for older mass spectral techniques but rather as a useful adjunct, which, if used in conjunction with other methods, may improve the quality of the final result or permit previously intractable problems to be successfully attacked.

The remaining area of NCI that merits discussion is its sensitivity compared to other methods of ionization. From the earliest reports, NCI was purported to offer substantial sensitivity advantages when compared to positive ion techniques [20]. However, Siegel [in 28] presented an elegant argument that no fundamental reason exists for either positive or negative ion formation to occur with greater likelihood. The relative sensitivity is determined simply by the relative rate constants for electron attachment and the particular ion–molecule reaction leading to positive ion formation from the sample. Thus, for selected compounds with large rate constants for the electron capture and low rate constants for positive ion formation, NCI exhibits sensitivity advantages over positive ion techniques. The converse is also true: Compounds with low rate constants for electron capture but high rate constants for proton transfer, for example, are more easily detected by positive ion techniques. The real advantage of NCI comes when one is attempting to determine species with high electron-capture cross section in a matrix consisting of compounds with low electron-capture rate constants. As has been pointed out by Dougherty [19], this is the case for the detection of halogenated aromatic compounds in many matrices of biological or environmental origin. Since the detection limits from such samples are limited by chemical rather than electronic noise, the selectivity of NCI offers very real improvements in sensitivity.

A variant of chemical ionization that merits discussion is atmospheric pressure ionization (API). In the simplest form of API, the effluent from a ^{63}Ni electron-capture detector is introduced into a very efficiently pumped mass spectrometer [29]. Refinements included the replacement of ^{63}Ni with a corona discharge as the source of energy. This ion source offers very high (and equivalent) sensitivities for either positive or negative ions. Since API is not fundamentally different from CI, it is not discussed in detail. It should be pointed out that the high and uniform sensitivity is one of the major limitations since no particular improvement in signal/chemical noise is offered by API.

3. Production of Ions from Low-Volatility Samples

All methods of sample ionization presented thus far required that the sample first be volatilized. Without question, great strides have been made from mass spectral data collected with this rather severe restriction. However, when one considers that 80–90% of the organics in water can be classified as nonvolatile, the importance of research into methods for the analysis of such materials becomes obvious.

There are two major approaches. The first is the chemical modification of the sample into a more volatile form amenable to mass spectral analysis. A thorough discussion of such derivatization reactions is best left to chapters dealing with chromatography. However, if mass spectrometry is to be used in conjunction with, for example, gas chromatography, the influence of the derivatization chemistry on the information content of the mass spectrum must be considered. Normally, derivatives are formed to replace active hydrogens or to reduce the effects of nonbonding electrons on volatility. A natural consequence of this is that the derivative is normally formed on or near a heteroatom. This frequently results in one of the more favored reactions being the loss of the added group. In so far as this fragmentation reaction reduces relative ion current from other, more informative reactions, it results in decreased knowledge about the sample. Therefore, one should choose reactions that lead to the most stable bonds being formed. Another consideration is the charge distribution in the fragments. The charge is preferentially retained by the fragment whose radical has the lower ionization energy. Clearly, it is more informative to see an ion resulting from the loss of the added group than to see that group as an ion in the spectrum. Thus, methyl esters are preferable to trimethylsilyl esters since the ester bond is stronger in the case of the methyl derivative and a trimethylsilyl group is more adept at retaining the charge.

The other approach is to develop new ionization techniques to permit the analysis of nonvolatile compounds than cannot be handled by direct probe or GC/MS techniques. This is the subject of the following discussion.

a. Field Desorption

In order for a sample to be analyzable by GC/MS, it must be suitable for analysis by GC as discussed in the chapter on GC. Analysis by direct-probe introduction requires that the sample have a vapor pressure of $\sim 10^{-7}$ torr at a temperature below which thermal decomposition is significant. Although this requirement is generally less stringent than that needed for GC analysis, only a small fraction of the organic compounds in the biosphere meet it. If one could in some way interfere with the intermo-lecular-bonding forces of such materials, then perhaps volatility could be enhanced. One possible means of achieving this is to perform the evapo-ration in the presence of high electrical fields. (The authors are aware of the raging controversies concerning the mechanisms of FD. This state-ment is not an attempt to become involved.) This method of sample volatilization–ionization is known as field desorption (FD). The use of FD for the ionization of organic materials began with the pioneering work of Beckey, Schulten and co-workers [30], and the subject has been reviewed thoroughly [31]. However, for the convenience of the present readers, FD is discussed briefly here.

The first step in an FD analysis is the preparation of the device on which the sample is introduced into the ion source, referred to as the emitter. In its usual form, the emitter consists of a 10-μm tungsten wire stretched between two supports, which has had its surface activated by the deposition of dendritic materials. These dendrites are most typically carbon [32], although silicon [33] and various metals [34,35] have been employed. After the sample has been coated on the surface, the emitter is placed in the FD source, which is essentially an EI source with the fila-ment removed and a counter electrode added. A voltage (typically 8–12 kV) is then applied between the emitter and counter electrode. Although the mechanism of FD is a subject of current debate [31,36], the authors have found the following simple model useful. The field gradient at an emitter approaches 10^8 V/cm. That is, a typical dipolar organic molecule aligned with the field experiences several volts potential difference along its length. Depending on the magnitude of the molecule's dipole moment, this could be equivalent to 100 kcal/mol or more of lattice energy without affecting the intramolecular vibrational energy (i.e., internal energy that might lead to pyrolysis). If this effect itself is insufficient to lead to sample volatilization, additional energy can be added by passing a small current (0–100 mA) through the emitter wire. Ionization can be thought of as occurring by field ionization as the sample leaves the emitter surface.

The primary advantage of FD is that a substantial number of polar compounds with limited thermal stability can be successfully analyzed by mass spectrometry. The major limitations are that the sample must be reasonably pure to reduce confusing condensed-phase reactions, the sen-sitivity is poor by MS standards with $\sim 10^{-7}$ g usually required for analy-sis, and little useful information other than molecular weight is readily available. These shortcomings not withstanding, FD has made valuable contributions in the biological sciences [37], and it may prove useful for many problems in water chemistry.

b. Surface Ionization

A different approach to the volatilization of low vapor-pressure materials involves the addition of energy in rapid, short bursts to localized areas of a surface coated with the sample. The basic idea is that adding energy suddenly to a small fraction of sample surface causes sample evaporation and ionization at a rate at least comparable with the rate of pyrolytic decomposition. These processes are termed desorption or sputtering. There have been a number of different sources of energy used to achieve this excitation.

The most logical starting place is use of a laser. This energy source is available in a wide range of power outputs, a selection of energies, is easily pulsed, and the position of excitation can be controlled with high precision. A variety of samples that were not amenable to classical direct-probe introduction have been successfully analyzed using the laser-desorption technique [38]. The major limitation of laser desorption probably results from the energy of the excitation source. The widely used 10.6-μm line from a CO_2 laser is in fact 948 cm^{-1}, which is of the proper energy to lead to intramolecular vibrational excitation of the sample. A method is needed that selectively excites the intermolecular bonds.

A method of achieving this might be to adapt the principles of secondary ion mass spectrometry (SIMS), which has enjoyed popularity for the determination of inorganic impurities on metal and semiconductor surfaces [39]. Grade and co-workers [40] investigated the conditions necessary for the sputtering of organics from metal surfaces. The SIMS technique actually involves two ion sources (Fig. 3). A primary ion beam, typically of Ar^+, is formed, accelerated, and focused to a point on the target, on which the sample has been coated. The kinetic energy of the Ar^+ beam is converted into excitation energy of the surface material. This leads to a number of processes, including volatilization and ionization of the sample, perhaps by condensed-phase processes analogous to chemical ionization. A recent review [41] covers the present state of this art. Briefly, SIMS frequently provides ions indicative of molecular weight,

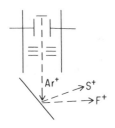

Fig. 3. Schematic diagram of a SIMS ion source. A primary-ion beam is generated, typically in a gas discharge, then directed to a surface where impact with the sample results in the formation of sputtered secondary ions. These may then be subjected to mass analysis to obtain the secondary-ion mass spectrum.

usually involving cationization with the supporting metal. A variety of surfaces can be used for sample introduction, including filter paper, which points to the possibility of conveniently interfacing a number of classical chromatographic techniques to mass spectrometry (see Section V). Molecular SIMS appears to be a very promising technique for obtaining mass spectra from compounds of a low to moderate molecular weight. The accessible molecular weight range is limited to ~1000 because quadrupole mass analyzers are typically used in present SIMS instruments. The commercial availability of SIMS with a high-performance analyzer should lead to an evaluation of this technique for higher molecular weight species.

At approximately the same time as the early investigations of molecular SIMS, another technique for adding energy to localized areas of a sample surface was being investigated. This involved the use of energetic fission products from the radioactive decay of ^{252}Cf [42]. Conceptually, the plasma desorption mass spectrometry (PDMS) experiment is quite simple. The sample is coated on a nickel foil and placed in the instrument in alignment with the ^{252}Cf source. Fission fragments pass through the sample foil producing desorption and ionization. The resulting ions are accelerated from the source and mass analyzed, usually in a time-of-flight system. MacFarlane and co-workers are actively pursuing the application of PDMS to a wide variety of previously intractable materials. Striking successes have been observed, including molecular weight measurement of species in the 10,000–12,000 range [43].

Reflection on the results obtained from laser desorption and PDMS leads to the postulation that the major function of the Ar^+ ions in the SIMS experiment is as an energy source. Precautions must be taken to dissipate the positive charge, which otherwise builds up on the target surface [41]. In addition, the interaction between a charged particle and a neutral species varies $1/r^4$ as opposed to $1/r^2$ for neutral–neutral interactions, where r is the separation. Thus, should SIMS result in initial ionization followed by volatilization of the charged species, this stronger interaction reduces effective volatility compared to a technique whose mode of operation involves either simultaneous vaporization–ionization or vaporization followed by ionization. Comparison of PDMS and SIMS results, although these are very limited, supports the idea that the charge associated with the Ar^+ beam is detrimental to the performance of the SIMS source. Barber et al. [44] developed a modification to the SIMS technique in which the Ar^+ beam is neutralized prior to impact on the target. This procedure has been termed "fast atom bombardment" (FAB). The technique has not been completely evaluated, but results (mostly on biomolecules) indicate that intense, stable ion beams can be obtained for samples that either could not be analyzed previously or that gave erratic ion beams when techniques such as FD were applied [45]. The position that FAB will hold in analytical mass spectrometry as applied to organics in water remains to be determined.

III. MASS ANALYSIS

The ideal mass analyzer would separate, without discrimination, all ions of a given mass from all ions of a different mass regardless of such initial conditions as angular dispersion, kinetic-energy spread within the ion beam, or mass, and display the entire mass spectrum simultaneously. Naturally, no single type of analyzer has achieved this ideal. However, different analyzers can perform very well in one or more aspects. The problem for the mass spectrometrist is to make the appropriate trade-offs so that the instrument purchased best meets the requirements of a particular laboratory. The purpose of the present section is to introduce the physical principles of the mass analyzers used in analytical mass spectrometry and compare their properties. Readers desiring a more in-depth treatment are referred to appropriate sources.

A. Quadrupole Mass Filters

The concurrent development of quadrupole mass filters, reasonably priced small computers, and adequate enrichment devices for GC/MS are the main technical factors contributing to the explosive growth of mass spectrometry in the 1970s. The quadrupole offers many features useful for computerized GC/MS. The systems are relatively inexpensive to build, easily interfaced to a computer and can be scanned rapidly with adequate precision for mass assignment to 0.1 dalton. The effects of their more prominent weaknesses of low resolving power, poor transmission at high mass-to-charge ratio (m/z) and susceptibility to contamination are somewhat mitigated by the gas chromatograph. Gas chromatography reduces the probability that two or more components are in the source simultaneously, generally restricts the molecular weight that is amenable to analysis, and reduces the amount of material necessary for an analysis. The reader interested in an in-depth treatment of the theory of quadrupole mass spectrometers is referred to the book by Dawson [46]. The following more general discussion of operating principles is a synopsis of a presentation by M. Siegel [47].

As one would deduce from its name, the quadrupole mass filter consists of four rods (Fig. 4) that are carefully machined to be smooth and

Fig. 4.   Schematic diagram of a quadrupole mass filter. The rods are of a diameter a_0 and are located a distance r_0 from the center of the filter.

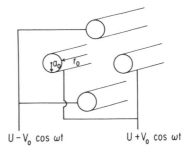

$$U - V_0 \cos \omega t \qquad\qquad U + V_0 \cos \omega t$$

aligned to be parallel. The ideal rod configuration is hyperbolic. However, such rods are more difficult to machine than round ones and in practical terms essentially equivalent performances can be attained with either form. Naturally, it is desirable to construct an instrument with round rods in a manner such that hyperbolic fields are approximated. The relationship between rod diameter, a_0, and the distance the rods are located from the center of the filter, r_0, to achieve this is $a_0 = 1.145\, r_0$.

Ions within an rf (radio frequency) quadrupole field follow complex trajectories. Depending on whether the amplitude of a particular ion's motion exceeds the dimensions of the filter, it can be classified as having either a "stable" or "unstable" trajectory. Within broad limits, the only relevant parameter determining whether a particular orbit is stable or not is m/z. That is, the quadrupole is relatively insensitive to factors such as kinetic-energy spread of the ion beam, angle of incidence, position of incidence, etc. Restrictions arise from the practical limitation that the quadrupole field must be finite. The mathematical concepts of stability and unstability are meaningful only in infinite quadrupole fields. In a practical instrument (with finite dimensions) some stable trajectories intercept the poles and some unstable trajectories pass through the exit aperture. This limitation manifests itself by a failure of the device to transmit ions whose incidence angle exceeds some critical value and/or the energy is below some critical value. Ions with large kinetic energies tend to be transmitted even though their trajectory was unstable. A detailed derivation of the equations governing a quadrupole mass filter is beyond the scope of this chapter. However, an examination of selected results serves to illustrate how various parameters interrelate to determine instrument performance. For a filter of length L and rf frequency f operated at resolution (i.e., peak width) Δm, the maximum component of kinetic energy along the long axis of the filter is

$$E_x < (1/25) f^2 L^2 \Delta m \qquad\qquad (24)$$

For the case of $f = 1$ MHz, $L = 25$ cm, and $\Delta m = 0.5$ daltons, the maximum ion kinetic energy is ~ 13 eV. Equation (24) states that other things being equal, the longer the filter and higher the operating frequency, the larger is the acceptable kinetic-energy spread on the ion beam. If this same quadrupole is constructed so that the rods are separated by $2\, r_0$, the maximum kinetic energy in the yz plane E is given by

$$E_{yz} < (1/2)f^2 r_0^2 \Delta m \tag{25}$$

If $r_0 = 0.5$ cm and other values are as before, then $E_{yz} < 0.06$ eV. The other interesting parameter is the maximum departure from an "on-axis" entry the filter tolerates. This is given by

$$r_a \cong (2/3)r_0(\Delta m/m)^{1/2} \tag{26}$$

At m/z 500, $r_a \cong 0.01$ cm. It should be noted that E_{xy} and r_a are interrelated in that E_{xy} decreases as $r \to r_a$ [i.e., Eq. (25) is valid only for ions that enter the quadrupole on axis]. In summary, the ability of a quadrupole to successfully analyze nonideal ion beams increases with rod length and diameter.

As stated, a rigorous treatment of ion motion within a quadrupole mass filter is beyond the scope of this chapter. However, the following heuristic treatment (again, from Siegel) was particularly helpful to the author in visualizing how such instruments function and is passed along in hopes the reader also finds it beneficial.

The quadrupole rods are "strong-focusing" devices. Referring to Fig. 5, we see that positive ions are subjected to a focusing force in the plane

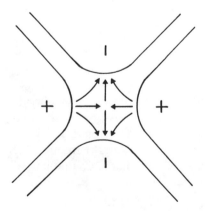

Fig. 5. Principle of strong focusing. The result of applying hyperbolic DC voltages is the focusing of ions in the plane of the axis with the + electrodes and defocusing of ions in the plane of the axis with the − electrodes.

of the positive rods and a defocusing force in the plane of the negative rods. It is clear that focusing cannot occur in both planes simultaneously since Gauss's law would require a charge (and hence an electrode) on the axis. This device is of little practical use since all positive ions would soon be lost to the negative rods. However, by placing a number of these devices in series with the relative position of the positive and negative rods interchanged between each device, the ions in a given plane are subjected alternately to focusing and defocusing forces (Fig. 6). The same effect can be accomplished in a mechanically simpler manner by alternating the voltage on a single set of rods so that spatial is replaced with temporal periodicity. To make this into a mass spectrometer, it is necessary only to add a positive and negative DC voltage to the appropriate poles (Fig. 4). For a DC voltage of U and rf amplitude of V at frequency ω

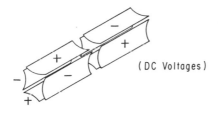

(DC Voltages)

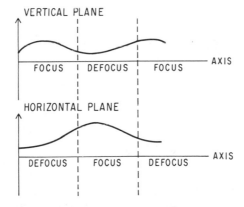

Fig. 6. Periodic strong focusing. In this case a series of quadrupoles are arranged with alternating positive and negative poles. An ion transversing the region is subjected to similarly alternating focusing and defocusing conditions as illustrated. If these alternating fields are repeated, the result is analogous to the actual situation when an alternating electric field is applied to a single set of quadrupole rods.

and time t within the filter, the ion beam is subjected to a quadrupole field described by

$$\phi(x,y) = \frac{\phi_0}{r_0^2}(x^2 - y^2) \tag{27}$$

where

$$\phi = U + V \cos \omega t \tag{28}$$

The z direction is along the axis of the filter, x and y directions are defined by the right-hand rule, and the centers of the respective rods are intercepted by the x and y axes. The distance from the z axis to the rod is r_0. By applying Newton's first law ($F = ma$),

$$m\frac{d^2x}{dt^2} = -\frac{2e}{r_0^2}(U + V \cos \omega t)x \tag{29}$$

and

$$m\frac{d^2y}{dt^2} = \frac{2e}{r_0^2}(U + V \cos \omega t)y \tag{30}$$

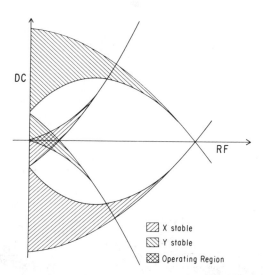

Fig. 7. Diagram showing regions of X and Y stability according to Mathieu functions. The areas of overlap correspond to regions of stable trajectories.

where e is the electronic charge. The solutions of these differential equations are known as Mathieu functions and have a complex form [46].

It is useful to make plots of U versus V from these equations. In Fig. 7 we see that for particular combinations of U and V, ions are stable in either the x or y direction. For those combinations in which the ion beam is stable in both directions, ions are transmitted through the filter. Note that the detailed shape of these curves depends on physical parameters of the filter. The vast majority of quadrupoles operate in the region nearest the origin in Fig. 7. An expanded view of this region (Fig. 8) drawn for two masses, m_1 and m_2, illustrates how a mass spectrum is obtained. As can be seen from the figure, the resolution (i.e., bandwidth) of the filter is determined by the ratio U/V, and the m/z transmitted is a function of V. Thus, by scanning the rf level while maintaining a constant ratio of rf to DC levels, a mass spectrum is obtained. Scan line (a) represents a higher resolution condition than scan line (b).

In this section the reader has been introduced to the fundamentals of quadrupole mass spectrometer operations. As should be apparent, the principles governing the operation of these "simple" devices are complex. This fact notwithstanding, quadrupoles are by far the most widely used type of mass spectrometer for organic analysis. This can be attributed to the ease with which they can be interfaced to chromatographs (gas or high performance liquid) and to computers, and also their relatively low construction costs. Offsetting these factors are susceptibility to contamination, ease of misuse, low resolution and poor high-mass transmission.

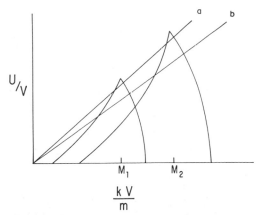

Fig. 8. More detailed view of the lower stability region in which most quadrupoles are operated. A mass scan is accomplished by scanning V and U so as to maintain a constant ratio. The upper scan line (a) corresponds to a higher resolution than the lower one (b).

B. Magnetic Sectors

Although instruments based on magnetic sectors are reputed to be more difficult to operate and maintain than quadrupoles, the authors find them easier to understand. Consider a beam of ions (of mass m) emanating from a point source after having been accelerated by voltage V. Then, for singly charged ions

$$eV = (1/2)mv^2 \tag{31}$$

where v is the velocity of the ions. If this ion beam is injected into a magnetic field of magnitude B arranged such that it is perpendicular to the original direction of ion motion, the ions experience a force perpendicular to both the direction of motion and B, whose magnitude is given by

$$F = Bev \tag{32}$$

From Newton's laws, this force is balanced by a centripetal force in the opposite direction whose magnitude is

$$F = m \left(\frac{v^2}{r} \right) \tag{33}$$

where r is the field radius. These may be set equal

$$m \left(\frac{v^2}{r} \right) = Bev \tag{34}$$

Recalling that $v = \left(\frac{2eV}{m} \right)^{1/2}$ from Eq. (31) and solving for m/e, we have the basic equation for a magnetic-sector mass spectrometer

$$m/e = \frac{B^2 r^2}{2V} \tag{35}$$

The mass spectrum can conveniently be scanned by varying either V or B.

Voltage scans offer the advantage of simplicity and a linear mass scale. Unfortunately, variations in acceleration voltage change the extraction efficiency of the ion source with concomitant changes in sensitivity. Therefore, the voltage is normally varied only when highly reproducible steps between points in the mass spectrum are required, as in selected ion monitoring. Thus in normal operations, the mass spectrum is scanned by varying the magnetic field. This approach gives uniform sensitivity throughout the mass range but requires a more complex procedure to calibrate the mass scale than for other types of mass analyzers.

In the absence of aberrations of the image, the resolution of a magnetic-sector mass spectrometer is determined by the radius. This is illustrated by rewriting Eq. (35) as

$$\frac{\Delta m}{m} = \frac{2\Delta B}{B} + \frac{2\Delta r}{r} - \frac{\Delta V}{V} \qquad (36)$$

By assuming ΔB and ΔV are zero,

$$\frac{\Delta m}{m} = \frac{2\Delta r}{r} \qquad (37)$$

Our ability to distinguish between two radii is a constant, so the resolving power $m/\Delta m$ depends linearly on r. Furthermore, it can be shown that

$$\frac{m}{\Delta m} = \frac{r}{S_1 + S_2} \qquad (38)$$

where S_1 and S_2 are the widths of the source and collector slits, respectively. Assuming that the ion source illuminates the slit uniformly, the transmission (i.e., sensitivity) of the sector depends on slit area and hence S_1 and S_2. Thus, at a given resolving power, sensitivity increases with r.

C.   Double-Focusing Mass Analyzers

Although not explicitly stated, Eq. (35) was derived on the assumption of a monoenergetic ion beam. This, of course, is never the case. The effects of energy spread can be examined by reconsideration of Eq. (36). Variations in kinetic energy can be treated as though they arise from variations in the acceleration voltage. Assuming ΔB and Δr to be zero,

$$\frac{\Delta m}{m} = \frac{\Delta V}{V} \qquad (39)$$

Thus, we see that with a constant energy spread ΔV, resolving power increases with V. However, mass depends inversely on V and problems from electrical discharges in the source increase with V. These considerations limit the voltage of commercial instruments to 3–10 kV. An alternative approach is to reduce ΔV. In early instruments, this was accomplished by placing an energy-dispersive device between the ion source and magnetic analyzer. The resulting loss of sensitivity made this unacceptable. Mass spectrometers with both high resolution and high sensitivity became possible when the direction focusing properties of radial electric fields were noted [48,49]. Ions moving in such a field experience a force perpendicular to the field and direction of motion, whose magnitude is eE. Balancing this is the centripetal force, as before

$$eE = \frac{mv^2}{r_e} \tag{40}$$

Recalling Eq. (31), we find that

$$r_e = \frac{2v}{eV} \tag{41}$$

Thus, the trajectory through the electric sector is independent of m/z and depends only on V and E. Further,

$$\frac{\Delta r_e}{r_e} = \frac{\Delta V}{V} - \frac{\Delta E}{E} \tag{42}$$

Assuming ΔE to be zero, ions that differ in kinetic energy by ΔV from the average follow a path that differs by Δr and can be resolved. Furthermore, it can be shown that the electric sector acts as a lens [50,51]. The optical equations are

$$l_e = \frac{r_e}{1.414 \sin 1.414\theta_e} \tag{43}$$

$$l_0 = \frac{(r_e^2/1.414 l_e) \sin 1.414\theta_e + r_e \cos 1.414\theta_e}{1.414 \sin 1.414\theta_e - (r_e/l_e) \cos 1.414\theta_e} \tag{44}$$

where l_e and l_0 are the image and object distances respectively, and θ_e is the sector angle. Note that one is free to chose two parameters from the set image length, object length and sector angle; the third is thus determined. Thus, direction focusing is possible in both magnetic and electric sectors; however, both are dispersive in energy. Mattauch and Herzog [48,49] derived the general equations for achieving first-order double focusing by making the energy dispersion of the electric sector equal in magnitude but of opposite direction to that of the magnetic sector. Thus, an ion beam of a given m/e, angular divergence, and energy spread from a

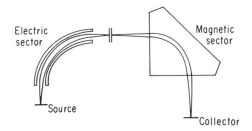

Fig. 9.   Schematic diagram of a double-focusing mass spectrometer.

source slit is brought to a focus at the collector slit. A schematic diagram
of a double-focusing mass spectrometer is shown in Fig. 9.

D. Time of Flight

In the time-of-flight (TOF) mass spectrometer, a pulse of ions is accelerated through a source-voltage difference V so that their kinetic energy
as given by Eq. (31) is $\frac{1}{2}(mv^2)$ and is constant. Thus, ions that have different m/z have different velocities. If they are allowed to drift a distance l,
then the flight time t is given by

$$t = l/v \tag{45}$$

$$t = \frac{l}{(2eV/m)^{1/2}} \tag{46}$$

$$t = \frac{(ml^2)^{1/2}}{2eV} \tag{47}$$

From this

$$\frac{m}{e} = \frac{2Vt^2}{l^2} \tag{48}$$

From Eq. (48), there exists a square relationship between m/z and t. As
a consequence of this, ions separated by 1 dalton arrive with less time
separation as m/z increases. For $l = 1$ m and $V = 2000$ V, the mean time
between successive pulses is shown below:

m_2	m_1	$t_2 - t_1$ (ns)
100	101	85
400	401	42
800	801	30
10,000	10,001	8.5

Thus at m/z 800, the ion pulse width must be short with respect to 30 ns in order to achieve unit resolution. Factors that contribute to pulse broadening include initial thermal energy, rise time of power supplies and amplifiers, and the position distribution within the source for ion formation. At present, resolutions of 500–600 can be easily obtained. TOF instruments share the advantage with sector instruments that resolution is independent of m/z. Thus TOF instruments are capable of separating low m/z doublets that are not resolved with "unit mass resolution" analysis, that is, quadrupoles. In addition, TOF analyzers offer the advantage of an unlimited mass range (at least in principle). Although the resolution is poor, Macfarlane and co-workers [43] reported ion currents in the m/z 10,000–12,000 range, pointing to new horizons for mass spectrometry.

IV. ION DETECTION

Almost every mass spectrometer involved in organic analysis is equipped with an electron multiplier as the ion-detection device. These come in two varieties, termed *venetian blind* (Fig. 10A) and *continuous dynode* (Fig. 10B). In both cases, the incident ion beam causes secondary electrons to be ejected from a surface. These electrons are accelerated, strike another surface causing further amplification, etc. Most electron

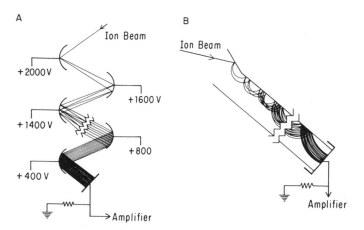

Fig. 10. (A) Schematic diagram of an electron multiplier. The incident ion beam generates secondary electrons, which are then accelerated by the voltage drop to the next stage where each electron generates a number of additional secondary electrons, and so forth. (B) Continuous-dynode electron multiplier. The incident ion beam leads to the formation of secondary electrons, which spiral down the tube generating additional secondary electrons with each collision with the surface.

multipliers have a maximum gain in excess of 10^6 when new (i.e., 10^6 electrons result per incident ion). With use, this is rapidly reduced to ~3×10^5 and remains at that value for several months to several years, depending on environmental conditions and the care with which the device is used. In the authors' laboratory, multipliers are routinely operated at a gain of approximately 40,000 to prolong operational lifetime. This is adequate to give a signal from a single ion arriving at the detector. Higher gain not only reduces lifetime but also increases noise and decreases linear dynamic range.

V.   CHROMATOGRAPHY/MASS SPECTROMETRY

A.   Gas Chromatography/Mass Spectrometry

As previously mentioned, the GC/MS/DS is one of the most powerful instrumental techniques available to the chemist for the qualitative and quantitative analysis of complex organic mixtures. The two instruments (GC and MS) are a good match with respect to sample size; however, the pressure requirements for packed-column GC are quite different from those of the mass spectrometer (1 atm versus 10^{-9} atm). Suitable interfaces are thus required to separate the bulk of the carrier gas from the column eluent. In capillary GC/MS, it is possible to couple the column directly to the ion source since carrier-gas flow rates are much lower than for packed columns (1 mL/min versus ~25 mL/min). In either case, pumping systems capable of maintaining proper source pressures with some carrier-gas flow must be included. A number of carrier-gas separators for packed-column gas chromatographs have been developed. These include the membrane, Watson–Bieman effusion, hydrogen–palladium, and jet separators. For a discussion of the design and operating principles of these, see McFadden [52].

B.   Liquid Chromatography/Mass Spectrometry

Although the gas chromatograph has historically been the most widely used mass spectrometer inlet system for water analysis, it is also possible to interface various liquid-chromatography systems [53]. These are potentially most useful when one is attempting to analyze compounds that are nonvolatile and require ionization techniques unsuitable for interfacing to a gas chromatograph, or are unsuitable for GC separation. An example of the former is provided by Day *et al.* [41] who described the use of paper chromatography as a separation method for organic SIMS. A paper chromatogram containing organics separated into spots was directly inserted into a SIMS source and the primary ion beam scanned along the axis of the chromatogram. Individual spots could then be identified by their secondary ion spectra. It should also be possible to utilize this same method with other ionization techniques such as fast atom bombardment. The ability to analyze nonvolatile organic constituents in water with this approach is attractive but has yet to be evaluated.

The analysis of unstable and other intractable organic compounds has been enhanced by the combination of the high performance liquid chromatograph (HPLC) with the mass spectrometer. In its simplest form, the chromatogram is developed off line and peaks collected in test tubes. Samples from these tubes can then be introduced into the mass spectrometer by a batch-inlet system. When the HPLC is directly coupled to the mass spectrometer, the technique is usually referred to as liquid chromatography/mass spectrometry (LC/MS). As with GC/MS, the ability to perform LC/MS rests on maintaining proper source pressure and conditions in the mass spectrometer by removal of the majority of mobile phase, in this case solvent. A number of interfaces have been developed but only three are widely commerically available.

One of these is depicted in Fig. 11 [54]. This device is usually referred to as the "moving belt" interface. Eluent from the LC column is first coated onto a rotating stainless steel or a polyimide belt, which then passes through various pumping chambers where the solvent is removed. The solute, which is presumably still on the belt, is then passed near a heater where it is flash-volatilized into the ion source. This system has

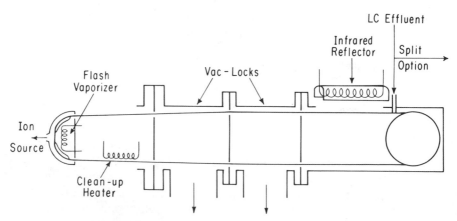

Fig. 11. Schematic diagram of a moving-belt interface for combined high-performance liquid chromatography/mass spectrometry.

several obvious limitations. First, the sample material must be sufficiently nonvolatile to remain on the belt when the solvent is pumped away. Second, it must be sufficiently volatile to be flash-vaporized into the ion source without thermal decomposition. Finally, the solvent must "wet" the belt. This places restrictions on solvent systems used and types of samples that can be analyzed.

Another device that somewhat circumvents these limitations is the "direct liquid introduction" interface [55]. In this system, a portion of the column eluent is split off (~1%) and injected directly into the ion source, solvent and all. The effect is analogous to a split injection in capillary gas chromatography. The solvent is utilized as a reagent gas for chemical ionization in the source, producing both positive and negative ions. The obvious limitation is that the majority of the sample is discarded (~99%), making sensitivity a problem in many cases. The choice of a solvent system in LC is usually based on optimum separation efficiency; however, in this case consideration must also be given to optimum ionization conditions. Insight into various processes that occur in the source that give rise to analyte ions, the degree of fragmentation observed for the analyte, and the identity of the many solvent cluster ions observed aid in optimization of this technique. These parameters have been investigated by Voyksner et al. [56,57] utilizing a combined LC/MS/MS instrument.

A recent addition to the family of interfaces for HPLC/MS involves an apparently new ionization process. The so-called thermospray interface [57a,57b,57c] involves the direct introduction of 1 to 2 mL/min liquid flow from the chromatograph into the vacuum system of the mass spectrometer under conditions of carefully controlled heating. Ionization is thought to occur as a result of slight charge imbalance in a small fraction of the droplets formed as the solvent enters the vacuum system. Although an extremely limited amount of information is available, this technique has been demonstrated to work much better with aqueous HPLC solvent systems than either of the other types of interface, and it appears to offer an advantage for highly polar or ionic analytes.

It is possible to interface microcolumn HPLC, which utilizes lower solvent flow than conventional LC columns, with a mass spectrometer. This is easier on the moving belt interface since there is less solvent involved. Also, splitting is unnecessary, so all the column eluent can be introduced into the ion source with the direct liquid-introduction interface. These apparent advantages are counterbalanced by the reduced sample-size range available on the microcolumn and the increased experimental complexity. The utility of microcolumn LC/MS is yet to be fully evaluated.

LC/MS coupling is an area of continuing development, with new interfaces certain to overcome the limitations presently encountered. Progress in LC/MS holds great promise for water analysis and should be followed closely by those in the field.

VI. TANDEM MASS SPECTROMETRY

Many environmental analyses involve the detection of single specific components (for example, 2,3,7,8-tetrachlorodibenzo-*p*-dioxin) in complex matrices. Such analyses can involve complicated and time-consuming extraction, cleanup, and derivatization procedures. It has therefore been desirable to investigate alternative mass spectral methods that offer increased specificity while minimizing the need for sample preparation. Such methods utilize batch-sample inlet systems to the mass spectrometer, such as a solids probe, and do not require preseparation. These techniques are collectively referred to as mass spectrometry/mass spectrometry (MS/MS) methods or, perhaps more appropriately, tandem mass spectrometry methods.

These terms, which imply separation and analysis within the mass spectrometer, have been used to refer to any instrumental method capable of providing information as to an unknown compound's identity or quantity through the use of metastable or collisionally induced daughter ions. A metastable ion is an ion sufficiently vibrationally excited to dissociate into a particular daughter ion and neutral species during the flight from the ion source to the detector. In sector mass spectrometers, this dissociation is most readily observed in one of the field-free regions of the instrument. Metastable ion transitions have long been used to aid in organic structural analysis [58]. Other stable ions can be caused to fragment in a field-free region by the addition of vibrational energy through a process known as collision-induced dissociation or collisional activation. In this process, ions from the source are allowed to collide with molecules of a neutral gas such as helium in defined regions known as collision cells and gain vibrational energy with accompanying increase in electronic excitation. This energy increase produces increased fragmentation and thus more daughter ions, termed collisionally induced daughter ions.

Connecting two mass spectrometers in series was first attempted for the study of mass-selected ion–molecule reactions [for a review of early literature, see ref. 59]. The analytical uses of tandem mass spectrometry can be traced to the development of what is termed mass-analyzed ion kinetic-energy (MIKE) spectrometry [60] or direct analysis of daughter ions (DADI) [61,62]. Such experiments employ a double-focusing mass spectrometer with a geometry such that the ion source is followed by a magnetic and then electric sector and finally the detector (Fig. 12). It is

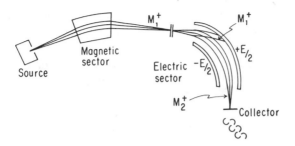

Fig. 12. Schematic diagram of a typical double-focusing mass spectrometer designed for mass-analyzed ion kinetic-energy measurements. The magnetic sector is adjusted to transmit the particular primary ion of interest (M_1^+, in this case). The electric sector can either be scanned to collect a complete daughter-ion spectrum or set at a preselected value so that the mass spectrometer is used in a selected-reaction monitoring mode.

convenient to have a detector at the intermediate focal point (second field-free region) so that the ion source–magnetic sector combination can be used as a single-focusing mass spectrometer. Thus, an ion beam of a particular m/z can be selected. If this ion has an m/z of m_1 and a kinetic energy of eV, then E_k, as expressed in Eq. (31), is

$$E_k = eV = (1/2)m_1v^2$$

For those ions that fragment after passage through the magnetic field to give ions of $m/z = m_2$, then

$$E_k^1 = \tfrac{1}{2}m_2v^2 \tag{49}$$

and

$$(m_1/m_2)E_k^1 = (m_1/m_2)(\tfrac{1}{2}m_2v^2) = E_k \tag{50}$$

Thus

$$m_1/m_2 = E_k/E_k^1 \tag{51}$$

It was noted that electric sectors such as those utilized in double-focusing mass spectrometers are effective kinetic-energy analyzers. By measuring the ion intensity as a function of applied electric-sector voltage, it is possible to measure the kinetic-energy spectrum of daughter ions produced from a mass-analyzed ion beam. Since the kinetic energy of the daughter ions is related to their m/z, we have a means of measuring what is effectively a mass spectrum of a given ion. As pointed out in the discussion of the QET in Section II,B, a rate constant for a given reaction is exponentially proportional to the internal energy in excess of the threshold for the reaction that is available to the ion. The converse of this also is true: The ions that fragment after passing through the magnet (i.e., metastable ions) are the ones that react slowly and hence generally have little energy above the minimum. Equation (1) shows that these slow fragmentation reactions may also be the result of a low-frequency factor; hence we can expect to see an increase in the important rearrangement reactions relative to simple cleavages. In summary, the QET predicts MIKE spectra to have a lower intensity than normal EI spectra and for rearrangement processes to dominate. This is, in general, the case.

One factor not directly apparent from the QET is the broadness of the peaks in a MIKE spectrum (Fig. 13). Consider an ion in transit between the sectors of a double-focusing instrument. For simplicity, let us restrict our attention to those m_1 ions whose velocity vectors are colinear with the path of the central ray and who happen to fragment at the focal point of the electric sector. As the parent ion breaks apart, the resulting product

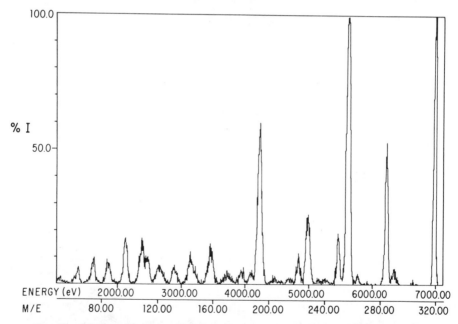

Fig. 13. Mass-analyzed ion kinetic-energy spectrum of 2,3,7,8-tetrachlorodibenzo-*p*-dioxin. The *m/z* 320 ion was selected by the magnetic sector, collisionally excited, and the daughter ions separated by means of an electric-sector scan. Prominent fragment ions correspond to those observed under electron-impact conditions.

ion and neutral molecule (neutral) impart a small velocity to one another as they partition the excess energy in the bond. If the ion–molecule (or ion–radical) reaction corresponding to the reverse of the fragmentation has an activation barrier, this "reverse activation energy" also contributes a component to the velocity. Consider those molecules whose fragmenting bond is colinear with the velocity of the center of mass. When these ions fragment, velocity is either added to or subtracted from the center-of-mass velocity depending on whether the ionic product is pointing in the positive or negative x direction. Obviously, all possible orientations of the fragmenting ions are possible and equally probable so that a Gaussian distribution of product-ion velocities is added to the center-of-mass velocity for a kinetic-energy release T. Since, in general, a given reaction involves a distribution of kinetic-energy release, a detailed analysis of the peak is required if one is interested in fundamental measurements. The interested reader is referred to an extensive treatment of this subject by Cooks *et al.* [58]. For our purposes, it is sufficient to realize that T is a characteristic of a given reaction and that by basing our T measurements on full width at half maximum, we can decide whether similar ions fragment via the same or different mechanisms. For example, the measurement of T for the $M^+ \rightarrow [M - Br]^+$ reaction permits one to distinguish bromomethylbenzene from bromotoluene [63].

The use of tandem mass spectrometry naturally increases the specificity in the detection step of an analysis when compared to low-resolution mass spectrometry. At the present, insufficient data are available to decide whether tandem mass spectrometry can compete with high-resolution techniques although the preliminary data indicate this may be the case, with the method of choice depending on the particulars of the problem. Naturally, the availability of high-resolution mass spectrometry with a magnetic–electric sector ($B–E$) configuration permits one to conveniently perform both measurements on a given sample.

An alternative means of studying the same ionic reactions is to apply a scan function that scans some combination of fields (i.e., accelerating voltage, electric sector voltage or magnetic field strength) simultaneously [64]. For example, if one scans the electric sector of a mass spectrometer with electric–magnetic sector ($E–B$) configuration down from that necessary to transmit the primary ion beam, of, for example, $m/z = m_1$, daughter ions of $m/z = m_2$ resulting from metastable fragmentation of the m_1 ions in the field-free region between the source and electric sector are transmitted when $E_2 = (m_2/m_1)E_1$. If the magnetic field is originally adjusted such that m_1 ions are transmitted, no m_2-ion current passes through the magnet. If the magnetic field is adjusted so as to satisfy the condition for the transmission of metastable m_2 daughter ions (i.e., $m^* = m_2^2/m_1$), the m_2 ions being transmitted through the electric sector are also transmitted by the magnet and hence detected. Molecular ions and fragment ions formed by reaction in the ion source are not transmitted by the electric sector under these conditions, because they do not have the required kinetic energy. A full spectrum of the daughter ions formed from a given parent ion by fragmentation in this field-free region can be conveniently measured by scanning the electric sector and magnetic sector simultaneously so that the ratio B/E remains constant [64]. Similar reasoning leads to the observation that, if the instrument is adjusted to transmit a given daughter ion under normal conditions, then a scan is obtained such that B^2/E remains constant resulting in the sequential detection of the relevant parent ions [64]. These scan modes, as well as that for a normal mass spectrum, are shown in Fig. 14. A number of authors have noted artifact peaks resulting from these scan modes and have proposed methods to handle this shortcoming [65–70]. Although a number of papers have been published noting the analytical potential of these techniques, a relatively small number of actual analytical applications have been re-

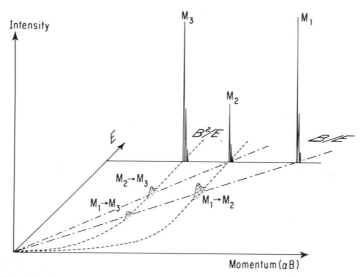

Fig. 14. The linked-scan plane. A momentum–energy surface for momentum at fixed energy (horizontal line) is just the ordinary mass spectrum and results in the collection of the ions formed in the source. As indicated, ions that fragment in the field-free region outside the source are deficient in momentum and energy. By adjusting the magnet to transmit M_1 and then scanning B with the ratio of B/E fixed, the daughter ions are collected (linear scan lines). By adjusting the magnet to transmit a daughter ion (e.g., M_3), then scanning B while maintaining the ratio of B^2/E constant, the various parents leading to the formation of that daughter are collected (curved scan lines).

ported [71–75, for example]. It should be noted that by implementing relatively complex linked-scan functions, it is possible to obtain spectra that result from the loss of some given neutral [76–78]. The spectrometer is transparent to all ions that do not undergo fragmentation by loss of this selected neutral. For example, a method specific for the detection of chlorinated compounds has been proposed that makes use of a constant neutral-loss scan for m/z 35 [78]. Obviously, many variations of this theme are possible.

Experimentally, perhaps the most convenient method of performing tandem mass spectrometry is through the use of the multiple quadrupole mass analyzer. The most common configuration is three quadrupoles in series with Q_1 and Q_3 used as primary and secondary ion mass filters, respectively, and the middle quadrupole Q_2 used as an rf-only collision cell (Fig. 15) [79]. Other configurations include the use of a clever system in which a material that is permeable to rf but impermeable to DC is used to define the rf-only region [80]. Thus, by locating a piece of this material within the mode of a two-quadrupole system, an rf-only collision cell is defined. (The relative merit of these systems is at a stage in which more heat than light has been cast on the subject.) The multiple quadrupole approach to tandem mass spectrometry offers, at least in principle, the advantage of simultaneous unit mass resolution on the primary and secondary ion beams.

Perhaps the most useful characteristic of quadrupole-based tandem mass spectrometers is the ease with which they may be scanned. It is convenient to set either spectrometer and scan the other to give either daughters from a selected parent ion or parents that lead to a selected daughter. By offsetting the mass of Q_1 by m_1 relative to Q_3, spectra can be collected selective for ions that fragment by loss of m_1, similar to the constant neutral-loss linked scan just mentioned. The capability of electrically adjustable resolution permits the analyst the option of operating either Q_1 or Q_3 as a high-pass filter so that ions of less than a selected mass

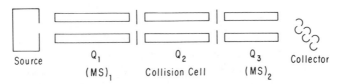

Fig. 15. Schematic diagram of multiple-quadrupole tandem mass spectrometer. The first quadrupole (Q_1), is operated as a mass spectrometer and used to select the primary ion(s). By operation of the second quadrupole in an rf mode (no DC), it is a strong-focusing lens and makes an effective collision cell when a gas is added. The third quadrupole is then operated as a mass spectrometer to analyze the daughter ions formed in the collision cell.

are not transmitted. This makes it convenient to use Q_1 to remove the reagent-ion current in CI experiments and for LC/MS/MS applications.

In comparing tandem quadrupole instruments with magnetic sector instruments of a $B–E$ configuration, the question facing the analytical chemist is whether a more satisfactory analysis can be achieved with unit-mass resolution on both primary and secondary ion beams or with moderate resolutions of 1000 to 3000 on the primary ion beam and a resolution of 50 to 200 on the secondary ion beam. Although a number of *a priori* statements have been made, insufficient data are available for an actual comparison. On the basis of experience in our laboratory, the authors' opinion is that MIKES are preferable for quantitative analysis when adequate resolution is available so that a relatively "pure" primary ion beam can be selected for the collision cell. Although secondary ions may be poorly resolved, the spectral pattern is easily recognized and secondary ion current is linear over several orders of magnitude of sample quantity. Adequate primary ion resolution is especially important when one performs the analogous experiment to selected-ion monitoring, namely, selected-reaction monitoring. Should an unexpected component be present substantially in excess of the material of interest, interferences may arise from low-probability processes that are unexpected and may not even be observed at the sensitivity one would normally use to examine the mass spectrum of a compound. On the other hand, if the problem is to identify an unknown compound, then a precise knowledge of daughter-ion mass becomes more important and the multiple quadrupole approach has an advantage.

One of the unique features of a quadrupole mass filter is the wide range of resolutions that is available. This can be exploited in a number of ways. For instance, Hunt and co-workers [81] have approached the sequencing of short-chained peptides by adjusting the resolution of Q_1 so that $\Delta M \cong$ 10 daltons and scanning Q_1 slowly while simultaneously scanning Q_2 rapidly at unit resolution. By using a 1:1 mixture of $CD_3CO : CH_3CO$ acetyl groups to derivatize the N terminus of the peptide, secondary ions from that end could easily be recognized by their appearance as a doublet separated by three daltons. If adequate sample purification is used prior to analysis, ions not appearing as a doublet can be assigned to the carboxyl terminus. Although sequencing polypeptides seems far removed from water analysis, the mass-spectral approach of operating Q_1 at substantially less than unit resolution should facilitate stable isotope tracer experiments in which one is interested in the transformations of a chemical in an aquatic environment. Another application of this wide range in possible resolution is to use Q_1 to filter out undesirable ions from a mass spectrum. The two most important limitations of the direct liquid introduction approach to on-line coupling of a high performance liquid chromatograph to a mass spectrometer are that relatively high-mass cluster ions result from the reagent plasma and that only a limited amount of fragmentation information is available as a result of restrictions imposed on the ionization technique. Voyksner *et al.* [56] reported a direct-coupled HPLC–tandem mass spectrometry method in which Q_1 of a two-quadrupole system is operated in the rf-only mode with the rf level adjusted so that cluster ions are filtered from the ion beam. By collisionally exciting the ion beam and using Q_2 to mass analyze the secondary ion beam, the desired fragmentation reaction information can be obtained.

The trade-off between primary and secondary ion resolution is avoided by the use of a sector mass spectrometer for primary ion selection and a quadrupole instrument for daughter-ion analysis. Since these instruments are just becoming commercially available, there is little information on which to base an evaluation of their suitability for routine analysis. However, within the mass-transmission limitations of quadrupoles, these should be powerful instruments. The present trend seems to be to use a double-focusing mass spectrometer for primary mass analysis with a quadrupolar collision cell and secondary ion mass filter. Thus the analyst can select the primary ion at high resolution (typically up to 10,000) which offers a significant advantage over multiple quadrupole instruments. Another consideration is that the double-focusing analyzer can be operated as a high-resolution mass spectrometer for exact mass measurements, etc. Therefore, any lab contemplating the addition of either a high-resolution mass spectrometer or a tandem mass spectrometer is advised to give serious consideration to these hybrid instruments.

What then are the limitations of these hybrid sector–quadrupole instruments and what other options might be considered for tandem mass spectrometry? The most severe limitations arise from the properties of the quadrupole mass filter. For low-to-moderate molecular weight samples, the resolution and transmission characteristics of the filter present no substantial limitation. However, if one is interested in high molecular weight compounds (>600), the mass discrimination becomes significant. At higher molecular weight (>1000–2000), the quadrupole is also unable to provide unit resolution. One approach to improve the high-mass performance in tandem mass spectrometry is to use triple- or quadruple-sector analyzers. Although this presents technical difficulty, sufficient progress has been made that commercial instruments are now being offered. The simpler of these has an $E–B–E$ configuration and permits the magnet to be used with either electric sector [82]. Thus the analyst has high resolution on the primary ion beam for MIKES. Other instruments incorporate four sectors as in the McLafferty design [83]. The commercially available version has four sectors in a $B-E-E-B$ configuration so that high resolution can be achieved simultaneously for primary and secondary ions. These more sophisticated instruments make practical tandem mass-spectrometric measurements on compounds with molecular weight up to 2000–5000. The reader interested in hybrid and other novel MS/MS instruments is referred to a recent review by Cooks [84]. It remains to be seen what new frontiers in water chemistry and environmental studies are possible with these newer techniques.

VII. QUALITATIVE ANALYSIS

This section contains the general principles of unknown identification from mass-spectral data. Those wishing to become more expert at identification are advised to read carefully McLafferty's book [13] and practice. The present discussion is limited to a brief description of spectrum interpretation and the use of computers for data enhancement and data searches.

A. Spectrum Interpretation

In the earlier discussion on gas-phase ion chemistry, it was emphasized that the ion–molecule and fragmentation reactions observed in a mass spectrometer are subject to chemical laws. Such basic principles of organic chemistry as relative stability of primary, secondary, and tertiary carbonium ions, the relative lability of C—Br and C—Cl, etc., also apply to gas-phase ions. Unfortunately, there is not always a direct correlation between solution-phase and gas-phase chemistry. Unusual behavior in the gas phase can be expected for reactions that are strongly influenced by the solvent. For example, proton affinities for gaseous alkoxide ions are in the reverse order of the proton affinities of the same ions measured in solution. Thus, the reader is advised to start with the premise that ions act in a manner that is sensible to an organic chemist and to remember the exceptions.

The identification of an unknown compound is normally a multistep process. For the particular case of organics in water, the analysis might involve isolation and perhaps a crude fractionation step. A thorough understanding of these procedures can provide valuable information concerning the nature of an unknown. When GC/MS is performed (as it usually is), the chemist automatically has available the GC properties associated with the unknown. By analyzing a relatively small number of reference compounds, the general characteristics of the GC system can be evaluated. Then, when the mass spectrum is being interpreted, possible structures can be eliminated on the basis of expected GC behavior. Naturally, any sample pretreatment must be taken into account when analyzing the sample [e.g., treatment with CH_2N_2, $(CH_3)_3Si—X$, etc.]. Finally, the chemist generally knows the source of the sample. However, care must be taken to insure that bias is not introduced by this knowledge.

For the novice, no better advice can be offered than that of McLafferty [13]: *Be systematic*. It is equally important to be skeptical. Always test the validity of each assumption as many ways as possible and, if feasible, gather ancillary data to go with the mass spectrum. For a positive identification, the unknown should give spectrum and retention characteristics that match that of a valid reference compound when *both* are analyzed by GC/MS using the same conditions.

Next, we consider the analytical procedure normally applied to organics in water samples in our laboratories. Experiments in development of preliminary isolation and pretreatment methods are monitored by GC using an appropriate detector, typically flame ionization. Once the chemist is satisfied that the isolation–separation methodology has been optimized, the sample with appropriate blanks and spiked samples (if quantitative analysis is desired) are submitted for GC/MS analysis.

The first step in the GC/MS analysis is data acquisition. In the initial phase, we acquire an exact mass-measured EI spectrum, typically operating the spectrometer at 1500 to 2000 resolving power and scanning at 1.0 to 3.0 s/decade while using an appropriate glass capillary column for sample introduction. Mass assignments are performed by the data system with C_2I_4 as a reference compound using the method originally proposed by Haddon [85]. The root-mean-square precision with which the higher mass peaks are assigned is typically 10–20 ppm. For cases requiring higher resolution and/or precision, we employ a method that effectively allows computer control of the peak-matching unit [86]. With this method, resolutions of 40,000 to 50,000 and accuracies of 1 to 5 ppm are available for a limited number of ions. The second phase of data acquisition is to reanalyze the sample using as precisely as possible the same GC conditions but performing positive ion CI with isobutane as the reagent gas. Although methane generally gives a more informative spectrum than butane, one rarely obtains any information that is not available from the combination of EI and butane CI. Because one must still be concerned with sample purity (i.e., GC-peak homogeneity) the simpler spectra given by butane are an advantage. For this analysis, the mass spectrometer is operated at 1000 resolving power and scanned at 1 s/decade. This results in a somewhat more faithful reproduction of the gas chromatogram than the conditions used for the EI analysis. By careful comparison of the two data sets, it is straightforward to correlate EI and CI spectra.

The next step is to interpret the data. First, the molecular weight is assigned from the CI spectrum if possible. The EI data are examined and appropriate elemental formulas are assigned to various ions based on the exact mass measurements. As an aside, we have not found the use of isotopic-peak ratios for elemental-composition assignment to be particularly useful except for compounds containing halogens and/or sulfurs. For CHNO compounds, the relevant ions are usually of low intensity so that the measurement of their exact intensity is unreliable. Once the elemental compositions are assigned, the functional groups present are deduced from the low-mass ion series and radicals lost from the molecular ion [13]. The loss of neutral molecules from the molecular ion is interpreted next. From these preliminary results, a list of possible structures is prepared. Any structure inconsistent with chemical or chromatographic information is eliminated. By this point, usually an unknown's structure has been limited to a small number of possible isomeric compounds. The remainder of the GC/MS data are examined for isomers or homologs and any tentative conclusions possible from elution orders are reached. By comparison of the mass-spectral and gas-chromatographic properties of the unknown with similar reference data obtained in our laboratory, it is frequently possible to make a positive identification. If the unknown is indeed a compound new to our experience, we return to the scientist requesting the analysis a list of possible structures and a statement that the positive identification requires appropriate reference compounds. At that point a decision is made concerning the amount of effort versus the benefit of the additional knowledge. If a positive identification is required, we obtain the necessary reference spectra and reanalyze the sample on a second GC column.

B. Computers and Mass Spectrometry

The availability of powerful computers at reasonable cost has been an essential factor in the widespread use of GC/MS in organic analysis. Originally intended to assist in the handling of the voluminous data generated during a GC/MS run, the processing programs have developed an increasing sophistication. Should one desire the most automated system possible, all work from tuning the mass spectrometer through unattended operations to generation of tables for the final report can be left to a computer. Of course, this implies a great deal of trust in the programmer and electronic devices in general. In this section, we discuss data system features common to most commercial systems that have been useful in our experience. No attempt is made to present a thorough survey, and no doubt oversights do exist.

All data systems have capabilities that might be termed data enchancement. These range from the simple subtraction of one spectrum from another to allow correction for background and partially resolved peaks, the ability to average any number of scans for the sample and background spectra, to sophisticated algorithms by which the computer decides which peaks belong to the sample and which to the background. It must be emphasized that these programs do nothing to improve the data. These manipulations simply allow the data to be presented to the chemist in a form that is more easily interpreted. In general, the quality of the data is reduced. For example, if one averages together three spectra, the effect on the GC resolution is the same as reducing the scan rate by a factor of three. Thus, averages should be performed only after GC peak homogeneity has been established.

The subtraction of a background spectrum is frequently a useful means by which the appearance of a low-intensity spectrum can be improved (Fig. 16). However, the indiscriminate use of such subtractions can lead to severe distortion of the system. Particular care must be exercised when

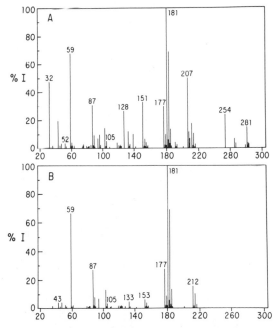

Fig. 16. Mass spectra before (A) and after (B) background subtraction (taken from a complex GC/MS run of ether-extractable chlorination products of aquatic fulvic acid [Millington, D. S., and Norwood, D. L., *Proc. 29th Annu. Conf. Mass Spectrom. Allied Top.* p. 59 (1981)].

background subtractions are used to "clean up" two partially resolved components from the GC. Before this can be done safely, it is necessary to demonstrate that the mass spectra share no common ions. This becomes less likely as the ratio of the quantities of the two components departs from unity, as the absolute intensity of the low-abundance ions of the major component becomes significant relative to the important ions of the minor component. When overlapping peaks are suspected, the most reliable procedure is to output several scans (corrected by a genuine background spectrum) across the region of interest and examine which ions are rising and falling together and from this deduce what ions belong in the same spectrum. The automatic enhancement algorithm that most nearly duplicates this procedure is that of Biller and Biemann [87].

One of the more useful ways in which computers can be employed for compound identification is for the systematic comparison of the experimental spectrum with those in a data base. In principle, the chemist has at his fingertips the collected knowledge (or at least the data) of mass spectrometry. At the present time, libraries with 30,000 to 50,000 mass spectra are available either on commercial data systems or through centralized facilities [88]. There are a number of reasons why this is not the panacea for which we hoped. The most severe limitation is the small size of the available libraries compared to the millions of known compounds. This limitation is being dealt with not only by efforts to increase the size of the data base in general but by emphasis on the addition of specific compounds likely to be of interest (pesticides, priority pollutants, drugs of abuse, common contaminants, etc.). A second major problem arises from the quantitative dependence of the mass spectrum on instrumental conditions. With care, a particular laboratory can reproduce ion-intensity measurements for a given compound within a few percentage points. In the case of magnetic sector mass spectrometers, reasonable attention to source temperature, sample pressure, electron-beam current, etc. results in data that can be reproduced reasonably well ($\sim 10\%$) by another laboratory equipped with a magnetic sector instrument. Quadrupoles, on the other hand, present a somewhat more difficult situation. To begin with, they suffer the disadvantage that relative transmission is mass dependent. For typical instruments of the same basic sensitivity at, for example, m/z 200, the sector instrument has a transmission at m/z 500 that is 5–10 times higher than the quadrupole. Efforts to compensate for this factor by decreasing the resolution at higher m/z for the quadrupole must be viewed with a little suspicion. Because the relative transmission of a quadrupole is influenced by contamination, it is necessary to establish a vigorous quality assurance and maintenance program if data of sufficient reproducibility for library searches, etc., are to be obtained [89]. If present trends continue, the problems relating to data distortion will solve themselves: laboratories making heavy use of library searches tend to have quadrupole instruments. Thus, the libraries should soon be predominantly of quadrupole-generated data. The final word of caution relates to the use of automatic data enhancement with library searches. Distortion of the spectrum can easily result in a misidentification or (more probable) a missed identification. The safest course of action is to include a visual examination of the data before results are reported.

An effective means of screening a particular sample for a set of targeted compounds is to enter the mass spectrum of each compound of interest into a library, acquire GC/MS for the sample, and treat the acquired data as though they were a data base and search the reference spectra for possible matches. With careful control of GC conditions, the GC retention time (or retention index) can readily be included. This can be relied on to eliminate compounds that are absent. It is still advisable to visually inspect mass spectra for presumed positives and to confirm their identity by analysis on a second column. This technique is referred to as "reversed searching" and is routinely employed as a method for priority pollutants in water [90].

VIII. QUANTITATIVE ANALYSIS

A. Introduction

The use of mass spectrometry for quantitative analysis dates to the earliest work of Thomson [1], in which the mass and abundance of isotopes were estimated. The earliest significant uses for chemical analysis were by the petroleum industry. However, the rising popularity of gas chromatography in the 1950s resulted in a decline in the use of mass spectrometry as a quantitative tool. The successful development of a GC/MS interface in the 1960s resulted in a rekindled interest in the use of mass spectrometry for quantitative studies. In this section, we attempt to give a general introduction to the area and pass along a few observations based on our own experience. Readers interested in a detailed treatment or specific applications are referred to more extensive reviews [91].

B. Advantages of Mass Spectrometry as a Quantitative Detector for Gas Chromatography

When one speaks of the advantages of a given technique, it is important to specify exactly what is being compared. Because mass spectrometry can be classified as a universal detector for GC, a reasonable reference point is the flame-ionization detector (FID) (at least for organic compounds). From a different perspective (e.g., that of a pesticide chemist) a more meaningful reference point is the flame-photometric detector or the electron-capture detector. Even this can be confusing, as there are a variety of operational modes for the mass spectrometer, each with different features. In the following discussion, we review these various modes of operation and compare them to the (in our opinion) appropriate GC detector.

1. Selected-Ion Monitoring

If one examines a typical mass spectrum, it is apparent that significant information is gathered for only a relatively small fraction of the time. A typical identification is based on the position and abundance of perhaps 10 or 20 ions. If the mass spectrometer is being scanned over a mass range of 400 to 500 daltons, even with instruments of unit resolution, no useful signal is received for 80 to 90% of the time. The situation is even more pronounced for higher resolution instruments. Obviously, if one is interested only in detecting a specific compound, a considerable gain in sensitivity can be achieved by simply stepping the mass spectrometer to monitor the intensity of ions selected as being characteristic of the component of interest. This experiment is known under a variety of names including selected-ion monitoring, selective-ion monitoring, single-ion monitoring, multiple-ion detection (MID), multiple-ion monitoring, etc. The accepted names are selected/single-ion monitoring (as appropriate) with the abbreviation of SIM.

The estimation of the gain in signal : noise ratio (S : N) can be made from a consideration of the change in sampling time per ion. For example, if one is scanning at 500 daltons/s and compares this with a SIM assay in which four ions are monitored for 0.25 s each, one would expect the S : N to increase by a factor of $[0.25/(1/500)]^{1/2}$ or ~ 10. However, even larger apparent improvements in sensitivity can be achieved, depending on the source of the noise. The above estimation assumes the sensitivity is limited by amplifier noise of the mass spectrometer. This is rarely the case. More frequently, the sensitivity limitation arises from a specific chemical interference. For those cases in which there are differences in the mass spectrum of the analyte and the interfering substances, one has only to select the ions to be monitored such that major ions from the interfering substances are avoided and the detection system can be operated at high gain. Thus, SIM typically results in sensitivity gains of approximately 100 when compared to scanning full spectra.

The above reasoning concerning the relatively small fraction of time that useful information is collected is not strictly valid. In many cases, the absence of a peak at a particular m/z is informative. For example, Phillipson and co-workers use the absence of an M-15 ion as part of their identification criteria for tetrachlorodibenzo-p-dioxins [92]. The trade-off for assays based on SIM techniques then is to collect sufficient information to assure the identification of the analyte while monitoring the minimum number of ions so that the sensitivity can be maximized. Unfortunately, as the detection limit is lowered, the probability of interference increases and, ideally, the degree of specificity of the analysis should also increase.

Another means of improving the specificity of the analysis is to increase the resolution of the mass spectrometer. This is illustrated in Fig. 17 where it is assumed that two components differing in mass by one dalton have not been separated by the GC. It is further assumed that the mass defects of the two components differ by 100 ppm. In Fig. 17A, we see the mass spectrum for each component scanned over the region of interest. If the higher mass component is the component of interest, the ^{13}C-isotope peak from the lower mass component could interfere with the measurement, depending on the relative amounts of the two components. Figure 17B illustrates the case when equal amounts of the two are present. In this case, a positive interference is observed. For the purpose of discussion, we assume the interfering component has 10 carbons, which amounts to 10–12% overlap and is not significant in most practical cases. When the lower mass component is present in greater abundance, the problem can become severe. In Fig. 17C, the case for a tenfold excess of the lower mass component is illustrated. Here, we see that the amount of analyte is overestimated by slightly more than a factor of two. The benefit to be realized from using higher resolution can be seen by examination of Fig. 17D, which is similar to Fig. 17B, except the resolution has been increased from unity to 10,000. By monitoring the center of the molecular ion of the higher mass peak, no positive interference occurs from the ^{13}C-isotope peak. Under these higher resolution conditions, even a 10-fold excess of the lower mass component does not present a significant interference (Fig. 17E). A significant problem can exist even at high-resolution SIM when the relative amount of the interfering substance becomes large as is illustrated for the case of a 100-fold excess (Fig. 17F). This problem can be mitigated somewhat by scanning the mass spectrometer over a narrow mass range (e.g., 200 ppm) and storing the peak profiles [86], permitting the spectrometrist to recognize the presence of the undesirable signal and provide some correction. In this discussion, the interfering signal was assumed to arise from the ^{13}C-isotope peak of a material giving a peak 1 dalton below the peak of interest. This was to illustrate that in determining analytes at very low concentrations, one must consider interferences to include peaks of less than 1% relative abundance. Mass spectral libraries frequently omit them. The use of such

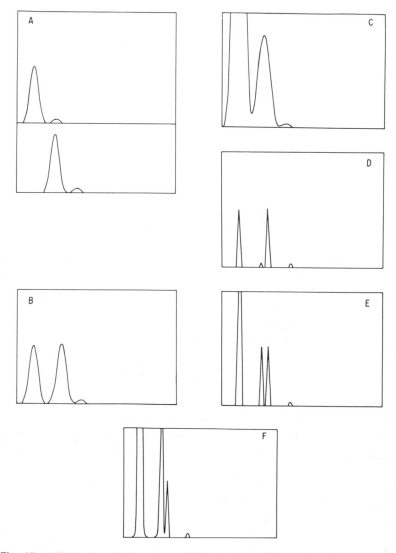

Fig. 17. Effect of resolution on specificity in selected-ion monitor based assays. (A) Two components separated in mass by one dalton with the mass spectrometer at "unit" resolution. (B) The hypothetical mass spectrum resulting from the same two components in the ion source simultaneously at the same concentrations. (C) The effect of a ten-fold excess of the lower m/z component. (D) Conditions as in (B) except resolution increased to 10,000. (E) Conditions as in (C) except resolution increased to 10,000. (F) Conditions as in (E) except 100-fold excess of lower m/z component.

tabulations to deduce the probability that a given combination of ionic masses and intensities are subject to nonspecific interference can lead to a false sense of security on the part of the analytical chemist.

Thus, the various SIM techniques can be viewed as analogous to selective detectors such as the electron-capture or flame-photometric detectors in that sensitivity and specificity have been emphasized at the expense of general information about the sample. In practice, SIM has significant advantages in flexibility, versatility, and ultimate sensitivity over other GC detectors.

2. Total Ion Current Measurement

At the opposite extreme in GC/MS is the use of total ion current as a measurement of the amount of sample present. It is rare that the actual total ion current from the source is monitored, rather, the individual ion currents collected in a given scan are summed and this summed-ion current is then plotted as a function of time (scan number) to produce the chromatogram. Since this requires the use of a computer-based data system, it is convenient (and usual) to modify this total by, for example, excluding ions outside a certain range (e.g., those below m/z 50) so that the signals from background gases are ignored. The use of full spectra obviously improves the confidence in the identification of the material being determined. The concomitant reduction in signal : noise ratio can be important if high precision is required or when working near the detection limit. For those cases in which adequate amounts of material are available, the use of full mass spectra make it possible to (at least in principle) perform a complete qualitative and quantitative analysis from a single GC/MS run. This is particularly attractive in areas requiring a high sample throughput, such as the monitoring of water for priority pollutants. A second important advantage is that the availability of complete mass spectra greatly enhances the probability that unexpected events in the assay are recognized.

The actual estimation of the quantity of a substance present is based, ideally, on a comparison of the chromatographic peak area to a calibration curve generated from the same compound. Many chemists find it impractical to generate the required calibration curve for each component present in a complicated sample and use relative response factors in a manner analogous to assays based on FID. In many cases, suitable accuracy (within a factor of two) can be achieved by assuming all compounds have the same sensitivity when analyzed by EI techniques. It must be emphasized that this applies only to EI (and then only under conditions of relatively high electron energy, e.g., 70 eV) and emphatically does not apply to techniques such as negative CI.

Full-scanned mass spectra provide the analytical chemist with the most information possible from a single experiment. Thus, this is comparable to FID.

3. Extracted Ion Current Profiles

The phrase "extracted ion current profile" (EICP) refers to a data manipulation procedure in which a GC/MS data file, stored on a computer, is searched for the occurrence of a given ion or set of ions. The resultant data, when plotted as a GC trace, have the appearance of a SIM analysis; the only advantage that SIM has over the EICP is its S : N ratio. This technique is in routine use in virtually every laboratory involved in trace organic analysis by GC/MS. The most important use of this technique is the rapid screening of a complex GC/MS data set for the candidate chromatographic peaks for a targeted compound or set of compounds. For example, by taking advantage of the knowledge that essentially all molecules containing trimethylsilyl give an intense peak at m/z 73, a simple search for m/z 73 permits the analyst to rapidly select the MS scans that can possibly come from this class of compounds.

The EICP can be used to advantage for quantitative analysis. As discussed above, the total ion current may be used to estimate the quantity of sample leading to a chromatographic peak. Two advantages can be realized by using the sum of the areas of a carefully selected subset of ions for this measurement. The more important improvement to the results is that the possibility of interference by a coeluting component has been reduced from any compound that has similar chromatographic properties to only those that also give significant ion currents at those m/z chosen for analysis. A less significant enhancement arises from the exclusion of low-intensity peaks (and hence their adverse influence on S : N) from the area measurements.

C. Internal Standards

The use of internal standards is recommended any time that a precision of greater than 30% relative standard deviation is required. Depending on the nature of the internal standard and the point at which it is added to the sample, it can serve a number of different functions. If one is concerned with correcting for imprecision in sample injection into the GC, any well-behaved compound known not to be in the sample suffices. However, the ability of the mass spectrometer to differentiate between isotopic variants of a given compound permits much more elegant approaches to standardization. In particular, by using an isotopic variant of the molecule it is possible to use the mass spectrometer to monitor sample loss during workup procedures, at least in principle. This technique is referred to as isotope-dilution mass spectrometry. The appropriate way of doing this is to prepare a series of spiked solutions containing a fixed amount of the isotopically modified compound and variable amounts of the analyte, to cover the range of concentrations of interest for the analysis. The medium used for the preparation of these solutions should be chosen to duplicate as nearly as possible the matrix to be encountered with the real samples. Each solution should be subjected to the complete analytical scheme to generate the data for the calibration curve that are a plot of the signal ratio versus amount ratio. The signal ratio is calculated as follows [93]:

$$\text{Signal ratio} = \frac{(R_y - R_m)(R_x + 1)}{(R_m - R_x)(R_y + 1)} \tag{52}$$

where $R = \dfrac{\text{EICP area of unlabeled compound quantification mass}}{\text{EICP area of labeled compound quantification mass}}$

m = ratio for the quantification standard (or unknown) analysis mixture

x = ratio for the pure unlabeled (naturally abundant) compound

y = ratio for the pure labeled compound

This equation should only be used when

$$2R_y < R_m < \tfrac{1}{2}R_x$$

Sufficient standards and/or replicates must be measured to give the desired precision in this curve. The actual samples are then spiked with the isotopic standard at the same concentration as the standard solution, the samples analyzed, and the signal ratios measured. The amount ratios can then be determined from the standard curve and hence the concentration of the unknown. Properly applied, such isotopic-dilution assays are capable of very high precision. With care, the accuracy can also be high. The most common deficiency in experimental design is the provision of convincing evidence that the isotopic spike has equilibrated with the analyte in the matrix.

This problem and other common shortcomings in the design and validation of extraction–cleanup procedures have been discussed in detail by Albro [94]. Assuming that a valid sample has been delivered to the mass spectrometry laboratory, there are still possible sources for substantial error. The most neglected of these is the possibility of interference in the measurement of the internal standard. It is common to see assays in which the abundances of several ions are measured and their relative intensities determined to add assurance that the signal being measured is coming from the targeted compound, whereas the internal standard is identified on the basis of only one ion appearing at the appropriate time in the chromatogram. If there is an undetected interference in the internal standard measurement, the result is an overestimation of the recoveries and hence erroneous results. The possibility of this can be minimized by including at least as many ions for measurement of the internal standard as are used for the analyte.

A second common shortcoming in the application of internal standards is the requirement that none of the internal standard be in the native sample. A glaring violation of this principle is the use of ^{37}Cl-labeled tetrachlorodibenzo-p-dioxin as an internal standard, because this is a naturally occurring constituent of the analyte. The converse situation is also to be avoided. That is, the internal standard must not contain any of the analyte. This obvious point can present some difficulty when labile hydrogens have been exchanged to form a deterium-labeled analog to the analyte. In addition to the more apparent problem of incomplete exchange, the analyst must also demonstrate the absence of the reverse exchange reaction during sample workup and storage.

The final common misuse of internal standards is the addition of the internal standard at inappropriate levels to the sample. In order to give meaningful measurement of recoveries, the standard should be near the concentration of the analyte.

D. Instrumental Considerations

The properties of the ideal detector for GC analysis were discussed in Chapter 4. Among the more important are linear dynamic range and specificity. The use of high-resolution selected-ion monitoring is probably the most specific general purpose means of detecting a particular trace component in a mixture of organic chemicals that can be analyzed by GC. It must be borne in mind that even this sophisticated technique has limitations, as were discussed in Section VIII,B. The discussion of possible interferences thus far has centered around factors that lead to an overestimation of the amount of analyte present. There are also instrumental factors that can lead to an underestimation of the analyte.

Mass spectrometers are generally assumed to give a linear response to varying concentrations of material within the ion source that covers five or six orders of magnitude and hence should offer no practical limitation for quantitative analysis. However, this large dynamic range is possible only if changes in the gain of the amplifier or detector (electron multiplier) are made. If a computer is included in the system, further restrictions to the dynamic range result. These arise from the necessity of converting the analog signal from the amplifier to a digital signal (i.e., a count) for the computer. The most commonly used analog-to-digital converters (ADC) have 12 bits and thus a maximum dynamic range of $1 : (2^{12} - 1)$ or $1 : 4095$. This dynamic range can be achieved only if there is no zero offset between the amplifier and the ADC. Normally, this is not the case so that the least significant bit is lost as noise. Thus, in well-adjusted instruments, the actual dynamic range is more nearly $1 : 2000$. Even this number is misleading since it refers to the ratio of the smallest signal : largest signal that can be measured. In order to achieve the same ratio for components of a mixture, it is necessary to inject precisely the proper amount of at least one analyte to either give a signal at the detection limit or a signal that exactly saturates the ADC. From practical considerations, the dynamic range is reduced further by a factor of two (or more). Finally, the S : N ratio at the detection limit is $3 : 1$. Obviously, this is unacceptable for quantitative measurements. In our laboratory, we strive for a S : N >10 on the least abundant component measured. The final conclusion is that a dynamic range of 100 to 500 is all that one can practically expect when an instrument with a 12-bit ADC is used for the measurement, unless the gain of the amplifier is altered during the course of the analysis. This limitation is being addressed in the current data systems by use of 16-bit ADCs and/ or autoranging preamplifiers. The result of undetected saturation of the ADC is, obviously, an underestimation of the quantity of material present.

A second limitation is the dynamic range of the ion source. Contrary to the belief that the ion source has a linear dynamic range of five or six orders of magnitude, especially when the mass spectrometer is being operated at high resolution, changes in the ion density within the source result in changes in the ion energy and hence detuning. It must be emphasized that these changes in tuning result from changes in ion density, no matter the compound from which the ions are coming. Thus, if one is monitoring for some particular compound and a different material coelutes that is present at orders-of-magnitude greater concentration, an erroneously low value for the analyte results. One can easily monitor for this source detuning by including a background ion in the sampling sequence.

IX. SELECTED APPLICATIONS

Now that the basic principles of organic mass spectrometry have been discussed and procedures have been put forward for their utilization in qualitative and quantitative analysis, it is useful to consider applications of these methods to specific problems in the field of water analysis. Over the past decade or more, the number of studies reported in the scientific literature in this particular field has increased rapidly. The authors feel that it would not be profitable for the reader if large numbers of these were simply listed, because excellent reviews, such as the *Analytical Chemistry Fundamental Reviews* [e.g., 95], are readily available and regularly updated. Rather, some representative examples of qualitative and quantitative studies are listed and several specific papers that the authors feel are most appropriate are discussed in some detail. The reader should recognize that the choices are somewhat arbitrary and reflects the authors' interests. It is hoped that this discussion gives novice mass spectrometrists as well as more experienced workers from other fields some idea how mass spectrometry can be applied to water quality problems.

Qualitative trace organic analysis of aqueous samples is an area of study that has received much attention in the literature. Many papers have been published concerning identification of micropollutants in raw and finished drinking waters [e.g., 96], industrial and municipal wastewaters [e.g., 97], large bodies of surface water [e.g., 98], and groundwater [e.g., 99]. One study that incorporates the general procedure for qualitative analysis discussed in Section VII is reported by Millington *et al.* [100].

Concern about the presence of toxic and carcinogenic micropollutants in public drinking waters has prompted interest in treatment methods capable of removing these substances. One of the most studied and utilized of these is adsorption onto activated carbon, either granular or powdered. In the United States, over 40 drinking-water treatment plants have granular activated-carbon filter beds in place. In order to evaluate the performance of these filters, it is necessary to establish the nature of the organic compounds accumulating on the carbon. This study reports on the development of a complete analytical scheme for desorbing organic material from spent carbon filter samples and identifying the many molecular structures present.

First, three desorption techniques were evaluated: Soxhlet extraction, ultrasonic agitation with organic solvents, and direct thermal desorption of the carbon into a capillary GC/MS. It was determined that a sequential extraction using solvents of different polarity was superior to any single-solvent extraction, the most effective series attempted being acetone, followed by methylene chloride, and then toluene. It was further found that ultrasonic agitation was almost as effective as Soxhlet extraction while being much simpler, faster, and less error prone. This is demonstrated by the reconstructed total ion chromatograms shown in Fig. 18. Direct thermal desorption was found to be the simplest and fastest method, but problems were observed including loss of high-boiling components and thermally induced chemical reactions.

With the extraction techniques evaluated, attention then focused on identification of the various components in the extracts. The criteria utilized for this were developed from earlier studies [101] and follow those published in a recent editorial in *Environmental Science and Technology* by Christman [102]. For each component, the data acquired included an EI mass spectrum for manual interpretation and comparison with available spectral libraries, a CI mass spectrum for molecular weight determination, and accurate mass measurements for elemental compositions of ions in the EI spectrum. No identification was considered complete until comparisons were made (both GC and MS) with authentic pure standards using the same equipment and conditions.

An example of a complete set of data is shown in Fig. 19 for tris(chloroethylphosphate) (see also Fig. 18). The EI spectrum matches that of the authentic compound, the CI spectrum indicates the molecular weight, and the accurate mass measurements indicate reasonable elemental compositions for two of the higher mass ions observed in EI. A coeluting component of molecular weight 215 may also be observed. Interpreting the accurate mass measurements requires a bit of chemical intuition. For example, the computer listed five possible choices for m/z 249 within certain specified parameters. Of these, $C_7H_5O_{10}$, $C_{18}H_2P$, and $C_{12}H_7O_2ClP$ can be immediately dismissed since the isotope pattern shows the ion to have two chlorines [13]. Because it is not chemically possible for any of the m/z 205 choices to come from $C_{13}H_7OCl_2$, it can also be ruled out, leaving $C_6H_{12}O_4Cl_2P$. The Kovat's Retention Index of the unknown was matched with that of the authentic compound on the same capillary GC column [103]. It is important to observe that this structure could not be found in any available spectral library. In all, over 50 different structural identifications were reported in the study.

Tandem mass spectrometry (MS/MS) may also be utilized for qualitative and semiquantitative analysis of water samples. One promising approach was proposed by Hunt *et al.* [104], who described the potential of a triple analyzer quadrupole mass spectrometer for the direct analysis (without chromatographic separation) of EPA priority pollutants in environmental matrices. The proposed procedure was to volatilize the entire sample into the ion source under CI conditions and perform functional

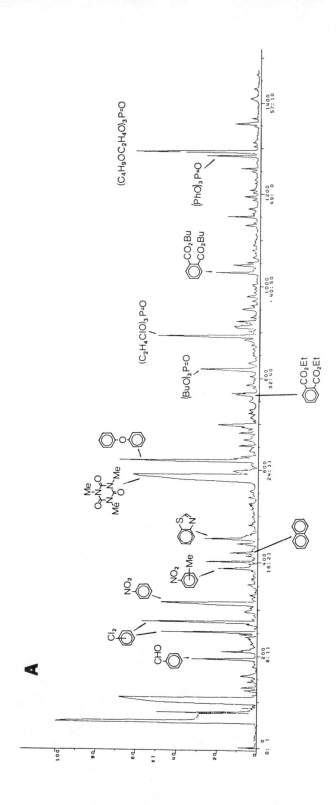

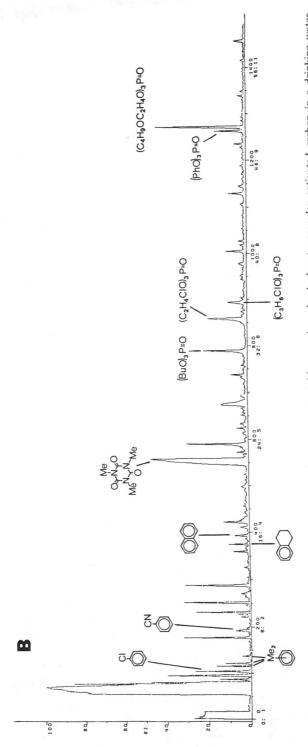

Fig. 18. Reconstructed total-ion chromatogram of acetone extractable organics adsorbed onto granular activated carbon in a drinking water treatment plant [100]. (A) Soxhlet; (B) ultrasonic.

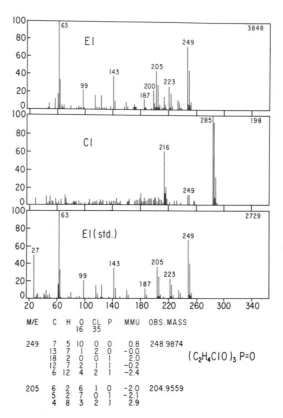

Fig. 19. Mass spectral data set for tris(chloroethylphosphate) derived from the GAC acetone extract in Fig. 18.

group analysis utilizing collision-induced dissociation and either a parent ion or constant neutral-loss scan. Although no analyses of aqueous samples were reported, it was stated that preliminary data indicated that the MS/MS method correlated well with GC/MS methods while providing a more rapid sample handling capability. The future of MS/MS methods in environmental analysis appears promising but needs careful evaluation.

The second major research area, quantitative analysis, is also best illustrated with an application to the priority-pollutant list. In the *Federal Register* one may find a number of methods that the EPA has proposed for use in such studies. Two of these (Methods 624 and 625) give the analyst the option of choosing external or internal standard techniques [105]. Others (Methods 1624 and 1625) specify the use of stable isotopically labeled internal standards and isotope-dilution mass spectrometry [106]. This is the method accepted by most mass spectrometrists and described in Section VIII,C. Colby and Rosencrance reported an evaluation of these approaches to quantitation of priority pollutants in aqueous samples and applied isotope dilution to a series of wastewater samples [93].

In order to compare the three methods, pure water and soap solutions were spiked with varying concentrations of four compounds (benzene, toluene, chloroform, and 1,2-dichloroethane) and analyzed by purge-and-trap GC/MS. The concentration range covered was 1–1000 μg/L. Each solution was also spiked with 20 μg/L of chlorobromomethane and 1,4-dichlorobutane (nonstructurally related internal standards) as well as stable isotopically labeled analogs of the four compounds. After analysis, calibration curves were generated using the external standard, nonstructurally related internal standard and isotope dilution methods. The curves produced may be seen in Fig. 20. In these cases, maximum error was observed with the external standard approach and least with isotope dilution. From the earlier discussion of internal standards, this is the anticipated result. The nonstructurally related internal standards did decrease the error somewhat but not as dramatically as the labeled standards. Standard addition experiments also failed to completely compensate for matrix effects.

After concluding that isotope dilution produced the best calibration curves, Colby and Rosencrance [93] demonstrated that the recoveries of various priority pollutants and their labeled analogs were identical to those from environmental matrices. This may be described as an attempt at method validation. They accomplished this by spiking 13 base–neutral compounds and their labeled analogs into four different industrial wastewaters, extracting by method 625, and quantitating each using an external standard GC/MS procedure. In each case, the recovery ratio (unlabeled : labeled) was near 1. The accuracy and precision of the method was further established by spiking 20 different types of industrial wastewaters with low-μg/L levels of 36 priority pollutants along with their labeled analogs and computing the accuracy of recovery-corrected concentrations for each compound. In each case, near 100% accuracy was achieved.

A body of evidence is accumulating, of which the Colby and Rosencrance study is but one example, which indicates that when the highest accuracy and precision are desired the use of stable isotope-dilution analysis is indicated. This is demonstrated for the priority pollutants but is also probably true for other compounds. In view of the commercial avail-

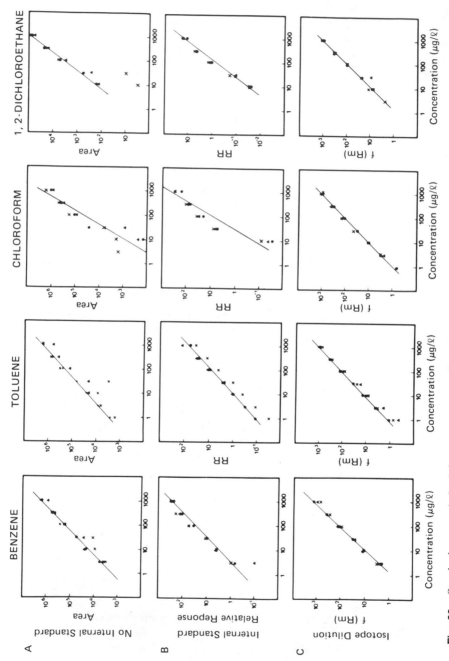

Fig. 20. Standard curves derived from pure water spiked with four priority pollutants using (A) no internal standard, (B) a nonstructurally related internal standard, and (C) isotope dilution [93].

ability of stable, isotopically labeled internal standards, the use of other methods is not recommended unless they can be convincingly validated and are accompanied by adequate quality assurance procedures. The reader is cautioned to take a skeptical view of any quantitative studies in which no attempt at validation has been made.

REFERENCES

1. Thomson, J. J., "Rays of Positive Electricity and their Application to Chemical Analysis." Longmans, Green, New York, 1913.
2. McLafferty, F. W., *Anal. Chem.* **28**(3), 306 (1956).
3. Beynon, J. H., *Microchim. Acta* **1**(6), 437 (1956).
4. Ryhage, R., *Anal. Chem.* **36**(4), 759 (1964).
5. Finnigan, M. A. T., *Mass Spectrom. Syst.* p. 6 (1981).
6. Williams, D. H., and Howe, I., "Principles of Organic Mass Spectrometry." McGraw-Hill, New York, 1972.
7. Bowers, M. T., "Gas Phase Ion Chemistry," Vols. 1 and 2. Academic Press, New York, 1979.
8. Kiser, R. W., "Introduction to Mass Spectrometry and its Applications." Prentice-Hall, Englewood Cliffs, New Jersey, 1965.
9. Roboz, J., "Introduction to Mass Spectrometry Instrumentation and Techniques." Wiley (Interscience), New York, 1968.
10. Marcus, R. A., and Rice, O. K., *J. Phys. Colloid Chem.* **55**, 894 (1951).
11. Marcus, R. A., *J. Chem. Phys.* **20**(3), 359 (1952).
12. Rosenstock, H. M., Wallenstein, M. B., Wahrhaftig, A. L., and Eyring, H., *Proc. Natl. Acad. Sci. U.S.A.* **38**, 667 (1952).
13. McLafferty, F. W., "Interpretation of Mass Spectra," 2nd ed. Benjamin, Reading, Massachusetts, 1973.
14. Munson, M. S. B., and Field, F. H., *J. Am. Chem. Soc.* **88**(2), 2621 (1966).
15. Hunt, D. F., *Adv. Mass Spectrom.* **6**, 517 (1974).
16. Bursey, M. M., Elwood, T. A., Hoffman, M. K., Lehman, T. A., and Tasersk, J. M., *Anal. Chem.* **42**(12), 1370 (1970).
17. Hass, J. R., Nixon, W. B., and Bursey, M. M., *Anal. Chem.* **49**(7), 1071 (1977).
18. Hunt, D. F., McEwen, C. N., and Upham, R. A., *Anal. Chem.* **44**(7), 1292 (1972).
19. Dougherty, R. C., and Weisenburger, C. R., *J. Am. Chem. Soc.* **90**(23), 6570 (1968).
20. Hunt, D. F., Harvey, T. M., and Russell, J. W., *J. Chem. Soc., Chem. Commun.* No. 5, p. 151 (1975).
21. Hass, J. R., Friesen, M. D., Harvan, D. J., and Parker, C. E., *Anal. Chem.* **50**(11), 1474 (1978).
22. Smit, A. L. C., and Field, F. H., *J. Am. Chem. Soc.* **99**(20), 6471 (1977).
23. Hass, J. R., and Bursey, M. M., eds., *Environ. Health Perspect.* **36**, 1 (1980).
24. Jennings, K. R., *Philos. Trans. R. Soc. London, Ser. A* **293**, 125 (1979).
25. Hass, J. R., Friesen, M. D., and Hoffman, M. K., *Org. Mass Spectrom.* **14**(1), 9 (1979).
26. Buser, H. R., and Rappe, C., *Chemosphere* **1**, 199 (1978).
27. Hunt, D. F., Stafford, G. C., Crow, F. W., and Russell, J. W., *Anal. Chem.* **48**(14), 2098 (1976).
28. Hass, J. R., *Pract. Spectrosc.* **3**, Chapter 4 (1979).

29. Horning, E. C., Horning, M. S., Carroll, D. I., Dzidic, I., and Stillwell, R. N., *Anal. Chem.* **45**(6), 936 (1973).
30. Beckey, H. D., *Int. J. Mass Spectrom. Ion Phys.* **2**(6), 500 (1969).
31. Beckey, H. D., "Principles of Field Ionization and Field Desorption Mass Spectrometry." Pergamon, Oxford, 1977.
32. Schulten, H. R., and Beckey, H. D., *Org. Mass Spectrom.* **6**(8), 885 (1972).
33. Matsuo, T., Matsudo, H., and Katakuse, I., *Anal. Chem.* **51**(1), 69 (1979).
34. Bursey, M. M., Rechsteiner, C. E., Sammons, M. C., Hinton, D. M., Colpitts, T. S., and Tvaronas, K. M., *J. Phys. E* **9**, 1405 (1976).
35. Wightman, R. M., Hinton, D. M., Sammons, M. C., and Bursey, M. M., *Int. J. Mass Spectrom. Ion Phys.* **17**(2), 208 (1975).
36. Holland, J. F., Soltmann, B., and Sweeley, C. C., *Biomed. Mass Spectrom.* **3**(6), 340 (1976).
37. Schulten, H. F., and Beckey, H. D., *Adv. Mass Spectrom.* **6**, 499 (1974).
38. Stoll, R., and Rollgen, F. W., *Org. Mass Spectrom.* **14**(12), 642 (1979).
39. McHugh, J. A., "Methods of Surface Analysis," Chapter 6. Elsevier, Amsterdam, 1975.
40. Grade, H., Winograd, N., and Cooks, R. G., *J. Am. Chem. Soc.* **99**(23), 7725 (1977).
41. Day, R. J., Unger, S. E., and Cooks, R. G., *Anal. Chem.* **52**(4), 557A (1980).
42. Macfarlane, R. D., and Torgerson, D. F., *Science* **191,** 920 (1976).
43. McNeal, C. J., and Macfarlane, R. D., *J. Am. Chem. Soc.* **103**(6), 1609 (1981).
44. Barber, M., Bordoli, R. S., Sedgwick, R. D., and Tyler, A. N., *J. Chem. Soc., Chem. Commun.* No. 7, p. 325 (1981).
45. Barber, M., Bordoli, R. S., Elliot, G. J., Sedgwick, R. D., and Tyler, A. N., *Anal. Chem.* **54**(4), 645A (1982).
46. Dawson, P. H., "Quadrupole Mass Spectrometry and its Applications." Am. Elsevier, New York, 1976.
47. Siegel, M., "Fundamentals of Quadrupoles With Applications to MS/MS," an invited lecture presented to the Research Triangle Area Mass Spectrometry Discussion Group, Research Triangle Park, North Carolina, 1981.
48. Herzog, R. F. K., *Z. Phys.* **89,** 447 (1934).
49. Mattauch, J., and Herzog, R. F. K., *Z. Phys.* **89,** 768 (1934).
50. Hintenberger, H., *Natl. Bur. Stand. (U.S.), Circ.* **522,** 95 (1953).
51. Hintenberger, H., Wende, H., and Konig, L. A., *Z. Naturforsch., A* **10A,** 605 (1955).
52. McFadden, W. H., "Techniques of Combined Gas Chromatography/Mass Spectrometry: Applications in Organic Analysis." Wiley (Interscience), New York, 1973.
53. Arpino, P. J., and Guiochon, G., *Anal. Chem.* **51**(7), 682A (1979).
54. McFadden, W. H., Schwartz, H. L., and Evans, S., *J. Chromatogr.* **122,** 389 (1976).
55. Baldwin, M. A., and McLafferty, F. W., *Org. Mass Spectrom.* **1,** 1355 (1973).
56. Voyksner, R. D., Hass, J. R., and Bursey, M. M., *Anal. Chem.* **54**(14), 2465 (1982).
57. Voyksner, R. D., Parker, C. E., Hass, J. R., and Bursey, M. M., *Anal. Chem.* **54**(14), 2583 (1982).
57a. Blakely, C. R., Carmody, J. J., and Vestal, M. L., *Anal. Chem.* **52,** 1636 (1980).
57b. Blakely, C. R., Carmody, J. J., and Vestal, M. L., *J. Am. Chem. Soc.* **102,** 5933 (1980).
57c. Blakely, C. R., and Vestal, M. L., *Anal. Chem.* **55,** 750 (1983).
58. Cooks, R. G., Beynon, J. H., Caprioli, R. M., and Lester, G. R., "Metastable Ions." Elsevier, Amsterdam, 1973.
59. Lindholm, E., *Adv. Chem. Ser.* **58,** 1 (1966).
60. Beynon, J. H., Cooks, R. G., Amy, J. W., Baitinger, W. E., and Ridley, T. Y., *Anal. Chem.* **45**(12), 1023A (1973).

61. Maurer, K. H., Brunce, C., Koppus, G., Habfast, K., Schroder, U., and Schulze, P., *Proc. 19th Annu. Conf. Mass Spectrom. Allied Top.* Paper K-9 (1971).
62. Schlunegger, U. P., *Angew. Chem., Int. Ed. Engl.* **14,** 679 (1975).
63. Hass, J. R., Tondeur, Y., and Voyksner, R. D., *Anal. Chem.* **55,** 295 (1983).
64. Boyd, R. K., and Beynon, J. H., *Org. Mass Spectrom.* **12**(3), 163 (1977).
65. Lacey, M. J., and Macdonald, C. G., *Org. Mass Spectrom.* **12**(9), 587 (1977).
66. Lacey, M. J., and Macdonald, C. G., *Aust. J. Chem.* **31,** 2161 (1978).
67. Lacey, M. J., and Macdonald, C. G., *Org. Mass Spectrom.* **13**(5), 243 (1978).
68. Lacey, M. J., and Macdonald, C. G., *Org. Mass Spectrom.* **13**(5), 284 (1978).
69. Lacey, M. J., and Macdonald, C. G., *Org. Mass Spectrom.* **15**(3), 134 (1980).
70. Shushan, B., and Boyd, R. K., *Int. J. Mass Spectrom. Ion Phys.* **34,** 37 (1980).
71. Haddon, W. F., *Anal. Chem.* **51**(7), 983 (1979).
72. Haddon, W. F., and Molyneux, R. J., *Proc. 28th Annu. Conf. Mass Spectrom. Allied Top.* p. 472 (1980).
73. Duholke, W. K., and Fox, L. W., *Proc. 28th Annu. Conf. Mass Spectrom. Allied Top.* p. 353 (1980).
74. Games, D. E., Eskers, C., Swann, B. P., and D. N. B. Mollen, *Proc. 29th Annu. Conf. Mass Spectrom. Allied Top.* p. 484 (1981).
75. Shushan, B., Safe, S. H., and Boyd, R. K., *Anal. Chem.* **51**(1), 156 (1978).
76. Lacey, M. J., and Macdonald, C. G., *Anal. Chem.* **51**(6), 691 (1979).
77. Haddon, W. F., *Org. Mass Spectrom.* **15**(10), 539 (1980).
78. Shushan, B., and Boyd, R. K., *Anal. Chem.* **53**(3), 421 (1981).
79. Yost, R. A., and Enke, C. G., *J. Am. Chem. Soc.* **100**(7), 2274 (1978).
80. Siegel, M. W., *Anal. Chem.* **52**(11), 1790 (1980).
81. Hunt, D. F., Buko, A. M., Ballard, J. M., Shabanowitz, J., and Giordani, A. B., *Biomed. Mass Spectrom.* **8**(9), 397 (1981).
82. Gross, M. L., Lyon, P. A., Crow, F. W., and Evans, S., *Proc. 28th Annu. Conf. Mass Spectrom. Allied Top.* p. 107 (1980).
83. McLafferty, F. W., Todd, P. J., McGilvery, D. C., and Baldwin, M. A., *J. Am. Chem. Soc.* **102,** 3360 (1980).
84. Cooks, R. G., and Glish, G. L., *Chem. Eng. News* **59**(48), 40 (1981).
85. Haddon, W. F., and Luken, H. C., *Proc. 22nd Annu. Conf. Mass Spectrom. Allied Top.* p. 436 (1974).
86. Harvan, D. J., Hass, J. R., and Wood, D., *Anal. Chem.* **54**(2), 332 (1982).
87. Biller, J. E., and Biemann, K., *Anal. Lett.* **7**(7), 515 (1974).
88. National Bureau of Standards, *Natl. Stand. Ref. Data Ser. (U.S., Natl. Bur. Stand.)* **NSRDS-NBS 63** (1978).
89. Eichelberger, J. W., Harris, L. E., and Budde, W. L., *Anal. Chem.* **47**(7), 995 (1975).
90. Slayback, J. R. B., and Kan, M. N., *Finnigan Appl. Rep.* No. AR8014 (1980).
91. Millard, B. J., "Quantitative Mass Spectrometry." Heyden, London, 1978.
92. Phillipson, D. W., and Puma, B. J., *Anal. Chem.* **52**(14), 2228 (1980).
93. Colby, B. N., and Rosencrance, A. E., *in* "Advances in the Identification and Analysis of Organic Pollutants in Water," Vol. 1, Chapter 13. Ann Arbor Sci. Publ., Ann Arbor, Michigan, 1981.
94. Albro, P. W., *Ann N.Y. Acad. Sci.* **320,** 19 (1979).
95. Burlingame, A. L., Dell, A., and Russell, D. H., *Anal. Chem.* **54**(5), 363R (1982).
96. Lin, D. C. K., Melton, R. G., Kopfler, F. C., and Lucas, S. V., *in* "Advances in the Identification and Analysis of Organic Pollutants in Water," Vol. 2, Chapter 46. Ann Arbor Sci. Publ., Ann Arbor, Michigan, 1981.
97. Linstrom, K., Nordin, J., and Osterberg, F., *in* "Advances in the Identification and

Analysis of Organic Pollutants in Water," Vol. 2, Chapter 52. Ann Arbor Sci. Publ., Ann Arbor, Michigan, 1981.

98. Proctor, B. L., Elder, V. A., and Hites, R. A., *in* "Advances in the Identification and Analysis of Organic Pollutants in Water," Vol. 2, Chapter 51. Ann Arbor Sci. Publ., Ann Arbor, Michigan, 1981.

99. Steurmer, D. H., Ng, D. J., and Morris, C. J., *Environ. Sci. Technol.* **16**(9), 582 (1982).

100. Millington, D. S., Bertino, D. L., Kamei, T., and Christman, R. F., *in* "Water Chlorination: Environmental Impact and Health Effects," Vol. 4, Chapter 31. Ann Arbor Sci. Publ., Ann Arbor, Michigan, 1983.

101. Millington, D. S., and Norwood, D. L., *Proc. 29th Annu. Conf. Mass Spectrom. Allied Top.* p. 59 (1981).

102. Christman, R. F., *Environ. Sci. Technol.* **16**(3), 143A (1982).

103. Millington, D. S., "Extraction and Analysis of Organic Compounds Adsorbed on GAC Filters used in Treatment Plants," final report submitted to U.S. Environ. Prot. Agency, Washington, D.C., 1982.

104. Hunt, D. F., Shabanowitz, J., Harvey, T. M., and Coates, M., *Proc. 30th Annu. Conf. Mass Spectrom. Allied Top.* p. 800 (1982).

105. *Federal Register,* Vol. 44, No. 233, Methods 624 and 625, pp. 69532–69552 (1979).

106. U.S. Environmental Protection Agency, "Methods 1624 and 1625." USEPA, Washington, D.C., 1981.

6 THE USE OF HPLC FOR WATER ANALYSIS

Karl J. Bombaugh

Radian Corporation
Austin, Texas

I. INTRODUCTION

High pressure/performance liquid chromatography (HPLC) is a valuable technique for analyzing nonvolatile substances. Consequently, it has great potential for use in water analyses where many dissolved substances are either ionic or polar and therefore exhibit low volatility. The technique of HPLC is not limited to nonvolatile components and should be used wherever advantage is gained by performing separations in a liquid state.

HPLC is a form of elution chromatography in which a narrow band of sample is pumped through a column of sorbent in a liquid carrier. As the sample travels through the column, its components distribute between a carrier and a sorbent and emerge from the column in separate bands. The emerging bands are monitored by a detector and recorded as a chromatogram. The technique is practiced with a liquid chromatograph. Complete systems can be obtained commercially or they can be assembled from components.

As commonly practiced, there are four types of liquid chromatography (LC):

1. Partition: using bonded phases and a polar carrier (reversed phase)
2. Adsorption: using silica gel and a nonpolar carrier (normal phase)
3. Ion exchange: using ion-exchange resins and aqueous buffers as carriers
4. Size exclusion: using porous resins and sorbable carriers

Many organic substances found in water can be separated successfully by partitioning on bonded-phase columns. Less polar materials can be separated by adsorption in the normal-phase mode. Ionic substances can be separated by ion exchange on properly charged resins, or in many cases they can be separated as ion pairs by partitioning on bonded phases. Polymeric substances having small differences in structures or composition but having measurable differences in molecular size require size separation using exclusion chromatography.

Major advances in column-packing technology have enabled vastly improved separations. Highly efficient columns packed with microparticles (5–10 μm) offer a significant advantage over older type columns prepared from larger particles (>30 μm) [1–5].

As water-borne components may be separated by microparticle packings in either the normal-phase or the reversed-phase mode, the choice of mode depends primarily on problem need. When the compounds to be separated are structurally related, the choice of mode may be based on which mode provides the best means of discrimination. Reversed-phase partition chromatography is usually preferred for separating homologous series; silica gel is preferred for separating structural isomers; ion-exchange is preferred for strongly dissociated compounds; and size-exclusion is preferred for separating macromolecules with small functional differences.

HPLC in some respects is analogous to gas chromatography (GC), yet it is uniquely different. HPLC is more complex than GC because the carrier enters into the separation process through its interaction with the solute–sorbent equilibrium. This interaction provides an extra degree of freedom which, although complicating the practice, markedly increases the versatility and utility of the technique.

The technology for HPLC has been derived from a broad range of scientific disciplines involving fluid dynamics, mechanics, chemical equilibrium, polymer chemistry, photometry, refractometry, gas chromatography, and classical column chromatography. It is therefore impossible to provide in a single chapter a comprehensive explanation of the many facets of HPLC. Nor is it possible to develop justifications for the many practices that have come into use through empirical research. The technology involved in the development and application of LC pumps, solvent programmers, sample injectors, column packings, and eluent detectors are each rooted in different areas of specialization. Those readers wishing to delve more deeply into any of these areas should first consult a comprehensive text on HPLC and then the specialized technical literature that deals with each subject of interest.

The purpose of this chapter is to acquaint the analyst with HPLC and its application to water analysis. As there is virtually no limit to the types of organic compounds that may contaminate water, the application of HPLC to the determination of organics in water is, in reality, a matter of organic analyses by HPLC. As far as inorganic molecules are concerned, they may be included in the discussion of ion chromatography, which is a segment of HPLC [6].

This chapter includes a discussion of both theory and practice. A brief historical background of LC is included to put current technology in perspective with the vast treasure of published literature on classical liquid column chromatography that can be used to advantage in solving separation problems. Chromatographic instrumentation and equipment are discussed from a functional point of view with emphasis on the principles of operation. No attempt is made to list equipment vendors or to describe in detail the relative merits of the various brands of equipment that are available in the marketplace. Space limitations do not permit the inclusion of much useful information that is available in numerous equipment vendors' brochures. Those wishing for such information should consult any of several equipment buyers' guides, such as the one published by *Analytical Chemistry*. Instead of comparing commercial equipment, the principles by which chromatographic equipment is governed are discussed. This discussion provides a background on how the equipment works and what each component is expected to do. From this information, the reader can make judgments in terms of his needs. Chromatographic theory is presented in practical terms so that it can be applied on the job. Selected applications have been included to illustrate chromatographic principles and to demonstrate their utility. Finally, a partial list of publications is provided in a bibliography of water analyses by HPLC.

Historically, three different chromatographic approaches have been used: frontal analyses, displacement chromatography, and elution chromatography. In frontal-analysis chromatography, the neat sample mixture is passed through the retardant bed until each of the respective component fractions emerges in a sequence of steps in order of their increasing affinity for the retardant bed. In displacement chromatography, the sample mixture is loaded onto an activated retardant after which adsorbed species are displaced from the active sites by a more strongly adsorbed displacement fluid. In elution chromatography, the sample is injected into a carrier fluid as a discrete band and transferred through the fixed retardant. When the carrier fluid is a liquid, the technique is known as liquid chromatography. With the introduction of small-particle column packings to improve separation efficiency, higher pressures have been required to achieve practical flow rates, and the technique has become known as HPLC.

The term HPLC is used interchangeably to represent either high pressure or high performance liquid chromatography, and no specifications have been put forth to designate at which pressure or at what performance classical liquid column chromatography becomes HPLC. Fundamentally, the principles involved in either technique are the same. The difference between the two lies in the approach. HPLC is performed with an engineered system that normally consists of a carrier pumping system, a sample injection system, a pressurized column, and a quantitative detector having a data readout system. The primary function of the HPLC system is to produce a chromatogram. In contrast, column chromatography normally is practiced in a preparative mode, using a gravity-propelled carrier and relatively low-pressure-drop columns. The technique generally is practiced without a recording detector, and its primary function usually is to produce fractions. In this chapter, HPLC is treated as an analytical technique, and although the differences between HPLC and classical LC are recognized, preparative-scale HPLC is treated in this chapter as an extended application of the analytical HPLC system.

HPLC is a product of both gas and liquid chromatography. Gas chromatography established the expectations in terms of speed, resolution, and system performance, whereas column chromatography provided background experience in LC. It demonstrated the significance of the carriers' involvement in the separation process. The additional degree of freedom provided by this involvement increases technique complexity but also greatly increases its versatility, and therefore its utility.

The major innovation that enabled classical LC to become HPLC was the elimination of band spreading that normally occurs in a moving liquid when it is transferred through open tubes, tube fittings, injectors, packed beds, or detector cells. By reducing particle diameter of column packings from 75 to 5 μm, reducing transport-tube diameter from 0.040 to 0.010 in. ID, and eliminating mixing volumes in transport couplings, all current advances in speed and performance were realized. An awareness of the problem created by longitudinal mixing must be acquired by all putative chromatographers who couple and uncouple chromatography columns and components, if they wish to avoid reverting to the performance of the pre-HPLC era.

II.   EQUIPMENT FOR HPLC

Equipment for HPLC can be obtained from commercial suppliers either as fully integrated systems, as working modules, or in many cases as components. This discussion explains the function of the elements that make up an HPLC system and describes the underlying principles of operation of each unit so that the potential user has a foundation with which to begin HPLC. Equipped with vendors' literature or instruction manuals, a prospective user may start the equipment and begin operation. Perhaps, if no equipment is available, he may select components and assemble them into a working unit.

A typical HPLC consists of the following elements: a carrier delivery system, a sample introduction system, a column with an appropriate packing and housing, a detector or a combination of detectors, and information recording and processing equipment. Each element is discussed in the paragraphs that follow.

A.   Carrier Delivery Systems

The carrier delivery system for HPLC, at a minimum, should be capable of providing a steady, reproducible flow of carrier through the LC column; head pressure should be adequate to provide a carrier velocity of at least 1 cm/s through a 30-cm column packed with microparticles. Carrier delivery is usually accomplished by a mechanical pump that was either designed for or adapted for use in HPLC. As various designs have been used in commercial and user-built systems, some discussion of this subject is in order.

Historically, two types of pumps were available for chromatography: constant pressure and constant displacement. Of the two, the constant-displacement type is generally preferred for HPLC because it provides a flow rate that is not altered by a change in solvent viscosity or system permeability.

Two types of constant-displacement pumps have been available: large displacement per stroke (syringe pumps) and small displacement per stroke (reciprocating pumps). Both types offer advantages and disadvantages. The syringe pump, which has given way to the reciprocating pump, normally relies on a mechanical screw to drive a plunger that produces a smooth flow with relatively low mechanical wear. At the end of each stroke, the pump must be refilled, but refilling cannot be done rapidly or automatically; stroke volumes range from 250 to 1000 mL so that several runs can be made between refills. Solvent change, like solvent refill, is not done rapidly. Because of the large solvent volume that is held under pressure in the syringe, solvent compressibility must be considered in maintaining uniform flow rates. For accurate work, it is necessary to stabilize the system at operating pressure before beginning a chromatogram and to inject the sample into a flowing system.

Small-displacement reciprocating pumps have gained wide acceptance in HPLC. Having low internal volume and continuous operation, they afford rapid solvent change, easy flow-rate adjustment, and a choice of solvent programming methods. They also permit recycling. As disadvantages, they produce a pulsed flow whose delivery depends on the reliability of check valves as well as piston displacement. Additionally, system pressure is reached slowly. Because small displacement requires higher stroke frequencies, component wear is greater. Pulsing, intrinsic with reciprocating pumps, is objectionable to HPLC because it produces detector baseline noise and added mechanical stress of the pump's packing. Several methods have been used to eliminate pulsing. The earliest was use of an inertial filter that dampened out the pulse. The filter consisted of a restrictor and a spring-loaded reservoir. Such a filter can be constructed from a 3-ft length of $\frac{3}{8}$-in. stainless steel tubing, pressed flat between the end coupling and coupled to a length of 10/1000 ID stainless steel capillary tubing. The length of capillary needed to achieve the desired degree of dampening can be determined by trial and error. The objective is to apply the minimum restriction needed to achieve dampening but avoid limiting flow rate unnecessarily. Pulse dampeners are available commercially from suppliers of HPLC accessories.

Although pulse dampeners were widely used to adapt reciprocating pumps to HPLC during the early development of HPLC and are still in use, they have been to a large extent replaced by modern systems that were especially designed to meet the needs of HPLC. Most modern HPLC pumps use multiple pistons or diaphragms running in phased opposition so as to minimize pulsing. In some cases, carefully designed overlapping cams are used to match the outputs of separate deliveries. In others, pump delivery speed is controlled by a flow sensor working through a rapid-response feedback loop to minimize time of the return stroke to achieve nearly uniform delivery. In the more sophisticated systems, a combination of flow feedback and dual pistons is used to gain pulse-free flow.

A constant-pressure syringe pump that has been used widely in HPLC is the pneumatic amplifier pump. The syringe delivers 50 mL per stroke and refills automatically in less than 3 s at the end of each stroke. The brief pulse introduced by the refill stroke does not interfere with chromatography. The pump affords a wide flow-rate range, and solvent change-over can be done rapidly. Liquid pressures achievable with the pneumatic amplifier pump are determined by the air pressure applied to the pneumatic piston. Pumps are available commercially with 22:1 and 44:1 ratios with a 100-psi limit on the pneumatic side. A 44:1 ratio provides maximum liquid pressure of 4400 psi.

The major disadvantage of the pneumatic amplifier pump is that it is intrinsically a constant-pressure pump whose output changes with alterations in either solvent viscosity or flow-system permeability. The addition of a feedback flow-control system is essential to achieving constant flow. Because of its large piston volume, recycle chromatography through the pump is impractical.

Because HPLC can be performed with any pump that produces a continuous pulse-free flow, one of the simpler HPLC pumps is a "pressured vessel" consisting of a coiled length of $\frac{1}{2}$-in. stainless-steel tubing attached to a cylinder of either nitrogen or helium that is equipped with a high-pressure regulator. This system offers the advantage of low cost but has the disadvantage of reduced versatility and convenience. It is quite suitable for isocratic work but is severely limited for generating solvent gradients, as discussed in the next section. Its chief advantage is that it can be put together, from materials found in most laboratories, as an answer to short-term needs. When using this system, it is customary to refill the reservoir at the one-half to three-quarters empty stage to prevent gas bubbles from forming in the column.

No single pump design offers all the advantages or is free from disadvantages, so the user must select which disadvantages he is willing to accept in order to gain the advantages preferred.

A state-of-the-art carrier supply system consists of a constant-displacement pump with a small internal volume and the capability of delivering a uniform flow of carrier at pressures up to 6000 psi. It is capable of producing either convex, linear, or concave gradients from mixtures of as many as three solvents. It provides capability for blending a solvent modifier at a 1% level and for an easy change in solvent. Such systems are available from several different LC manufacturers. If HPLC equipment is needed, but modern equipment is not available, the principles just described can be used to guide the chromatographer in fabricating a carrier-delivery system from available substitute components. Such a system may not be as convenient as the state-of-the-art system, but when properly applied, it is capable of high performance liquid chromatography.

B. Carrier Gradient Systems

In the early days of HPLC, the major discussion among practitioners concerned the best type of pump; constant flow versus constant pressure; reciprocating pistons versus a syringe-type piston. That question seems to be settled, and all major manufacturers offer some version of a reciprocating piston pump. The present consideration concerns what type of gradient system is best, and this question probably will not be resolved for several years until enough experience has been gained with the various apparati and the disadvantages of each can be distinguished from the claimed advantages.

Gradient systems (actually solvent-composition programmers) fall into two major categories: prepump and postpump solvent mixing systems. Prepump systems can only be used with small-volume reciprocating piston pumps. The prepump system feeds solvents to a small-volume mixing device on the suction side of the pump. The solvent mixture is then pumped into the column by a single pump. Feed rates of the respective solvents are governed by time-proportioned solenoid valves that are controlled by a microprocessor. Sophisticated logic provided by the manufacturer permits a wide choice of gradient shapes, both simple and complex. Prepump mixing also enables ternary mixtures to be programmed without premixing of components as is typical with dual-pump gradient systems. Programmers of this type may be used to select isocratic mixtures or to vary mixtures automatically. Since solvents are premixed before entering the constant-displacement pump, carrier flow rate does not change during the gradient because of solvent association while mixing.

Postpump mixing is generally accomplished using two electrically controlled LC pumps whose feed rates are varied to produce any one of several (up to 10) predetermined gradient shapes. The more simple logic of the two-pump system normally does not allow complex gradient shapes. Most systems of this type provide for a linear plus several concave and convex gradient shapes; included is an initial and final isocratic "hold."

With the two-pump system, ternary solvent mixtures can be achieved only by switching from solvent B to C stepwise rather than as an A–B–C blend. The effect of this difference on LC separations has not been well demonstrated. Advocates of postpump mixing claim that it produces fewer problems from solvent outgassing during mixing.

One weakness of a multipump mixing system is that at the extremes of the gradient (e.g., 1% A and 99% B) one pump is required to deliver very small volumes. Because delivery is based on a fixed-volume increment, long pauses occur between deliveries of the minor component. At these conditions, carrier composition varies in pulses. It is therefore preferred to operate a two-pump system at >10% of the minor component. When lower concentrations are needed, the low-feed solvent should be blended to permit a higher feed rate of the low-concentration component.

For the benefit of those who do not have an automatic solvent programmer available, some words are in order about manual gradient systems. Solvent programmers have a long history in pH control and a theoretical treatment by Drake [7] is worthy of notation. Manual solvent programmers can be either closed-vessel, open-vessel, or shaped-vessel type; vessels may be used in series or in parallel. For shaped vessels, the program is determined by the vessel shapes. The simplest shape, the complimentary cone, produces a linear gradient [8]. In an emergency, a gradient generator of this type can be improvised from two Erlenmeyer flasks (one inverted and equipped with a breathing tube) joined by small-diameter transport tubing that is fed to the suction of a chromatographic pump. If the vessels are constructed to withstand pressure, the device can be used as the delivery pump with helium displacement gas as the pressure source.

A closed-vessel gradient generator, based on an exponential-dilution flask, is used on the high-pressure side of the LC pump [9]. This system yields a gradient described by

$$C_B = B_0(1 - e^{-v/v})$$

(1)

where B_0 is the actual concentration of the modifying solvent, C_B is the instantaneous concentration of modifier in the effluent, v is the volume eluted from the mixing chamber, and v is the volume in the mixing chamber. Gradient shape from this system is highly reproducible because it is fixed by the concentration of the modifier and the volume of the mixing chamber. The system can be used with any source of pressure, such as an LC pump or a pressurized gas cylinder. Its chief limitation is that it is somewhat cumbersome to operate and it wastes solvent.

An open-vessel gradient generator consists of a series of cascaded vessels with different solvents in each [10]. Solvents moved from one vessel to the next are mixed in each to form a continuous gradient. In a multichamber gradient system, solvent levels are normally held constant to avoid spillover between vessels. In a two-chambered system, shapes can be varied simply by controlling the feed of B into A. When the flow-rate ratio of $B : A = 1$, the gradient shape is identical to that produced by the closed-vessel programmer. When $B : A > 1$, the slope of the gradient is greater than that from the closed-vessel system; when <1, the slope is less than that of the closed-vessel system. To obtain a concave gradient, the feed rate of B must be varied during the program. Changing the ratio of $B : A$ from 0.2 to 0.7 produces a useful concave gradient shape. To use this approach, the initial volume of solvent A must be determined before starting.

A simple open-chamber gradient generator can be improvised by using a small laboratory metering pump to pump solvent B, at a measured rate, into a predetermined volume of solvent A in a stirred vessel, that serves as the solvent reservoir for the HPLC pump [11].

C. Sample Inlet Systems

Sample introduction in HPLC can be done at either elevated or atmospheric pressure. Injection at atmospheric pressure entails stopping carrier flow and opening the system to insert the sample. Elevated-pressure injection into the flowing carrier is accomplished with a loop valve or a self-sealing diaphragm that is capable of withstanding system pressure. Most diaphragms (septa) are limited to pressures of less than 1000 psi and must either be replaced or reinforced for operations above this limit.

A convenient and widely used method of sample injection in HPLC is by syringe injection into a sample valve loop that is isolated by means of a pressure lock from the system's pressure (Fig. 1). This system permits sample injection volumes to be varied without changing loops. It provides the advantage of injecting into the carrier stream without objectionable head pressure. It omits the need to stop carrier flow, and it permits injection in very small volumes (1–2 μL). This system can be used with any type of pump. In this method of injection, sampling precision is limited by the sampling precision of the syringe.

The most reproducible method of introducing samples into a flowing carrier stream is by means of a fixed-volume loop valve (Fig. 2). Sampling volume is determined by the loop volume; sampling precision is independent of the syringe injection that fills the loop. Changes in sampling volume are made by interchanging sample loops. This method of sample injection is preferred for routine analyses where uniform sample volumes are practical and sample quantity is adequate to flush the loop. The sam-

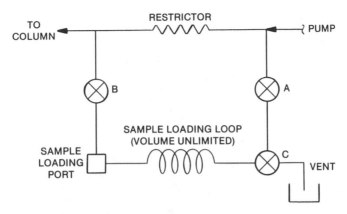

Fig. 1. Schematic diagram of waters' septumless valve injector. (Illustration courtesy of Waters Associates.)

ple volume of fixed-loop injection valves may also be varied by partially filling the loop. Sample–solvent mixing is prevented by the inclusion of an air bubble between the solvent and sample volumes [12,13].

The stop-flow technique is normally used with constant-pressure pumping systems that are capable of returning to system pressure quickly. To make an injection, the carrier flow is stopped by means of a quick-close valve placed before the injector. The sample is then inserted into the injector by means of a syringe. Small injections may be made without a septum enclosure since the sample solution simply displaces a small volume of carrier from the inlet volume. The system is then closed and the carrier flow valve opened to resume operation.

Large or prep-scale samples can be injected into a stop-flow system if the injection is made from a large syringe through a septum that withstands the back pressure of the injection. By injecting slowly, injections of several milliliters can be made by displacing carrier through the column.

D. Column Hardware and Housing

The column for HPLC is usually made from $\frac{1}{4}$-in. stainless steel tubing (0.45 mm I.D.); normally a 30-cm length is used with microparticle packing, and 50- to 120-cm length is used with large-particle packing ($\sim$37 μm). Column length may be increased by joining sections by means of low dead-volume couplings. For ultramicro packings, microcolumns constructed from $\frac{1}{16}$-in. tubing 3–5 cm in length are used.

Most HPLC is done at room temperature with columns exposed to the

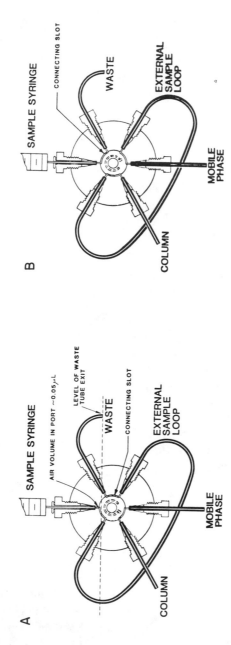

Fig. 2. Valco injection valve with exchangeable sample loop. (A) Load position; (B) inject position. Valve is viewed from spring end. Load to inject is a counterclockwise rotation. (Illustration courtesy of Valco Instruments.)

laboratory atmosphere. For precise work, the column should be placed in an enclosure to protect it from variations in ambient temperature. The enclosure can be as simple as an insulated cover or as complex as a precisely controlled oven. Where separations can be enhanced by operation at elevated temperatures, a temperature-controlled oven is beneficial. Most commercial instruments provide a sheet metal enclosure into which temperature controls can be added. If a commercial oven is not available when needed, a suitable one can be improvised from commercially available laboratory equipment.

III. DETECTORS FOR HPLC

A variety of measurement principles are used to monitor HPLC column effluents. Included in these measurement principles are photometry (UV–visible, fluorescence, and infrared), refractometry, electrochemistry, and flame ionization. The most widely used detectors for HPLC use "light" (photons) as the basis of measurement: light absorption (photometry), light bending (refractometry), and light shifting (fluoremetry). Other detectors, based on electrometric principles (e.g., electrical conductivity and voltametry), are also used extensively but have applications primarily with aqueous systems.

A. LC Detectors Using the Properties of Photons

1. Photometric Detectors

Photometric detectors are applicable to all substances that absorb energy in the ultraviolet–visible range of the spectrum, provided that the compounds of interest absorb energy in the region where the carrier is transparent. Detector response is determined by Beer's law, which states that absorbance $A = abc$; where a is the component's absorptivity, b the cell length, and c the component's concentration in the cell. To gain maximum sensitivity, LC flow cells are designed to use maximum path length and minimum (hold-up) volume relative to the energy throughput of the cell.

One problem experienced with UV/LC detectors is interference from refractive index changes produced by the liquid lens formed by the flowing liquid that results in flow-rate sensitivity. This interference is minimized by use of cell designs that compensate for the contributing effects. For example, one design uses a tapered light path in its flow cell; others split the flow stream longitudinally in the light path.

Two types of photometers are used: a fixed-wavelength filter photometer that uses a mercury vapor lamp as a UV source and a variable-wavelength spectrophotometer that uses a deuterium lamp as a source. Each photometer type offers advantages and suffers limitations. The fixed-wavelength detector, with its mercury vapor source, offers greater energy throughput and greater wavelength stability at lower cost than the spectrophotometric detector. The mercury-vapor source used in the fixed-wavelength detector offers high sensitivity for compounds absorbing at 254 nm; the purity of the major emission line provides a linear response over a wide dynamic range. Instruments with a maximum sensitivity range of 0.0025 absorbance units and a dynamic range of five orders of magnitude [2.56 absorbance units (au), full scale] are available commercially. These instruments are simple to operate and maintain. Properly designed units are not sensitive to either pressure or flow. They provide low baseline noise ($\pm 2.5 \times 10^{-5}$ au with carrier flowing) and low drift ($\pm 2.0 \times 10^{-4}$ au/h). Cells and transport tubing in these units are constructed of materials that are resistant to attack from most carrier solvents and sample materials.

For circumstances in which detection at 254 nm is not adequate, some manufacturers offer a capability for detection at other wavelengths. As a modification to the fixed-wavelength detector, they use other mercury emission lines in conjunction with narrow-bandpass filters to detect at any of the following wavelengths: 280, 313, 334, and 365 nm. By using other source lamps to gain other emission lines (e.g., Cd at 229 and Zn at 214 nm), additional long-wavelength detections are added. In circumstances where they can be used advantageously, selective wavelength detectors of this type, although limited, provide a lower cost option to the spectrometric detector.

2. Spectrophotometric Detectors

The spectrometer detector normally uses a broadband emission source and a grating monochrometer to provide a detector that is continuously variable across the entire UV–visible spectrum from 190 to 650 nm. Although prism instruments have been used for LC detectors, a grating instrument with all reflective optics provides uniform resolution over the entire optical range of the instrument. The source of the UV emission is normally provided by a deuterium lamp and either a tungsten or a quartz iodine lamp is used to provide energy in the visible region. Some detectors include a wavelength-scanning capability as an option. Spectral scans are done by stopping carrier flow while the component of interest is trapped in the cell. In some instruments, spectral scanning can be done automatically; in most, it must be done manually. Other options available with variable-wavelength detectors include a choice of slit width that influences resolution. As the width of the slit determines the spectral band width, one would normally use a narrow slit for scanning the spectrum and a wider slit for monitoring the column effluent.

A remarkable advance in spectrophotometric detection of HPLC effluents was demonstrated by Denton and co-workers, who used a computer-controlled rapid-scan UV spectrophotometer as an HPLC detector [14]. With a scan time of <1 s (0.94 s), they were able to scan the flowing effluent and obtain a spectrum every 20 s producing a three-dimensional chromatogram on time, quantity, and wavelength axes. The three-dimensional chromatogram showed each component peak as a composite of spectra whose maxima were at different wavelengths. The computer also provided a two-dimensional chromatogram on time–quantity coordinates with each component peak recorded at its own wavelength maximum. An advantage of having a three-dimensional plot is that it is able to show mixed component peaks on the wavelength peak magnitude plane with greater clarity. Commercial instruments offering these capabilities are available.

3. Refractive Index Detector

When used with a properly chosen carrier, a differential refractometer is virtually a universal detector for LC. The differential refractometer measures the bending or refraction of a light beam by the sample relative to a reference that may be either fixed or flowing. Normally the reference is the chromatographic carrier. The most sensitive operation is achieved when a large difference in refractive index (RI) exists between the measured component and carrier. Common LC carriers (such as water, acetonitrile, isopropanol, and hexane) have refractive indices in the vicinity of 1.3 RI units; aromatics (such as benzene and toluene) have RI units of ~1.5.

In order to be detected by a differential refractometer, the RI of a component must differ from that of the carrier to produce a signal that is 4× noise. Since the noise level of the differential refractometer is, at a minimum, about 5×10^{-8} RI units, the limiting detectable signal is 2×10^{-7} RI units. Productive operation is achieved at much greater differentials—perhaps 1×10^{-5} RI units between cells. In practical terms, the detectable limit of a differential refractometer is ~1 ppm concentration in the flow cell for materials that differ from the carrier by at least 0.1 RI units. To achieve a 1-ppm concentration in a column effluent requires a higher concentration (perhaps 30 ppm) in the sample. With components having a RI difference <0.1, the sensitivity would be lower.

Substances having RIs less than the carrier produce a negative deflection, whereas those having greater RIs would produce a positive deflection. Because the refractive index of a substance changes in proportion to changes in density, the temperature must be maintained constant for all components interacting with the light beam, including the measuring cell and the sample solution. Solute pressure must also be held constant to prevent spurious response.

Commercial refractometers are of two types, Fresnel and deflection. Fresnel's law of reflection states that light reflected from an interface is proportional to the angle of light striking the surface and the difference in RI across the interface; in this case, the interface is a prism wetted by the chromatographic effluent. To cover the RI working range requires prisms of two different densities: a low-density prism for the range of 1.31 to 1.45 and a high-density prism for 1.40 to 1.55. Since the measurement is made at the interface, the glass must be kept clean to achieve reliable performance. The detector's response is proportional to the light reflected from a prism wetted by the chromatographic effluent relative to a reference solution located in an adjacent area of the prism. When used as an LC monitor, a static reference solution is normally used and response is measured from a baseline that was established when the RI of the sample and reference were the same. Since only the liquid wetting the prism is involved in the measurement, very small-volume cells (2 μL) are possible.

The deflection-type instrument measures the deviation of a columnated beam of light passing through a triangular cell containing the sample solution. A triangular reference cell also in the light beam converts the triangle to a rectangle such that when the RI of the liquid in the two cells are the same, no deflection occurs. The deflection is directly proportional to the difference in refractive index between the two complimentary triangular cells. As with the reflective instrument, this refractometer is normally operated with a reference cell, and response proportional to component concentrations is measured from baseline. There is no need, however, to change component parts to cover the RI range. Because the light beam passes through the liquid, this instrument is not as sensitive to contaminant buildup on the cell surface, but it is affected by highly colored or opaque solutions. It is extremely sensitive to its optical components, such as the light source, cells, mirrors, and photodetectors. Reliable performance requires that all components be held mechanically and thermally stable on an optical bench. Its cell volume of 8–11 μL, although larger than that of the reflective-type instrument, is comparable to those used in UV detection.

Because the refractive index of any solution changes with a change in density, the temperature in both cells of either type of instrument must be the same to avoid drift. This is accomplished by means of a heat exchanger using either a temperature-controlled water bath or by an efficient temperature-controlled enclosure. Because solution density is affected by pressure, pump pulsation (if present) is visible on the baseline of either instrument at high sensitivity. If baseline pulses interfere with the chromatogram, additional pulse dampening should be added to the pumping system.

Refractive-index detectors are useful for samples that do not absorb UV or visible radiation. They can be used with UV-absorbing carriers that interfere with absorbance detectors. For use in gradient elution, they can only be used with a combination of isorefractive carriers, that is, carriers of differing chromatographic strength whose refractive indices are about the same [15]. For example, hexane can be used with isopropanol at low sensitivity. If higher sensitivity is required, ethanol can be added to the isopropanol to match the RI of the mixture to that of hexane.

4.   Fluorometric Detector

Fluorometric detectors measure the energy emitted from a solution that is excited by UV radiation. Many compounds absorb energy that they then fluoresce at longer wavelengths; often absorbing UV radiation and emitting it in the visible region of the spectrum. At low concentration and low absorption, fluorescence F approaches linearity with concentration as defined by

$$F = I_0\phi abc = I_0\phi A \qquad (2)$$

where I_0 is the intensity of the incident radiation, ϕ is the fluorescence yield characteristic of the component, a is the absorptivity of the incident radiator, b is the cell path length, c is the solute concentration, and A is the absorbance, $= abc$. At higher concentrations, where absorbance exceeds 0.01, response becomes nonlinear and measurement should be handled by absorbance rather than fluorescence.

Fluorometric detectors may be either filter or spectrofluorometric photometers. Excitation may be accomplished in a variety of ways. Normally, lamps with phosphor-coated envelopes are used to provide a choice of exitation lines across the spectrum from 254 to 550 nm. Excitation energy is achieved with mercury vapor, Zenon-arc, tungsten–halide, and other types of lamps.

In the filter-type fluorometer, source energy is passed through an optical filter and focused on a cell that can be either a right-angle or an in-line configuration. The incident radiation is blocked with a filter so that mainly the emitted radiation strikes the photodetector. A reference path is used to compensate background radiation that escapes through the filter.

Spectrofluorescence detectors used monochrometers to provide a continuously selectable excitation energy so that components of interest can be detected at their fluorescence maximum. Source energy is provided by a deuterium/tungsten–halogen lamp combination to cover the entire UV–visible range.

Fluorescence-detectable compounds include polynuclear aromatics, vitamins, certain amino acids, many biological compounds, and many drugs and drug metabolites. In addition, fluorescing derivatives can be made from many nonfluorescing compounds to facilitate their detection with a fluorescence detector. This detection method offers high sensitivity but limited linearity. Its greatest asset is that it enables one to monitor, selectively, a fluorescent compound in a background matrix that might prohibit its detection by other techniques. When using a fluorescent detector, the carrier should not absorb radiation at either the exiting or the emitting wavelength and it must not quench the emission. Further, the component of interest must be separated from any substance that might interfere with its fluorescent response. Although the absolute limits of fluorescence detection are generally comparable to those of UV detection at low wavelengths (<210 nm), the superior spectral selectivity of fluorescence detection provides better analytical discrimination and permits greater freedom in selecting carrier solvents.

5. Photoconductivity Detector

The photoconductivity detector is relatively new in HPLC [16]. Although it relies on photons, it is a variation of the electrochemical detector. This detector combines photo-induced ionization with electrochemical detection to gain high selectivity for certain types of compounds. This detector shows selective sensitivity to certain heteroatoms, certain chlorinated hydrocarbons, nitrosamines, sulfonamides, etc. It has a reported linearity range of 10^7–10^4 and sensitivity of 1–2 ng of compounds such as atrazine. Because electrolytes interfere with the detector, it must be used with buffer-free phase systems.

B. LC Detectors Using the Properties of Electrons

All detectors described in the previous section rely on the effects of photons. Another useful group of LC detectors is based on the effects of electrons. Electrometric detectors (which include conductometric, potentiametric, and electrochemical) have all been used in HPLC, although their application has been limited, their potential for use in water analyses is significant.

1. Electrical Conductivity Detectors

Electrical conductivity detectors measure the change in electrical resistance across a flow cell to detect ionic components in the column effluent. The detector is capable of detecting any ionizable substance including acids and salts. Electrical conductivity can be measured with either direct current (DC) or alternating current (AC). Normally, AC is used to avoid electrode polarization, which can cause drift and nonlinear response. Phase-sensitive electronic circuitry is normally used to eliminate dielectric interferences normally associated with AC measurement.

Although the electrical conductivity detector has limited use in HPLC, it is used extensively in ion chromatography where a guard column is used to remove electrolyte and buffer from the eluent prior to detection. Electrical conductivity detectors are temperature sensitive and should, therefore, be either thermostated or temperature compensated. Electrical compensation has been used to reduce the need for a guard column.

2. Electrochemical Detectors

There has been a resurgence of interest in electrochemical detectors after the development of a practical flow cell using thin-film electrodes [17–19]. Electrochemical detectors, although offering high sensitivity (0.5 ng) for some compounds (e.g., biogenic amines), are restricted to a fairly narrow linear dynamic range. These detectors utilize redox reactions as the basis for measurement and are useful for detecting substances that enter into either oxidation or reduction reactions within the range of the applied potential. An electrochemical cell that carries a reaction to completion is a coulometric cell; one in which the reaction is diffusion limited is an amperometric cell. Amperometric cells are more popular in HPLC.

A typical amperometric cell (Fig. 3) consists of a thin-layer glassy carbon electrode and a silver chloride reference electrode. An amperometric controller maintains a fixed potential difference across the reference and working electrodes that serves as a driving force for the redox reaction. A

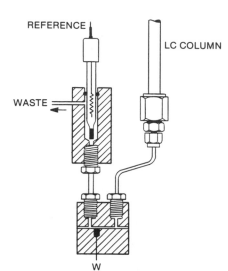

Fig. 3. Thin-layer amperometric detector. (Illustration courtesy of Bioanalytical Systems, Inc.)

sensitive ammeter measures the minute diffusion current (pico- to nano-amp range) by converting it to a voltage suitable for input to a strip-chart recorder. The device may be used either in an oxidative or reductive mode. Potentials between 0 and 2 V/mequiv are applied to the working electrode.

In normal operation, an applied voltage equivalent to the potential of the voltametric plateau [i.e., the hydrodynamic voltagram (HDV)] of the species is used. The voltage is set and a baseline established for the background current of the cell with carrier flowing. An offset voltage is used to displace the background current to zero baseline and the cell is ready for use. As compounds elute from the HPLC column, those which undergo electrochemical reaction invoke an amperometric current proportional to the electron transfer and are detected quantitatively.

The magnitude of the signal, however, is not straightforward. An electrometric cell operating in the coulometric mode is governed by the Faraday law, which states that during a 1 g-eq chemical change, 96,500 couloms of electricity must transfer. However, when operating amperometrically, the energy transfer is diffusion limited and therefore dependent on conversion efficiencies. Therefore, the energy transfer Q is defined by

$$Q = fFnN \tag{3}$$

where f is the energy conversion efficiency, F is Faraday's constant, n is the electron change, and N is the number of moles of analyte. The response, therefore, is dependent not only on the number of equivalents reacted but also on the conversion efficiency. In many cases, with the cell operating on the voltage plateau, similar functional groups have similar conversion efficiencies and therefore produce similar responses. In those cases where the detector cell is not on the voltage plateau, however, the responses may not be similar and a sensitivity factor is required for each component [20]. Since calibration for detector response is a common practice in quantitative HPLC, operation off the plateau should not limit the usefulness of this detector. Because of its high sensitivity and high specificity for a variety of noxious pollutants, this detector should find increasingly broad application in water analyses.

C. Considerations for Choosing an HPLC Detector

Among the detection principles used for general purpose HPLC, photometric detection with a fixed wavelength detector is the most popular. In water analyses, where severe background interferences can be expected, serious consideration should be given to spectrophotometric detection for use at the net wavelength maxima of the compounds of interest. Even greater consideration should be given to electrochemical detection for its high sensitivity to specific classes of compounds. For quantitative analyses, detection by any of these principles requires calibration using either internal or external standards. When used for quantification at high precision and high accuracy, vigorous control of carrier flow rate is mandatory since HPLC detectors using these principles are concentration dependent. A change in flow rate produces a proportionate change in the detector's area response.

IV. DESCRIPTION OF THE CHROMATOGRAPHIC PROCESS

HPLC is a form of elution chromatography in which a narrow band of sample is transported by a liquid carrier through a fixed bed of sorbent. As the sample travels through the bed, its components distribute between the carrier and the sorbent. Its migration through the bed is governed by the distribution of its individual components between the mobile and fixed phases. Components having similar distribution ratios travel through the bed together in a band that gradually increases in width as it advances through the column. A chromatographic separation is achieved only when the migration rate of a band is sufficiently different from that of others that it is eluted from the column separately, that is, free of contamination by other bands. For the separation to occur, the differential migration rates must be able to compensate for the injection band width plus the band spreading that occurred during transport through the bed. To achieve the needed differential migration rate and overcome the band spreading, it is necessary for the sorbent–carrier system to provide the needed differential band retention. Thus, it becomes evident that *band retention* and *band spreading* are the two most basic problems in elution chromatography.

A. Solute Band Retention

Band retention can be measured as either retention time (t_r) or retention volume (V_r). The relationship between the two depends on carrier flow rate (F)

$$V_r = t_r F \tag{4}$$

Solute retention in a chromatographic column can be described as

$$t_r = t_m + t_s \tag{5}$$

where t_m is the time in the moving phase, t_s is the time in stationary phase (which is equivalent to t', the net retention), and t_s is determined by

$$t_s = \frac{K_D V_s}{F}, \tag{6}$$

where K_D, the solute distribution coefficient, is defined as

$$K_D = \frac{[\text{solute}]_s}{[\text{solute}]_m}, \tag{7}$$

where $[\text{solute}]_s$ and $[\text{solute}]_m$ are the concentrations in the stationary (s) and moving (m) phases, respectively.

K_D is a thermodynamic property that defines an intrinsic relationship between the phase combination being used and the solute; it is the basis for retention in chromatography. For a separation to occur, K_D of a component must be different from that of others, for example, $K_{D2} > K_{D1}$. V_s, a column property, is determined by the phase load and volume of the particular column being used.

Because t_r is dependent on phase load, column volume, and flow rate, the dimensionless term K' or capacity factor is generally preferred as the expression for solute retention. K' is defined in Eq. (8) in both fundamental and chromatographic terms:

$$K' = \frac{t_s}{t_m} = \frac{t_r - t_m}{t_m} = \frac{V_r - V_m}{V_m} = \overbrace{\frac{KV_s}{V_m} = \frac{M_s}{M_m}}^{\text{Fundamental terms}} \tag{8}$$

where t' is the net retention time $t_r - t_m$ and M is the mass of solute in the indicated phases.

The relationship M_s/M_m in Eq. (8) shows that a solute's retention is a function of its mass distribution between the stationary and mobile phases. K', known also as phase ratio, is an indication of the extent to which the sorbent is being used. It is, therefore, a most basic parameter of the chromatographic process. K' is related to t_r by Eq. (9) and to V_r by Eq. (10):

$$t_r = t_m(1 + K') \tag{9}$$

$$V_r = V_m(1 + K') \tag{10}$$

B. Chromatographic Band Spreading

Band spreading is intrinsic to the chromatographic process. As a narrow band of solute consisting of a single species is transported through an LC column, the band width increases in proportion to the distance traveled. Several factors contribute to the spreading. These contributions are additive as variances:

$$W_t = (W_0^2 + W_1^2 + W_2^2 ... W_n^2)^{1/2} \tag{11}$$

Some of the contributions to the total band width are

$W_0 =$ initial band width of the sample injection
$W_1 =$ spreading in the injector
$W_2 =$ spreading in pre- and postcolumn transport tubes
$W_3 =$ spreading in couplings
$W_4 =$ spreading in the detector
$W_5 =$ spreading in the column

W_5 can be divided further into "spreading in the mobile phase" and "spreading in the stationary phase." All contributions affect chromatographic peak shape, and all contributions increase the demand placed on the column to produce a greater differential retention in order to achieve a separation. W_1 through W_4 are related to chromatographic equipment and normally are set by the equipment manufacturers. However, the careless selection of a replacement fitting or a piece of transport tubing can produce devastating results that are beyond the control of the equipment manufacturer.

In a properly designed HPLC system, extra column void volume is negligible so that V_r is determined by the column as

$$V_r = V_m + K_D V_s \tag{12}$$

Chromatographic band spreading is directly related to travel through the column so that the ratio $V_r : W_v$ may be used as a simple indication of band broadening during transport through the column; W_v is the chromatographic band width expressed in volume units.

Several concepts have been used to define band spreading. Perhaps the oldest and still most widely used is the Theoretical Plate concept, which envisions the chromatographic column as a series of equal sections in which the solute establishes an equilibrium between the two phases at each section or Theoretical Plate. Using the probability function

$$\frac{1}{1 + t_s} \tag{13}$$

where t_s is the time the solute spends in the stationary phase, several expressions have been derived for calculating the Theoretical Plate Number (N) of a chromatographic column. The commonly accepted method is

$$N = 16(V_r/W_v)^2 = 16(t_r/W_t)^2 \tag{14}$$

where W, expressed as either t or V, represents peak width as measured between tangents drawn from the peak's sides to its baseline, as illustrated in Fig. 4.

N signifies the number of theoretical equilibration stages provided by the column. Consequently, a large N signifies a good column with minimal band spreading during transport. Conversely, a small N signifies a poor column with a high degree of nonselective dispersion or band spreading during transport. N describes column efficiency rather than band spreading. Therefore, the reciprocal relationship, Height Equivalent to a Theo-

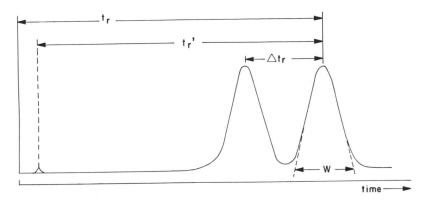

Fig. 4. Diagramatic representation of the expanded resolution equation expressed in terms of retention time and peak width [24]:

$$R = \left[\frac{\Delta t_r}{t_r'}\right]\left[\frac{t_r'}{t_r}\right]\left[\frac{t_r}{W}\right] = \left[\frac{\Delta t_r}{W}\right]$$

(Reprinted with permission of copyright owner.)

retical Plate H, defined in Eq. (15), is used to define band spreading:

$$H = L/N \qquad\qquad (15)$$

where L is the column's length.

Several theoretical models have been proposed to describe band spreading in a chromatographic column; an excellent review of the subject has been presented by Grushka *et al.* [21]. One of the earliest models, proposed by Van Deemter, is shown in simplified form in Eq. (16):

$$H = A + B/U + CU \qquad\qquad (16)$$

A, B, and C are column constants and U is carrier velocity. The A term relates to bed uniformity and includes consideration of particle diameter and the time for a molecule to pass a particle. It is commonly called the multiple-path term. The B term deals with molecular diffusion in which the constant B includes provision for a tortuous path and for solute diffusion in the moving phase. The C term deals with resistance to mass transfer in the stationary phase in which C includes provision for the thickness of the stationary phase, the solute diffusion rate in the stationary phase, and the number of transfers between phases as a function of retention. The effect of retention on H is defined as $K_D/(K_D + 1)^2$, which reaches a maximum when $K_D = 1$. At this condition, the solute spends equal time in each phase, and therefore makes the maximum number of transfers.

This historic equation was used for many years as the band-spreading model for gas chromatography and in later years, as the foundation model for liquid chromatography. A plot of the relationship of H versus U produces a hyperbola with a minimum value that designates an optimum carrier velocity. Any further increase in carrier velocity beyond the optimum requires a greater increase in column length to regain lost resolution, which results in a net increase in analysis time.

When the Van Deemter model was applied to liquid chromatography, it was discovered that in most data no minimum was observed. It was deduced that, because of the slow rate of solute diffusion in liquids, the B term was virtually insignificant and the minimum occurred so close to zero velocity that it was of no practical significance. Consequently, an empirical relationship between H and U, as defined in Eq. (17), was developed and had a profound effect on the development of HPLC:

$$H = DU^n \tag{17}$$

where D is the column constant with reported values ranging from 0.008 to 0.5 cm, U is the carrier velocity, and n is an exponent with values of less than unity and typically 0.3–0.4.

When n is less than unity, the carrier velocity can be increased without producing an equivalent increase in H. This means that theoretically carrier velocity can be increased to the limits of the pressure capability of the system; column length can be increased to recover the resolution lost by the carrier-velocity increase, thereby gaining a net reduction in analysis time. This is the basis of high performance liquid chromatography.

The magnitudes of both D and n appear to be influenced by the particle diameter of the HPLC column packing D_p. Snyder showed a relationship between n and D_p [22], and Majors found the dominant relationship to be between D and D_p, in which $D = D_p^{-1.8}$ [23]. Overall results vary, probably from factors such as packing quality and uniformity. The significance to the analyst is that, with a properly packed microparticle column, carrier velocities of the order of 1 to 2 cm/s can be used without seriously increasing band spreading. The dominant theme of current models for predicting band dispersion in HPLC is that a principal source of spreading is from static pools of carrier trapped between large particles of packing; when this is eliminated, mass-transfer resistance is reduced, and modern HPLC is possible. In practice, if peaks broaden excessively as carrier flow rate is doubled, it can be concluded that something is wrong in the system that needs correction. If a correction is not practical at the time, carrier flow rate should be reduced to a practical minimum where band spreading is less severe until the malfunction can be isolated and corrected.

C. Resolution

The extent to which chromatographic peaks are separated is called resolution. The term is defined by Eq. (18) and is illustrated in Fig. 4.

$$R = \frac{V_2 - V_1}{\bar{W}_V} = \frac{t_2 - t_1}{\bar{W}_t} = \frac{\Delta t}{\bar{W}} \tag{18}$$

where V_1 and V_2 are the retention volumes of components 1 and 2, respectively, and $\bar{W}$ is the average peak width. A separation is complete when one solute is eluted from the column in a different volume of carrier from the other, that is, when $R = 1.5$. In many cases, complete resolution of peaks is not necessary for reliable analytical work. However, the amount of resolution needed varies with the relative amounts of overlapping components present in the sample. For example, when components are at equal concentration, a resolution of 0.6 may still be adequate; at concentrations of 100:1, however, a resolution of 1 is still inadequate as is illustrated in Fig. 5.

A systematic method of estimating R by visually comparing a chromatogram with a reference chart has been reported by Snyder [25]. The reference chart contains an array of 12 theoretical chromatograms representing three concentration levels and four resolution values. The estimated resolution can provide a basis for making judgments about separation. Such comparisons are extremely valuable when an unresolved component is seen only as a shoulder or a misshaped peak.

If separation is not adequate, it becomes necessary to either increase band separation or reduce band spreading. The effect of each option is illustrated diagrammatically in Fig. 6. The parameters addressing both options are found in the expanded resolution equation:

$$R = \left(\frac{\alpha - 1}{\alpha}\right)\left(\frac{K'}{1 + K'}\right)\left(\frac{1}{4}\sqrt{N}\right) \tag{19}$$
$$\quad\quad\quad A \quad\quad\quad B \quad\quad\quad C$$

where N is the number of theoretical plates, α is $\frac{K_2}{K_1}$, K_1 and K_2 are K_D of individual components, and A, B, and C designate selectivity, capacity,

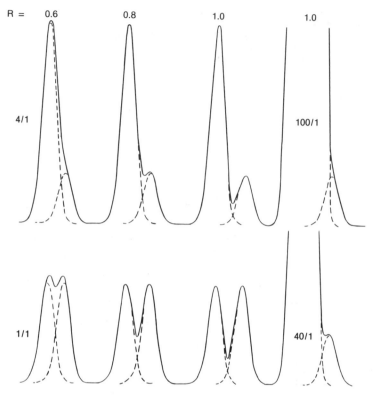

Fig. 5. Effect of relative peak size on peak overlap for values of R as indicated. Peaks are defined by $y = ae^{-x}$ and $a = 5, 20, 200,$ and 500 [24]. (Reprinted with permission of copyright owner.)

and column efficiency terms, respectively. Each term in this equation is addressed separately. The equation may be represented in empirical terms using simple measurements taken directly from the chromatogram, as illustrated in Fig. 4 and stated in Eq. (20):

$$R = \left(\frac{\Delta t}{t'}\right)\left(\frac{t'}{t_r}\right)\left(\frac{t_r}{\overline{W}}\right) = \frac{\Delta t}{\overline{W}} \qquad (20)$$
$$\quad\quad A \quad\quad B \quad\quad C$$

A chromatogram evaluation can be done either by computation using Eq. (19) or by inspection using Eq. (20). If term A, which defines selectivity, is large ($A > 0.5$), the column is providing selectivity; whereas, if it is very small ($A < 0.1$), a different phase may be needed. If the capacity,

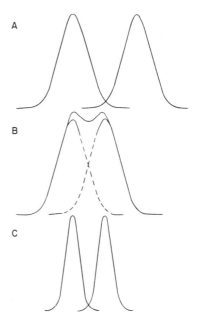

Fig. 6. Effect of band width and band position on resolution of Gaussian peaks [24]. (A) Increase relative retention, $R = 1.2$, $W = 2$; (B) initial separation, $R = 0.6$, $W = 2$; (C) decrease band width, $R = 1.2$, $W = 1$. (Reprinted with permission of copyright owner.)

term B, is very small ($B < 0.1$), the stationary phase is not being used and resolution suffers. To increase resolution, increase B by decreasing solvent strength. A practical working range for B is $0.9 > B > 0.5$. Figure 7 provides an illustration of the effect of the B term on resolution when the A and C terms are held constant. In practice, it is not uncommon for all three terms of this equation to be influenced somewhat by changing sol-

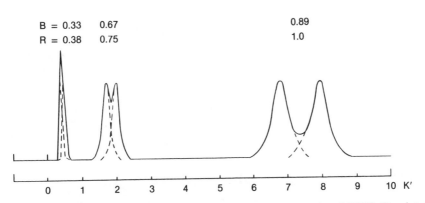

Fig. 7.   Illustration of the effect of B on resolution at constants A and C [26]. (Reprinted with permission of copyright owner.)

vent strength; however, this illustration demonstrates, theoretically, how resolution is lost with poorly retained components.

The C term, which relates to band spreading, was discussed in detail in the previous section. Equation (19) shows the relationship of band spreading to peak resolution. In the previous section, it was indicated that a simple measurement (t_r/W) is a reasonable indication of column performance. As seen in Eq. (20) that relationship is the C term in the resolution equation. It can be seen that resolution is increased by increasing relative retention (column selectivity), increasing retention (capacity factor), and reducing band spreading (column efficiency).

V. SOLUTE RETENTION IN HPLC

Solutes are retained in HPLC either by sorbing them on the surface of a phase (adsorption) or by solvating them in the bulk of the phase (absorption). The retention may be the result of surface charge, hydrogen bonding, van der Waals forces, complexation, or, as is often the case, a combination of several forces. Rather than attempt to classify HPLC in terms of these uncertain mechanisms, an empirical classification has developed that is based on the types of columns and solvent systems that are used for the separations. The classifications, listed in Table I, are described in the sections that follow.

A. Normal-Phase Chromatography

Normal-phase chromatography (NPC) is practiced with a porous sorbant (e.g., silica gel) of small uniform particle size and a nonpolar carrier

TABLE I

Empirical Classifications

Type of chromatography	Column packing	Carrier
Normal phase	Silica gel, aluminum, or polar-banded phases	Nonpolar
Reversed phase	Bonded polymers	Aqueous/organics
Ion exchange	Ion exchanges	Aqueous buffers
Size exclusion	Porous gels	Strong solvent
Paired ion	Normally practiced in the reversed-phase mode but can be practiced in a normal-phase mode as well	

(e.g., hexane), whose eluting strength is adjusted by adding polar modifiers (e.g., alcohol). Solute retardation is achieved by sorption on the polar sites that are rigidly fixed on the surface of the sorbant particle. The surface area of the fully porous sorbants are usually 250 m^2/g or greater.

A solute's retention on a polar sorbant is strongly related to the number and nature of its functional groups. In simplistic terms, it can be stated that on either silica gel or alumina, the retention of COOH > OH > COOR > ROR > RCHO > aromatics > paraffins, and that the retention of any species increases as the number of polar groups is increased.

Adsorptive retention in a chromatographic column is a competitive process in which the solute molecules and the carrier molecules compete for the active sites of the packing so that the net retention of a solute is determined by the activity of the sorbant and the net polarity of the solute in the carrier. Spatial factors are a consideration also. Although solute molecules in the liquid phase are free to align themselves with solvent molecules for maximum interaction between their functional groups, they lack this freedom in the adsorbed state where the spatially fixed sites vary with the geometry of the solute molecule. Consequently, a uniqueness in selectivity can arise from the interactions between functional groups on the solute and those on the rigid sorbent. As a result, spatially oriented geometric isomers having similar total polarity can often be separated on solid adsorbents when other methods fail. In contrast, the weak interactions of alkyl groups with the active sites on solid adsorbents render liquid adsorption chromatography relatively ineffective for separating a homologous series of molecules having a similar polarity.

The most commonly used adsorbants for normal-phase chromatography are silica gel and alumina, but other materials such as aluminosilicate clay, florisil, and charcoal have been used in selected applications. For use in HPLC, adsorbant particles must be small and of a narrow particle size range.

Commercial columns are prepared from either 5- or 10-μm particles with a narrow size-distribution range. Particles may be either spherical or irregular; both are available from suppliers. Higher efficiencies are claimed for spherical particles although irregular particles continue to be in wide use and are providing high-quality separations.

Previously, both fully porous and pellicular or superficially porous particles in the 35-μm size range were in common use. The latter, to a large extent, have given way to the fully porous microparticles, whose particle diameter (d_p) is 5–10 μm.

Bonded phases, originally developed for use in the reversed-phase mode, have come into use in normal-phase chromatography also. Commercial packings having either amino or cyano groups are being used successfully with nonpolar carriers to which polar modifiers are added to increase migration. The most popular polar-bonded phases are based on silane reactions, for example, aminopropyl, ethylamino, propylpropyl-amino, cyanopropyl, or cyanoethylpropylaminosilane. These packings offer the selectivity of their organic moiety, have excellent stability, respond rapidly to changes in mobile-phase compositions and are compatible with gradient elution.

As disadvantages, bonded amine columns are susceptible to attack by oxidation and/or reaction with certain sample constituents that can limit their useful life. Guard columns, available from chromatography suppliers, are generally used to protect the amine groups from such attack.

Nonpolar solvents are used in NPC. Solvent strength is increased (decreasing retention) by adding polar modifiers at levels ranging from a trace to several percent. Typical carriers are n-hexane, benzene, chloroform, and methylene chloride. Typical modifiers include isopropanol, methylene chloride, and tetrahydrofuran. Additional information on carriers and modifiers are included in the discussion of mobile-phase selection.

B. Reversed-Phase Chromatography

Reversed-phase chromatography (RPC) uses a polar mobile phase and a nonpolar stationary phase. As currently practiced, the mobile phase consists of water mixed with a polar organic solvent. The stationary phase consists of a hydrocarbon that is chemically bonded to a silica support. Although variations occur, such as using nonaqueous mobile phases or nonhydrocarbon stationary phases, the basic concept remains the same, that is, a nonpolar stationary phase used in conjunction with a polar mobile phase in which the solute is soluble but not miscible.

The retention mechanism in reversed-phase chromatography appears to be based on a net interaction that depends on a solute's insolubility in a carrier and a comparatively weak attraction exerted by the packing. Although the most commonly used models for RPC retention are solvophobic in nature, it appears that a more broadly based model is needed. Simplistically, reversed-phase chromatography works best with substances that are weakly soluble in water and have hydrocarbon moieties that can be attracted by the molecular forces of the hydrocarbon groups immobilized on the column packing.

Although relative solvophobic effects play a dominant role in RPC separations, the contribution of the stationary phase should not be ignored; surface wetting, solvent penetration into the bonded phase, and organic modifier penetration into the bonded phase all have the capability of influencing solute retention, and thus altering selectivity [27].

Packings for RPC generally consist of a silica particle to which an organic moiety has been attached by means of an Si–C bond. Most widely used additions consist of one of the following: C_{18}, C_8, phenyl, CN, or NH_2. Typically, the support particle's diameter is either 5 or 10 μm, having a surface area in a range between 90 and 400 m^2/g. Particle shape may be either spherical or irregular.

Phase thickness, governed by the reaction mechanism, may be grouped into two categories: monolayers and multilayers. Monolayers, offered by some vendors, are reputed to allow faster analysis, particularly with high molecular weight substances. Theoretically, monolayers should provide less resistance to mass transfer because of their shorter diffusion paths in the stationary phase. In practical HPLC, both speed of analysis and phase capacity must be considered when choosing between monolayered and multilayered reversed-phase packings.

Packing performance is affected by the nature of the support as well as the composition of bonded organic phases. Both surface area and pore diameter of the solid support influence a packing's properties, because these parameters affect phase load and availability. The separations compared in Fig. 8 illustrate the difference between two C_{18}-bonded packings: one having a high surface (400 m^2/g) and small pore diameter (80 Å); the other having a low surface (90 m^2/g) and a large pore diameter (300 Å). The large-pore packing material (TP) carries a higher phase density ($\times 2.3$) and a lower phase load ($\times 0.52$) than the small-pore material (HS). The large-pore material (TP) shows a higher total retention and a better performance with large hydrophobic molecules. The small-pore material, having a higher phase load, offers greater retention, higher capacity, and better overall performance with smaller hydrophylic molecules that have access to all the sites.

Carriers for RPC are prepared from mixtures of polar solvents such as water and acetonitrile or water and methanol. Solute migration rate is controlled by the proportion of the organic component in the mixture. If the organic solvent used in the original mixture does not provide an adequate migration rate for some strongly retained component, a more lipophilic solvent such as tetrahydrofuran (THF) may be used to further modify the carrier. The proportions of water-to-organic solvent used in the mixture depend on the sample composition and whether the separation is being made isocratically or by gradient elution. For isocratic operation with an unknown sample, a suitable starting concentration is 1 : 1 water : organic solvent, which can be modified as needed to obtain the desired retention. For gradient operation, a typical program might begin with a 70 : 30 mixture of water : acetonitrile and end with 100% acetonitrile. In a case where either greater carrier strength or solvating power is needed, the gradient can be extended with acetonitrile : THF to 100%

THF. The chromatogram shown in Fig. 9 was prepared with a carrier gradient that began with a 1:1 mixture of water:acetonitrile and programmed to acetonitrile, then ethylacetate, and finally methylene chloride. A reversed-phase carrier need not be limited to a water:organic–based system. Any group of solvents that provide the needed solubility gradient and the partitioning distribution for the components in the sample can be used.

Cascaded solvent systems are increasing in popularity as equipment manufacturers make provision for their use in state-of-the-art solvent supply systems. If existing equipment does not make provision for cascaded solvent programs, higher strength solvent blends (e.g., 30% THF in acetonitrile) can be used in conventional two-solvent gradient systems.

The water used in HPLC carriers should be free from organics to avoid problems with spurious peaks. Reversed-phase packings have a tendency to sorb traces of organics from the water and then release them as the carrier strength is increased. For this reason, pumping 100% water through the column when starting a run is discouraged.

C. Paired-Ion Chromatography

Paired-ion chromatography (PIC) is a useful variation of HPLC in which ionic substances are converted to lipophylic liquids that distribute

---

Fig. 8. Separations on two different types of C_{18} reversed-phase packing. (A) Larger pore, greater phase density. (1) Benzene; (2) naphthalene; (3) acenaphthylene; (4) acenaphthene; (5) fluorene; (6) phenanthrene; (7) anthracene; (8) fluoranthene; (9) pyrene; (10) benz(a)anthracene; (11) chrysene; (12) benzo(b)fluoranthene; (13) benzo(k)fluoranthene; (14) benz(a)pyrene; (15) dibenz(ah)anthracene; (16) benzo(ghi)perylene; (17) indeno (1,2,3,cd)pyrene.

Column:	Vydac 201TP5415
Gradient:	solvent No. 1, water; solvent No. 2, acetonitrile; start 50% solvent No. 2 for 3 min, then run linear gradient programmed over 7 min to 100% solvent No. 2
Flowrate:	1.5 mL/min
Detector:	UV at 254 nm

(B) Large surface area, higher phase load per gram of packing. (1) Solvent front; (2) glyoxalic acid; (3) tartaric acid; (4) malic acid; (5) lactic acid; (6) acetic acid; (7) citric acid; (8) succinic acid; (9) propionic acid; (10) maleic acid.

Column:	Vydac 201HS104
Solvent:	0.01 M triethylamine adjusted to pH 2.0 with phosphoric acid
Flowrate:	1.5 mL/min
Detector:	UV at 220 nm [28]

(Illustration courtesy of The Separations Group.)

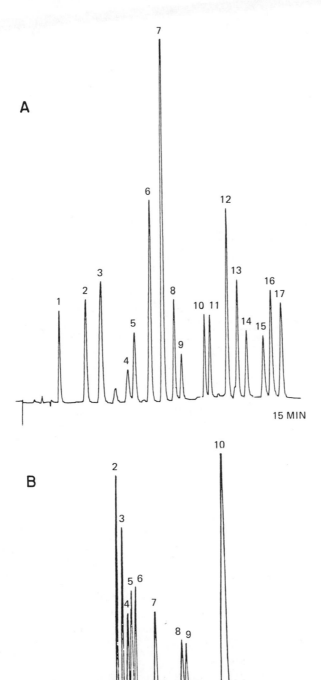

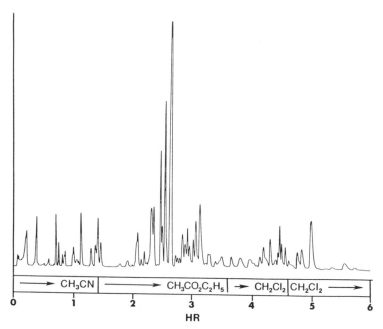

Fig. 9. Polyaromatic compounds extracted from carbon black and separated by HPLC using a ternary phase gradient. Eluent: (1) Initial 15 min with 1 : 1 water/acetonitrile; (2) next 70 min with a slightly concave gradient; (3) Waters program 5 to 100% acetonitrile; (4) next 130 min with a linear program to 100% ethylacetate; (5) final 60 min with a linear program to 100% methylene chloride. Compound range: (1), Pyrene = 202; 53, pyreno 4,3,2- a b c coronene; M = 448. [29] (Permission to reprint was granted by the American Chemical Society.)

between conventional LC phases. Typically, a water-soluble solute is combined with an ion-pairing reagent to produce a lipophylic liquid that partitions between a polar and a nonpolar phase; either phase may be held stationary. In the reversed-phase mode, the reagent is added to the aqueous carrier and used with a conventional bonded packing. In the normal-phase mode, the sample, dissolved in water, is combined with the ion-pairing reagent and then extracted into the organic mobile phase for separation. The aqueous stationary phase containing the ion-pairing reagent and pH buffer, added to minimize dissociation of the ion pair, is dispersed on silica packing.

By the normal-phase technique, Karger *et al.* [30,31] separated a variety of materials such as carboxylic acids, sulfa drugs, thyroid hormones, and biogenic amines. Either tetrabutylammonium hydrogen sulfate or perchloric acid–sodium perchlorate was used for ion pairing and as the stationary phase. With a system of 0.04 M TBA HSO_4–0.31 M Na_2SO_4 in

0.25 *M* borate buffer at pH 8.5, they separated 11 sulfa drugs in less than 12 min using 25% v/v *n*-butanol in heptane as the carrier.

Although satisfactory for isocratic operation, normal-phase PIC seems to be at a disadvantage where solvent programming is required because the aqueous stationary phase is unstable with changing carrier composition. As a result, operation in the reversed-phase mode has found greatest acceptance. Ion-pairing reagent kits for use with reverse-phase columns are available commercially.

The most commonly used reverse-phase ion-pairing reagents are alkyl sulfonates, for use with cationic solutes, and alkyl ammonium salts and quaternary amines, for use with anionic compounds. Many of the most useful ion-pairing reagents, especially used at the concentration needed for high efficiency, are extremely corrosive to HPLC columns and equipment. Consequently, care must be exercised to flush the system after use and to avoid allowing reagents to stand in the system when not in use.

There remains an on-going discussion about the actual retention mechanism of PIC. The discussion centers on whether the solute is separated as an ion pair that is formed in solution or by a dynamic exchange at the adsorption site. Regardless of which mechanism is correct, the model used here provides a basis for understanding the technique; a technique that provides a useful solution to otherwise difficult problems and that utilizes conventional HPLC columns.

As an alternative to PIC, ion suppression may be used with weakly dissociating compounds to increase their lipophylicity. Solute dissociation is accomplished by adding a common ion to the carrier. For example, the addition of a small quantity of one of the following acids: trifluoroacetic, trichloracetic, perchloric, acetic, or phosphoric, to the carrier increases retention and decrease tailing of carboxylic acids in a reverse-phase system. In a like manner, the addition of a basic substance performs a similar function on basic solutes. This technique is not new. In 1966, Kessner separated 15 Kreb's-cycle acids in the normal-phase mode using a column of silica gel coated with aqueous H_2SO_4 and a carrier programmed from chloroform to tertiary amyl alcohol via a concave gradient [32].

D. Ion-Exchange Chromatography

Ion-exchange chromatography (IEC) is practiced with pH-controlled aqueous carriers and ionically charged packings. The solute is dissolved in an aqueous carrier and pH is maintained at a value to promote dissociation. The packing, usually a synthetic resin, consists of an insoluble polymeric matrix whose surface contains fixed-charged groups and mobile counterions. The counterions can be exchanged for similarly charged ions in the carrier. The exchange is based on chemical equilibrium. The charge of the counterion identifies the type of resin. Resins that attract negative counterions are called anion exchangers; those attracting positively charged counterions are called cation exchangers.

Anion exchangers are produced as either strongly or weakly basic, while cation exchangers are produced as either strongly or weakly acidic. The strongly basic anion exchangers most commonly used in HPLC contain a quaternary ammonium exchange group $[CH_2N^+(CH_3)_3Cl^-]$, and weakly basic exchangers have protonated amines $(N^+HR_2Cl^-)$. The most commonly used strongly acidic cation exchangers contain a sulfonic acid group (SO_3H^+), and the weakly acidic exchangers contain a carboxylic acid group $(COOH^+)$. The respective counterions of the anionic and cationic exchangers $(Cl^-$ and $H^+)$ exchange with appropriately charged ions in the carrier.

The substrate for the charge groups is normally a porous resin particle prepared as a copolymer of polystyrene. Newer exchangers have been prepared by introducing charged groups into organic matrices that previously have been bonded to a porous silica substrate. The merits of each are discussed in vendors' literature.

Solute retention in ion-exchange chromatography is controlled by four factors: (1) charge strength of the solute ion; (2) ionic strength of the carrier; (3) charge strength of the counterion on the resin; and (4) pH of the carrier. Solute retention is decreased by increasing ionic strength of the carrier; by replacing the starting counterion with one of lower charge or by adjusting carrier pH to suppress dissociation of the involved ionic species.

Ion-exchange resins are received from vendors in a stable form, anionic resins as chlorides and cationic resins as sodium salts. Prior to use, they should be equilibrated in the carrier because column capacity (V_s) depends on degree of coverage by the active counterion. The counterion must be chosen to provide the desired activity. As a guideline, a cation exchanger is most active in the lithium form and least active in the silver form. Note that the resin, having the lowest affinity for lithium, is most active when it is bearing this weakly bonded counterion. The sequence of affinities for monovalent counterions is as follows:

$$Li^+ < H^+ < Na^+ < NH_4^+ < K^+ < Rb^+ < Cs^+ < Tl^+ < Ag^+$$

Affinity increases with molecular weight and valency. For bivalent ions, the affinity sequence is $Be^{2+} < Ca^{2+} < Sr^{2+} < Ba^{2+}$. The affinity sequence for some common anions is

$$F^- < HCO_3^- < OH^- < Cl^- < BrO_3^- < HSO_3^- < CN^- < Br^- < NO_3^- < I^-$$

Many separations of organic anions are done with the exchanger in the nitrate form, which is less active than the chloride form received from the vendor.

The greater the affinity of an ion for an exchanger, the more difficult it is to remove the ion by subsequent elution with a solvent. In fact, some rare earth elements are so strongly held that hot concentrated acid is needed to remove them from the exchanger. Precautions are needed when ions of great affinity are involved in the separation because, if they are not completely removed from the column, unsuspected contamination may affect later operation. Broadly speaking, the greater the diameter of the ion, the more firmly it is held by the resin. This suggests that affinity is related to the activity coefficient of the ion on the resin phase.

Aqueous solutions of buffers and neutral salts are used for carriers in ion-exchange chromatography. The salts provide the ionic strength that is needed to regulate solute migration, while the buffer maintains the carrier at the desired pH where retention is independent of solute concentrations. The effect of pH on retention is difficult to predict accurately, since several properties are influenced by pH, including solute dissociation, resin dissociation, and resin swelling, all of which affect resin capacity.

Useful information can be gained by running titration curves of an ion-exchange packing to determine the pH region in which exchange capacity is stable. Such information can be useful in cases where the preferred pH for the solute is different from that for the resin and where a compromise must be made. In general, the point of net maximum dissociation between the resin and the solute is prepared. Because salts of a weak acid experience greater dissociation than the acid itself, pH should be increased from the pK_a by one or two units as illustrated by the following treatment of the ionization equation:

$$K_a = \frac{[H^+][B^-]}{[HB]}, \tag{21}$$

which can be expressed as

$$pH = \log \frac{[B^-]}{[HB]} + pK_a \tag{22}$$

This equation indicates that with the carrier pH at the pK_a of the solute, dissociation is only 50%; at pH $= pK_a + 1$, dissociation is 91%; and at $pK_a + 2$, dissociation is 99% complete.

Weak bases can be protonated and treated as cations. When this is done, the same guidelines apply; that is, a reasonable carrier pH is at $pK_a + 2$.

E. Size-Exclusion Chromatography

In the various modes of chromatography just described, solute retention is based on an attraction between a packing and a solute where the attracting force is influenced by a solute's composition. There exists another very useful form of chromatography where a deliberate effort is made to suppress all forces that respond to a solute's compositional differences, so that molecules, diffusing freely, can be separated according to their size in solution. This method, known as size-exclusion chromatography (SEC), is applicable to virtually any soluble species of any size or molecular weight.

To make a separation by SEC, the sample is passed through a column of a porous permeable packing by a carrier that fills both the packing's pores and the interparticle spaces. Sample molecules are free to diffuse into any liquid available to them. Sufficiently small molecules can diffuse into all the available pores; those too large to enter any pores must remain in the interparticle space. Between these extremes, the intermediate-sized molecules can penetrate some fraction of the pores in accordance with their size in solution. As the carrier passes through the column, it carries with it only the molecules in the interparticle spaces. The molecules inside the pores are left behind until they diffuse out of the pore and into the moving carrier. Large molecules that cannot enter the pores are swept along with the carrier and eluted first. Small molecules that spend time in the pores are eluted later. Retention, therefore, is related to the time that species spend in the gel pores.

This relationship can be expressed as an equilibrium distribution between solute concentration in the solvent, trapped in the gel (C_p), and interparticle solvent (C_o) as

$$K_o = C_p/C_o \qquad (23)$$

when $K_o = 0$, that species is excluded. When $K_o = 1$, it is totally permeating and therefore retained to the maximum extent possible. For all cases where $1 > K_o > 0$, the molecules are retained in accordance with their size in solution and may be quantified by means of a calibration curve as illustrated in Fig. 10. If any peaks show $K_o > 1$, sorbant forces are in effect that distort the validity of the size measurement. Such effects may be of no consequence if molecular size measurement is not being made. When an analytical separation is sought, such effects may even serve as an advantage.

Because the solute distributes between trapped carrier in the pore (V_s) and a moving carrier (V_m), SEC may be considered a special form of

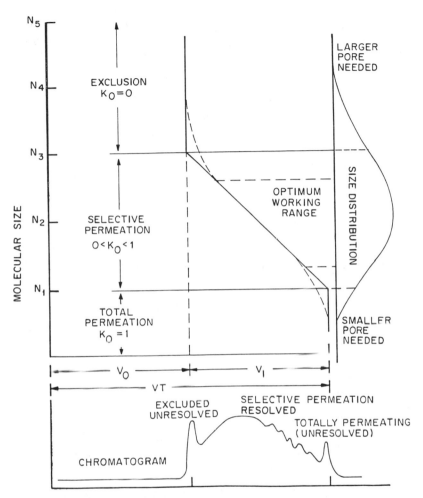

Fig. 10.   Diagramatic representation of the interrelationship between the molecular size distribution of the sample (top right), the calibration curve (top left) and the chromatogram (bottom) [24]. Solid lines, theoretical curve; dashed line, real curve. (Reprinted with the permission of copyright owner.)

liquid–liquid chromatography in which retention volume is controlled by the permeation coefficient defined as

$$K_o = \frac{[S]V_{pa}}{[M]V_{pv}} = K_D \frac{V_{pa}}{V_{pv}} = \frac{V_{pa}}{V_{pv}}, \tag{24}$$

where V_{pa} is the volume of pore available to any size species, V_{pv} is the pore volume of the packing (which is equal to the volume of stationary phase), [S] the solute concentration in the stationary phase available to a given size species, [M] is the concentration of solute or carrier, and K_D is the distribution coefficient.

When the solvent trapped in the pores is identical to the carrier, as it should be for SEC, the partition coefficient K_D for any species equals unity and vanishes from the equation so that the permeation coefficient for a species of any given size is determined by the fraction of the pore volume available to it, as shown in Eq. (24). Retention volume in SEC is described by the relationship

$$V_r = V_m + K_o V_{pv} = V_m + V_{pa} \tag{25}$$

Note that the retention volume V_r in SEC is simply V_{pv}.

Packings for SEC are composed of particles having pores of controlled size. Since the pore size distribution determines the working range of the packing, the permeation range of a column must be selected to accommodate the hydrodynamic size of the solute as is illustrated diagrammatically in Fig. 10. If the size distribution of the sample is outside the range covered by a given column, additional columns of different pore size may be coupled together to provide the total range needed. It has been a common practice to arrange coupled columns in descending order of pore size.

The permeation range of a column should be related to the expected molecular weight range of the sample. Permeation range is usually identified by the exclusion limit of the packing. However, different vendors use different methods of expressing the dimension. The following are three different methods used to describe the permeation of commercially produced SEC packings.

1. Polystyrene gels. Exclusion limit, called porosity, is expressed as the extended chain length of a polystyrene molecule that, by extrapolation, is excluded from the gel.
2. Polydextrans. Exclusion limit is expressed as molecular weight of the spherical molecules used in the determination.
3. Porous glass. Average pore diameter of the packing is determined by a physical measurement, for example, mercury porosity.

In most cases where the dimension does not relate to the hydrodynamic dimension, a calibration curve is required to relate the given dimension of the packing to the molecular size or molecular weight range of the sample. Fortunately, many vendors are now identifying their packings with a recommended molecular weight range as well as a size dimension.

TABLE II

Physical Properties of Solvents Commonly Used for SEC[a]

Solvent	Boiling point (°C)	Viscosity[b] (cp)	Refractive[b] index	UV cutoff (nm)
Chloroform	61.2	0.56 (20°C)	1.4476 (20°C)	245
Tetrahydrofuran	66.0	0.51 (25°C)	1.4070 (20°C)	220
Ethylene dichloride	84.0	0.84 (20°C)	1.4443 (20°C)	230
Toluene	110.6	0.52 (25°C)	1.4893 (24°C)	285
Tetrachloroethane[c]	146.5	1.84 (15°C)	1.4940 (20°C)	—
Dimethylformamide	153.0	0.90 (25°C)	1.4280 (25°C)	—
Dimethyl sulfoxide[c]	189.0	2.0 (25°C)	1.4773 (25°C)	—
Water[d]	100.0	1.0 (20°C)	1.3330 (20°C)	200

[a] From Weast and Astle (1978–1979) and Dean (1973).
[b] The experimental temperatures are given in parentheses.
[c] High-viscosity solvents are generally used at elevated temperatures.
[d] Buffers or electrolytes, when added to an aqueous carrier, alter the listed physical properties of water.

Packings for SEC are basically two types, rigid and nonrigid. Nonrigid gels (such as Sephadex and Biogel*) are used extensively in gel filtration chromatography but do not find much application in HPLC. Rigid packings (such as Styragel, Zobax, and Spherosil**) withstand the velocity flux associated with the high flow rates used in HPLC and are the principal candidates for modern HPSEC.

Carriers in SEC, technically do not enter into the equilibrium distribution in the same way as in sorption chromatography. The chief function of the carrier in SEC is to dissolve the sample and transport it through the column. To be a good carrier, the solvent should be sufficiently similar to the packing so as to suppress sorption. It should have a low viscosity and be compatible with the detector. Solvents commonly used as SEC carriers are listed in Table II along with physical properties [33, 34].

Although the carrier should not affect solute retention in exclusion chromatography, solvent effects can occur either by altering the size of the solvated molecule or by altering the pore dimension of the swollen gel. Molecular solvation and association, experienced with polyelectrolytes and biopolymers, require the addition of electrolytes to the carrier to prevent molecular relaxation in solution [35]. In the relaxed or extended

* Manufacturers for these trademark products are Pharmacia Uppsala and BioRad.
** Manufacturers for these trademark products are Waters Associates, E. I. DuPont, and Rhone-Poulene of France, respectively.

state, the hydrodynamic volume of these materials becomes concentration dependent and produces anomalous results. To obtain reproducible retention volumes that are independent of concentration, proper control of pH and of electrolytic strength is necessary.

VI. MAKING A SEPARATION

Although the operational procedure is simple, HPLC is a relatively complex technique because of the many variables that affect solute retention and because of the inexact composition of column packing materials. HPLC is at best an empirical technique that requires judgement, especially with samples of unknown or uncertain composition. In many cases, optimization may be required for a particular sample or for a new column even when well-defined procedures are being used.

When a sample of unknown or uncertain composition is encountered, the following steps provide a systematic approach to solving the separation problem.

1. Choose a separation mode
2. Choose a column for that mode
3. Choose a carrier of moderate strength for that mode
4. Make a trial separation
5. Optimize the separation

A. Choosing a Separation Mode

The choice of separation mode depends on the following:

1. The nature of the sample
2. The type of separation wanted
3. The selectivity needed
4. Experience with similar compounds

Important questions about the nature of the sample are

1. Does the sample contain polymeric material?
2. Does the sample contain organic extractable material?
3. Does the sample contain ionic material?
4. Is the sample acidic, basic, or neutral?

If the compound classes are known, their physical and chemical properties may be listed in the various handbooks and can be used in making decisions. Basic chemical and physical properties provide a basis for selecting a separation strategy.

The next consideration is "What type of analysis is wanted?"

1. Is the analysis of molecular size or composition?
2. Is the determination of selected compounds or of complete speciation of all components in the sample?
3. Must the components of interest be separated from a background?
4. What is the background matrix?
5. Is qualitative or quantitative information or both wanted?

The next consideration is information about the liquid chromatographic properties of the sample. Potential information sources are

1. Published methods for samples of like sample composition
2. Operating experience with similar samples
3. Previous experience with similar compounds in other matrices
4. Previous experience with analogous compounds

In many cases, the water analyst does not start from "scratch." Published methods on the compounds of interest or their analogs may be available from various sources and should be used. Adjustments may be required to compensate for the sample matrix, but the retention data on columns and carriers are usable for reference. If information on specific compounds of interest is not available, then data from analogous compounds can be used for making judgments about an HPLC approach. The diagram in Fig. 11 provides a relatively simple scheme that can be used as an aid in making selection.

The first decision in this scheme is based on the molecular weight of the sample. The decision value of $M = 2,000$ is, indeed, arbitrary since size separations have been performed on compounds having molecular weights of less than 100 and reversed-phase separation performed on compounds having molecular weights greater than 20,000. Note also that many compounds have alternate routes open to them so that other considerations may be factored into a choice of route. Among these considerations are

1. What columns and solvents are available at the time of need?
2. What is the solubility range of the compounds in a specific sample?
3. What range of solvent strengths is required in one separation mode versus another?

Most organic-soluble substances can be separated by reversed-phase HPLC. Nonpolar hydrocarbons, separable by either mode, require liquid solid chromatography (LSC, discussed in Section VI,B) for isomeric separations and RPC for homologous separations. Ionizable organics previously separated by ion exchange may be modified with ion-suppressing or

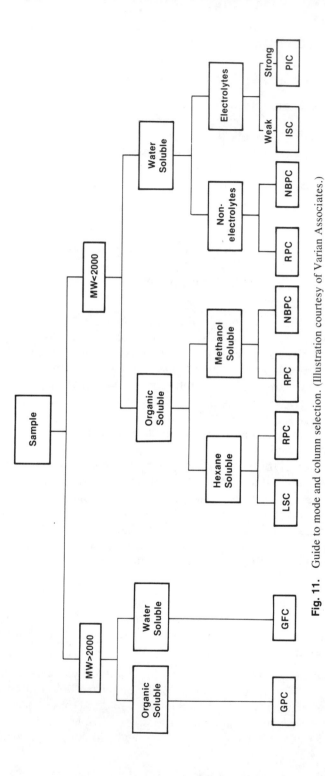

Fig. 11. Guide to mode and column selection. (Illustration courtesy of Varian Associates.)

ion-pairing agents to render them suitable for reversed-phase systems. However, inorganic ions of neutral salts must be separated by ion-exchange chromatography.

As a general rule, reversed-phase chromatography may be used to greatest advantage when the solute is soluble in hexane and slightly or moderately soluble in water. For many separation problems, a single 30-cm reversed-phase column is adequate. In some cases, the required resolution cannot be achieved from a single-phase system because selectivity cannot be optimized concurrently for all components in the mixture. In such cases, first efforts should be directed to separating components of greatest interest, then, if additional separation is needed, consideration should be given to the use of other phase systems. When used either concurrently (on the whole sample) or sequentially (collecting fractions from one system and injecting them into the other), RPC plus NPC provide a very powerful separation technique for mixtures of organic compounds that might be found in water.

Thin layer chromatography (TLC) and published TLC data can be useful when selecting an HPLC phase system. Using Eq. (26), R_f data can be converted to K' [Eq. (8)]:

$$K' = 1/R_f - 1 \qquad\qquad (26)$$

Although K' values in an HPLC column are not identical to those computed from R_f on TLC plates, the information can be used to advantage in several ways:

1. TLC indicates the range of migration rates of a sample's components.
2. TLC demonstrates the potential success of a normal- or a reversed-phase packing.
3. TLC data indicate a probable composition of a compound's eluting solvent and thereby aid in choosing a mobile-phase composition.

R_f values between 0.5 and 0.1, representing K' values 1–9, respectively, are in the working range of HPLC. Retention values at any given carrier strength are usually lower in HPLC than in TLC so that a correction factor may be needed to convert from TLC to HPLC. Generally, HPLC carrier strength should be slightly lower (3–5%) than that used to make the equivalent TLC separation [36]. If gradient elution is being used, the strength of the initial carrier should be ~20% weaker and range to a concentration that is 20% stronger than the TLC developer. For example, a TLC development at 60:40 water:methanol indicates an

HPLC starting carrier of 80 : 20, which is then programmed to at least 60 : 40 [37].

The approach just described can be used with either RP or NP chromatography. However, when working with LSC using silica or alumina, carrier strength relationships are quite different from those of reversed-phase chromatography (RPC), and the 20% rule does not apply.

B. Mobile-Phase Selection

The mobile phase plays a critical role in HPLC because solute migration is controlled by the mobile phase interacting with the stationary sorbent. The sorbent represents a comparatively fixed attracting force that is attenuated by the mobile phase thereby controlling solute migration. HPLC carriers are usually composed of binary or tertiary mixtures of solvents of differing strength whose properties are chosen to produce the desired solute migration rates. Solvents that differ greatly in polarity from a column packing are considered weak solvents because they are poorly retained by the column and do not produce solute migration. Strong solvents are retained by the packing and enhance solute migration. When the carrier solvent is maintained at a constant strength during an elution, it is an isocratic separation.

HPLC carrier solvents should be high in purity, low in viscosity, and compatible with both the sample and detector. Boiling points in the range between 40 and 110°C are preferred. Impure solvents should generally be avoided since they may cause operating difficulty. For example, impurities from a weak solvent can accumulate on the column and be eluted later by a stronger solvent producing anomalous peaks and a drifting baseline. Highly viscous solvents can also cause operating difficulty. In addition to causing high back pressure, they can, when used with a less viscous companion solvent, suffer viscous streaming during a solvent gradient program. Viscous streaming can produce detector noise and nonreproducible retention values.

When for some reason a viscous solvent is needed in a solvent system, it should be blended with another solvent of similar strength but of lower viscosity. Relatively viscous solvents can be used in blends with less viscous solvents at concentrations as great as ~50% without significantly increasing the blend's viscosity over that of the less viscous component.

1. Carriers for Normal-Phase Chromatography Using Solid Adsorbants

In liquid–solid chromatography (LSC) the carrier competes with the solute for the active sites on the adsorbant thereby causing solute migration. Carrier strength is increased either by selecting a more polar solvent or by blending a stronger and a weaker solvent to obtain the desired strength. Single-solvent carriers are advantageous when one with the desired strength is available. However, solvent blends are usually required to meet the needs of modern HPLC. The list in Table III, called an elutropic series, shows solvents in order of chromatographic strength [38]. This table, which provides solvent properties of chromatographic interest, can be helpful for LSC carrier preparation.

In HPLC, especially with gradient elution, solvent admixtures are used to produce nonviscous transparent carriers. It has been a common practice to add a polar modifier such as an alcohol to a nonpolar solvent such as a hydrocarbon to achieve a desired carrier strength in LSC. This practice allows a wider range of strength to be obtained from a single set of components, but it requires precise blending of components to achieve

TABLE III

Properties of Solvents for LSC Listed in Order of Increasing Affinity for Al_2O_3

Solvent	$E_o{}^a$	Boiling point (°C)	Viscosity (cp)	Refractive index $(n_D)^b$	UV cutoff (nm)
Fluoroalkane FC 78	−0.25	50	0.4	1.267	210
Hexane	0.01	69	0.32	1.3754	190
Cyclohexane	0.04	81	1.00	1.4264	200
Carbon tetrachloride	0.18	77	0.97	1.4607	265
Isopropyl ether	0.28	69	0.37	1.3678	220
Benzene	0.32	80	0.65	1.50142	280
Chloroform	0.40	61	0.57	1.4476	245
Methylene chloride	0.42	40	0.44	1.3348 (15°C)	245
Tetrahydrofuran	0.45	65	0.51	1.4040 (25°C)	220
Ethylene dichloride	0.49	84	0.79	1.4165	230
Methyl ethyl ketone	0.51	80	0.38	1.38071 (15°C)	330
Acetone	0.56	56	0.30	1.356	330
Ethyl acetate	0.58	77	0.45	1.3728	260
Acetonitrile	0.65	82	0.37	1.3460	210
Pyridine	0.71	115	0.88	1.507	330
Propanol	0.82	82	2.30	1.3854	210
Ethanol (anhydrous)	0.88	78	1.20	1.36212 (18°C)	210
Methanol	0.95	64	0.60	1.3276 (25°C)	210
Water	> Methanol	100	0.89	1.3330	200
Acetic acid	> Methanol	118	1.1	1.37182	210

^{a} E_o is the solvent-strength parameter based on Al_2O_3 [38]. Other physical properties are from Karger *et al.* (1974) and Peresson and Karger (1974).

^{b} n_D is the refractive index for D line of sodium at 20°C unless otherwise indicated.

reproducible results. Since the solvent strength scale is not a linear function of binary composition, some trial and error is generally required to achieve a precisely desired strength. As an indication of this nonlinearity, Snyder demonstrated that the addition of 20% of a polar modifier achieved 80% of its strength while only a 5% addition achieved 50% [39]. Nonlinearity is less pronounced with solvents of more similar strength, but the trend is the same.

A preferred method of adjusting LSC carriers to a precisely controlled strength is to use components of differing strengths so that the addition of small increments of a strong solvent is not necessary. Small adjustments in strength can be made with the intermediate-strength component. For example, the combination of hexane ($E_o = 0.01$), chloroform ($E_o = 0.40$), and propanol ($E_o = 0.88$) provides a wide range of solvent strengths and an ability to adjust carriers to a precise strength at any intermediate value.

When using silica as a column packing, it is beneficial to add ~5% water to the alcohol that is used as the polar modifier. Water is needed to block the high-activity sites on the silica. If these sites are not blocked, the fully active silica accumulates water from the carrier until an equilibrium is reached between the carrier and the sorbent. The equilibrium level is determined by the moisture level in the carrier. During moisture accumulation by the silica, retention time changes. To insure operational stability, it has been a common practice to pacify the silica using a carrier consisting of 95% hexane, 4.75% isopropanol, and 0.25% water, and to add about 5% water to the modifier as a method of maintaining the previously established equilibrium. However, high efficiency silica gel columns should not be exposed to high concentrations of either water (10%) or methanol (50%) since such exposure reduces efficiency irreversibly.

2. Carriers for Normal-Phase Chromatography with Bonded Packings

Carriers for normal-phase chromatography with bonded phases follow the same principles as outlined for LSC. Because the surfaces of these bonded-phase packings are much less active than those of silica gel, lower strength solvents and lesser quantities of polar modifiers are required to achieve effective solute migration rates. Typical carriers for these packings include hydrocarbons and chlorinated hydrocarbons with acetonitrile as the commonly used polar modifier.

3. Carriers for Reversed-Phase Chromatography

The retarding forces in reversed-phase chromatography are comparatively weak, so that retention is enhanced by the carrier through an inverse solubility relationship with the solute. A carrier's strength is increased by making it a better solvent for the solute. Water is the most commonly used weak solvent, and methanol and acetonitrile are the more commonly used strong solvents. In special cases, tetrahydrofuran is used in conjunction with either acetonitrile or methanol to improve solubility and to increase carrier strength beyond the capability of these less lipophylic solvents. Buffers and either strong acids or strong bases may be added to the aqueous component of the carrier to extend the range of lipophylic chromatographic exchange. Some typical additives and their function are summarized in Table IV.

Acetonitrile is probably the most widely used strong solvent [27]. In many cases, methanol and acetonitrile appear to be interchangeable but, in practice, they are not. As indicated by their solubility parameters (Table V), dispersion forces are about equal but they differ significantly in their bonding forces. If the C_{18} moiety on the packing can be represented by a C_{10} paraffin, the solubility parameter of the packing should be about 7.6 and be based entirely on London dispersion forces. The net result indicates that acetonitrile is a significantly stronger RP solvent than methanol.

The solubility parameter δ is related to its contributing components as follows

$$\delta_I^2 = \delta_D^2 + \delta_O^2 + \delta_H^2 \tag{27}$$

where D is the dispersion, O is the orientation forces (dipole), and H is hydrogen bonding. In RP chromatography, each parameter component influences the overall solubility of the solute in the carrier so that a con-

TABLE IV

Typical Additives to RPC Carriers

Typical additive	Function
Phosphate salts	Buffer to stabilize pH
Acetate salts	Buffer to stabilize pH
Formic acid	Suppresses dissociation of weak acids
Trifluoroacetic acid	Suppresses dissociation of weak acids
Ditertiary butylamine	Suppresses dissociation of weak bases
Heptane sulfonic acids[a]	Forms ion pairs with organic bases[a]
Tetramethyl ammonium hydroxide[a]	Forms ion pairs with organic acids[a]

[a] A range of alkyl acids and bases have been used for ion-pairing reagents (Bidlingmeyer, 1980).

TABLE V

Solubility Parameters of Solvents for Reversed-Phase HPLC[a]

Solvent	Molecular volume (cc)	Dispersion (D)	Dipole orientation (O)	Hydrogen bonding (H)	Total (T)
Tetrahydrofuran	81.7	8.2	2.8	3.9	9.5
Acetonitrile	52.6	7.5	8.8	3.0	11.9
Methanol	40.7	7.4	6.6	10.5	14.2
Water	18.0	7.6	7.8	20.7	23.4
n-Decane[b]	196	7.7	0	0	7.7

[a] From Hansen and Beerbower (1971).

[b] The values for n-decane are included as being indicative of the C_8 and C_{18} moieties bonded to the RP packing.

tributing parameter can be altered to gain selectivity in the separation process.

The solubility parameter relates to the solute distribution coefficient by Eq. (28):

$$\log K_D = \frac{\bar{V}(\delta_x - \delta_m)^2 - (\delta_s - \delta_x)^2}{2.3RT} \qquad (28)$$

where δ_x is the solubility parameter of the component, δ_m the solubility parameter of the mobile phase, δ_s is the solubility parameter of the stationary phase, $\bar{V}$ is the molar volume, R is 1.987 cal degree^{-1} mol^{-1}, and T is the absolute temperature. Although the moiety on a reversed-phase packing is not acting precisely as a liquid solvent, the relationship presented in Eq. (28) provides a means of visualizing the interacting forces that influence the reversed-phase separation and can be used for making first estimates.

4. Carriers for Ion-Exchange Chromatography

Aqueous solutions of buffers and neutral salts are normally used as carriers for ion-exchange chromatography. Because solute migration rate is directly proportional to ionic concentrations, neutral salts that produce properly charged ions are used to control ionic strength; buffers are used to maintain the carrier at a pH where dissociation, and therefore retention, is independent of solute concentration.

The effect of pH on retention is difficult to predict because both the resin and the solute are affected. In some cases, as with weakly basic ion exchangers, the effects of pH on resin exchange capacity and ionic dissociation run counter to each other so that the optimum pH range for the resin is limited. Maximum retention occurs at the point of net maximum dissociation between the solute ion and the resin counterion. Because salts of weak acids experience greater dissociation than the acid, pH should be increased beyond the pK_a by one or two units where dissociation is more favorable, as predicted by Eq. (22).

It is a common practice to hold pH at an optimum value when programming the ionic strength. However, when a sample's components cover a wide range of pK_a, it may be necessary to program pH also.

C. Gradient Elution

The technique of gradient elution permits components of widely differing polarities to be eluted from an HPLC column in a single pass. The technique is accomplished by beginning an elution with a carrier of low strength and gradually increasing carrier strength until all components have eluted. The justification for this practice follows.

When a sample's components are eluted isocratically, poorly retained components tend to be poorly resolved; whereas strongly retained components tend to be spread out in excessively broad peaks. Such strongly retained peaks are so severely diluted in the carrier that they cannot be seen by conventional HPLC detectors. Only the components eluting in a relatively narrow K' range are likely to be resolved and detected. As can be shown from chromatographic theory, the more efficient the column, the greater are the number of resolvable components and the wider the workable K' range.

The theoretical relationship between retention, resolution, and peak width are illustrated by the data in Table VI. The numerical values are based on a 0.45×30 cm HPLC column having a void volume of 3 mL and yielding 3000 plates at a flow rate of 1 mL/min. Peak heights were calculated on the assumption that peaks were of constant area and were isosceles triangles.

Several observations can be made from the data in Table VI. For example, as K' is reduced below 1, R diminishes rapidly, and as K' is increased beyond 5, resolution is increased modestly but both time and peak widths increase disproportionately. Peak height (signifying detector response) also diminishes significantly. This data summary shows numerically why HPLC works best in the region of $5 > K' > 2$.

To operate in the desired K' range, the strength of the mobile phase must be changed during the elution by blending a high-strength solvent (B) into a low-strength starting solvent (A) forming a gradient. As the carrier strength is increased, the component having the lowest K' begins to move

TABLE VI

Theoretical Relationships between Retention Resolutions
and Peak Dimensions in HPLC

K'	R	t (min)	Peak width[a] (min)	Relative peak height
0.10	0.01	3.3	0.24	4.17
0.25	0.20	3.75	0.27	3.70
0.50	0.33	4.5	0.33	3.03
1	1.00	6	0.43	2.33
2	1.33	9	0.65	1.54
5	1.67	18	1.31	0.76
10	1.82	33	2.4	0.41
15	1.88	48	3.5	0.28
20	1.90	63	4.6	0.22
25	1.92	78	5.7	0.18
50	1.96	153	11	0.09

[a] Assumptions for the data are as follows: V_o = column volume in mL, with a 30-cm column, N = 3000 plates, F = 1 mL/min, R = resolution, K' = phase ratio [Eq. (8) and (26)].

away from the more strongly retained components. During migration, the instantaneous K' values fall within the range 1–5 where separation is optimum. Since the concentration of the strong solvent at the front of the band is less than at the rear, the peaks are sharp and symmetrical. With the proper gradient shape, all compounds are eluted at optimum migration rates; in practice, the gradient can be shaped to space solute bands more or less uniformly.

Gradients are usually prepared as binary mixtures by adding a strong solvent (B) to a weak starting solvent (A). Tertiary mixtures can also be used where an elution is started with A, programmed to B, and then to C. Any of these solvents (A, B, or C) may themselves be solvent mixtures. In practice it is not uncommon to use a 70 : 30 water : methanol mixture as A and a 10 : 90 water : methanol mixture as B.

The rate at which B is added to A determines gradient shape, which influences solute migration rate and therefore peak spacing on the chromatogram. Commercial instruments offer a wide choice of shapes and complexities. In general, the choice includes a linear gradient ($dc/dt = C$) plus a selection of several convex and concave curvatures. The optimum shape of a gradient for a particular distribution of solutes is best determined by trial and error. However, some simple guidelines can be offered in advance:

1. A linear gradient should produce bands of constant width across the elution.
2. A positively curved (convexed) gradient should show early bands ($K' < 1$) to be narrower than later bands ($K' > 5$).
3. A concaved gradient should produce the reverse of the convexed gradient.
4. A convexed gradient may be used to advantage when interest is focused primarily on the high-retention components.
5. A concaved gradient is advantageous with a bimodal distribution of early and late eluting components.

In practice, a linear gradient does not always produce equal peak widths so optimization of gradient shape is necessary.

Column Regeneration

After completing the elution, it is necessary to restore the column to its starting activity by reverse programming to the starting carrier and then flushing the column with starting carrier until equilibrium is reached. It has been shown that by reverse programming, regeneration is achieved more quickly than by simply flushing with the A solvent [42]. Failure to allow adequate time for equilibration can result in nonreproducible retentions. The time required for equilibration varies. Bonded phase and pellicular packings require less time than porous silica or alumina. Columns that have been programmed to high strengths require more equilibration time than those programmed to lesser strengths. Where time is a factor, a gradient program need not be carried beyond the strength needed to elute the sample, because extending the gradient simply adds to the time needed to regenerate the column.

When choosing a chromatographic method, the advantages of gradient elution should be weighed against the disadvantages. Since analysis time normally includes both sample elution and column regeneration, there is a point (a sample distribution width) below which isocratic operation is preferred to gradient elution, particularly because samples can be injected sequentially without waiting for the column to be regenerated.

Isocratic operation provides greater accuracy than gradient operation; retention volumes and band widths are more reproducible from sample to sample over the entire elution range. Column–carrier equilibrium is more stable in an isocratic operation since the system is running at a steady rate. As an empirical observation, isocratic operation appears to be more applicable to LSC than to RPC with bonded phases.

D. The Trial Separation

The practice of HPLC usually begins with a trial separation. When working with an unknown sample, conditions for the trial separation should be selected to elute an unknown sample quickly. If using gradient elution, the carrier should be programmed from a low strength to a high strength in a relatively steep gradient to provide a survey of the component distribution in the sample. After the initial survey, the gradient slope can be reduced to achieve optimum resolution. If operating isocratically, it is preferred to start with a carrier of high strength that elutes all components quickly and then adjust the strength as needed to gain the resolution desired.

1. Sample Injection Volume

In practice, a sample injection volume of about 20 μL is typical for a 0.45 × 30 cm HPLC column. The normal working range is 10–50 μL. If a sample injection volume is too large for the system, it interferes with the separation. The extent of interference varies inversely with peak retention, that is, the band width of a sample injection interferes less with more highly retained peaks than with those that are not as highly retained. The influence of injection band width (W_i) on a solute peak width (W_p) can be calculated by:

$$W_F = [W_p^2 + (4/3)W_i^2]^{\frac{1}{2}} \tag{29}$$

Normally, W values are measured in volume. When $W_p/W_i = 3$, the final peak width, W_F, experiences a 9% increase over the undistorted solute peak width (W_p). Therefore, as a rule of thumb, if injection volumes are held to less than 30% of an eluting peak's volume, that peak experiences less than a 10% resolution loss to an adjacent peak. When using gradient elution, where the K' value of all symmetrically eluted peaks is ~2, the 30% injection volume is ~200 μL (based on a 0.45 × 30 cm column).

2. Sample Load

The sample load is determined by sample injection volume and sample concentration. Excessive sample concentrations overload the column and distort the separation process. Column overloading should normally be avoided; however, in special cases, overloading may be used advantageously. The primary reasons for using high sample concentrations are (a) to detect some components that are present at low concentrations, and (b) to increase the quantity of component that can be collected from a preparative separation. Because no stringent limits can be set for column loads, a discussion of the factors affecting column overloading is appropriate.

A column is overloaded when any solute does not have access to the stationary phase by virtue of its concentration in the migrating band. When the solute is prevented from entering the stationary phase, it migrates more rapidly and elutes earlier than under proper conditions. Column overload, therefore, causes an apparent reduction in retention time (t_r) with an increasing sample load.

When chromatographic bands are narrow, solute concentration at the sorption site is greater than when they are broad, so that sharp peaks (low K') have lower mass-load limits than wide peaks (high K'). Therefore, high-efficiency columns are more sensitive to overloading than those of lower efficiencies.

The relative concentration of a sample that can be loaded on a column without impairing column performance is determined by the solubility of the component in the carrier and the sorption capacity of the packing. Porous layer packings generally provide lower capacities than fully porous packings. In general, bonded packings provide lower capacities than LSC packings.

The mass limit for a sample that contains many widely distributed components is greater than it is for a sample containing only two or three narrowly distributed components. However, the load of a single component in a multicomponent mixture should not be treated individually unless the component is reasonably well retained, because all molecules in the mixture compete for the same sorption sites during the initial portion of their migration. As components separate into separate bands, this competition diminishes and the concern of overloading shifts from the total sample to the individual components.

Sample load for most analytical work is governed by the detector's sensitivity and linearity. Frequently, the detector overloads before the column packing. For preparative work, a practical load limit is 1×10^{-4} g of the component per gram of packing in the phase ratio range of 2 to 5. Thus, load limits may be increased significantly by eluting at a higher K'.

VII. OPTIMIZING THE SEPARATION

The trial separation is an important step in the application of HPLC to analytical problems. The chromatogram from the trial separation provides important information that is necessary to optimize the separation. A visual inspection of the chromatogram shows whether a minor modification of the carrier program is adequate or if major changes to the system are needed.

Some of the more important indicators of chromatographic performance are addressed in the following questions:

1. Are the peaks symmetrical or are they skewed?
2. Are the peaks spaced uniformly across the chromatogram?
3. Are the peaks separated or do some overlap?
4. Is overlapping expressed in partially resolved, discernible peaks, discernible shoulders, or virtually undetectable distortions (slight tails, etc.)?

The separation can be optimized using the three terms, A, B, and C of Eq. (20).

If the peaks are widely separated, solvent strength can be increased to reduce analysis time. If peaks are badly skewed, a modifier should be added to the carrier to reduce solute–sorbent interaction. The problem of optimization is most important when more resolution is needed than was attained in the trial separation.

When optimizing a separation, first consideration should be given to the efficiency term $C = t_r/W$ as described in Eq. (20). If the value of term C is 8 or greater, the column is performing adequately. If $C \leq 5$, the column should be replaced. When $8 > C > 5$, the column is substandard, but the decision to replace is left with the operator.

The next consideration is phase utilization as indicated by term $B = t'/t_r$. Values of B between 0.5 and 0.8 indicate proper phase utilization by an isocratic separation. When using gradient elution, peaks should not be run together at the beginning of the chromatogram nor should they be spread out excessively at the end of the chromatogram.

When both column efficiency and phase utilization are adequate and additional resolution is needed, either column length must be increased or column selectivity must be increased by means of a more selective phase system. Resolution R increases in proportion to the square of the increase in column length L as defined by Eq. (30):

$$R/R_0 = (L/L_0)^2 \tag{30}$$

In practice, this equation defines a minimum requirement for column length since significant efficiency loss occurs in the column couplings.

Increasing resolution by means of a more selective phase system is a matter of trial and error, because the composition of both the mobile and the stationary phase influence selectivity. Effects of mobile-phase composition on chemical selectivity have been described for reversed-phase [43–46], normal bonded-phase [47] and liquid–solid [48–49] chromatographic systems. A concept based on a selectivity triangle has been proposed as a basis for systemization [50] and computerization [51]. However, data on the retentivity and selectivity of the stationary phase are inadequate at this writing to facilitate practical systemizations. The many packings from the many vendors, particularly reversed- and normal bonded-phase packings, differ so widely in retention properties that an empirical approach is likely to be the most productive [52]. When a selective-phase system can be found, it can make the most significant contribution to resolution. If a search is being made for a phase that separates a poorly resolved pair of compounds, a selectivity $A = \Delta t_r/t_r'$ of ~0.17 is needed to separate the components on a commercial column. When values for term A fall between 0.16 and 0.10, separations (if possible at all) can be made only with difficulty.

The guidelines presented here apply to commercially available columns with reputed plate numbers between 1,000 and 3,000. Some vendors advertise column efficiencies of 10,000 plates/m. Unfortunately, these efficiencies apply to 30-cm columns.

A. Quantification in HPLC

Quantification in HPLC, as in many other instrumental techniques, is accomplished by use of a calibration standard to obtain either response factors or a calibration curve. As in gas chromatography, the standard may be external (i.e., injected separately from the sample) or internal (i.e., added to and injected with the sample). An external standard is more convenient to use but it does not compensate for errors in injection volume. When using a fixed-volume injection value, external standards are more dependable than when using syringe injection. Internal standards are desirable when the standard peak can be eluted without interfering with sample components.

Quantification validity depends on reproducible sample injection, reproducible detector responses, and uniform carrier flow rate. When detector response is nonlinear, the calibration standard should be at approximately the same concentration as the components of interest. For accurate work, a response factor S should be determined for each component,

$$S_g = \frac{\text{Response/g compared}}{\text{Response/g of standard}} \tag{31}$$

and used to convert either peak height or peak area to mass of component. Molar response may be used in place of mass response by the appropriate conversions.

Quantification based on peak height for HPLC requires precise control of carrier strength since a slight change in K' alters peak height. Conversely, quantification based on peak area requires uniform and reproducible flow rate since any given peak width in the chromatogram is affected inversely by the carrier's flow rate.

B. Sampling by Trace-Enrichment HPLC

In some cases, the water sample may be injected as the neat liquid directly into the chromatograph. In most cases, however, this practice is not acceptable because the component concentrations are too low to be detected and/or the background interferences are too high to permit valid analyses. When direct injection is not feasible, the organics of interest must be concentrated by one of several available techniques:

1. Liquid–liquid extraction and concentrations by solvent evaporation
2. Absorption onto a sorbing resin followed by desorption and concentration in organic solvents
3. Freeze-drying supplemented with liquid–liquid extraction
4. Membrane ultra filtration
5. Trace enrichment onto a reversed-phase HPLC column

Of the methods listed above, 1 through 4 are in general use in analytical chemistry and are not discussed further in this chapter (see methodology in ref. 27). However, the method known as trace enrichment (Table VII) is, in reality, a variation in HPLC operating procedures. In its simplest form, a measured quantity of the water sample is pumped through the RP column, which retains most of the organics at the head of the column. The carrier program is then initiated to produce a separation. If little or no

TABLE VII

Water Analysis Using HPLC Trace-Enrichment Techniques

Type of determination	Reference
Trace organics in waste water, tap water, and seawater	53
18 Pesticides at 20 ppb level	54
Water-miscible solvents	55
Oil-contaminated seawater	56
Phthalate esters	57
Chlorophenols	58
Fluoronated organic acids	59
Toxic amines	60
Trace metal ions in nuclear-reactor coolant	61

migration occurs during deposition, a relatively large volume of sample can be extracted and a significant degree of concentration achieved without additional sample handling or exposure to other sources of potential contamination.

Trace enrichment may be performed directly on the analytical column but at the risk of fouling the column with sample contaminants. A preferred practice is to use a replaceable precolumn a few centimeters in length that attaches directly to the analytical column without causing band spreading. The precolumn can be packed in the user's laboratory and replaced as needed.

ACKNOWLEDGMENT

The author expresses his deep appreciation to Ms. Melody De Moss for typing this manuscript and to Radian Corporation for their support in this effort.

REFERENCES

1. Piel, E. V., *Anal. Chem.* **38,** 670 (1966).
2. Huber, J. F. K., and Hulsman, J. A. R., *Anal. Chim. Acta* **38,** 305 (1967).
3. Snyder, L. R., *Anal. Chem.* **39,** 698 (1967).
4. Scott, C. D., *Anal. Biochem.* **24,** 292 (1968).
5. Majors, R. E., *Anal. Chem.* **11,** 1722 (1972).
6. Small, H., Stevens, T. S., and Bauman, W. C., *Anal. Chem.* **47,** 1801 (1975).
7. Drake, B., *Ark. Kemi* **8,** 1 (1954).
8. Scott, C. D., Jolley, R. L., Pitt, W. W., and Johnson, N. F., *Am. J. Clin. Pathol.* **53**(5), 701–712 (1970).
9. Bombaugh, K. J., *J. Chromatogr.* **107,** 201–206 (1975).
10. Peterson, E. A., and Sober, H. A., *Anal. Chem.* **31,** 857 (1959).
11. Bombaugh, K. J., *Am. Lab.* **5,** 69 (1973).
12. Harvey, M. C., and Stearns, S. D., *J. Chromatogr. Sci.* **20,** 487 (1982).
13. Schmidt, A., *Chromatographia* **12,** 825–831 (1979).
14. Denton, M. S., De Angelis, T. P., Yacynych, A. M., Heineman, W. H., and Gilbert, T. W., *Anal. Chem.* **48,** 20 (1976).
15. Bombaugh, K. J., King, R. N., and Cohen, A. J., *J. Chromatogr. Sci.* **43,** 332–338 (1969).
16. Popovich, D. J., Dixon, J. B., and Ehrlich, B. J., *J. Chromatogr.* **17,** 643–650 (1979).
17. Kissinger, P. T., *Anal. Chem.* **49,** 447A (1977).
18. Kissinger, P. T., Bratin, K., King, W. P., and Rue, J. R., *ACS Symp. Ser.* **136,** 57–58 (1981).
19. Bratin, K., and Kissinger, P. T., *J. Liq. Chromatogr.* **4**(10), 1777 (1981).
20. Bratin, K., and Kissinger, P. T., *Talanta* **29,** 365–370 (1982).
21. Grushka, E., Snyder, L. R., and Knox, J. H., *J. Chromatogr. Sci.* **13,** 25 (1975).
22. Snyder, L. R., *J. Chromatogr. Sci.* **7,** 352 (1969).
23. Majors, R. E., *J. Chromatogr. Sci.* **11,** 88 (1973).
24. Bombaugh, K. J., *in* "GLC and HPLC Determination of Therapeutic Agents" (K. Tsuji and W. Morozowich, eds.), p. 126. Dekker, New York, 1978.

25. Snyder, L. R., *J. Chromatog. Sci.* **10**, 200 (1972).
26. Bombaugh, K. J., *Prog. Anal. Chem.* **6**, 207–208 (1973).
27. Majors, R. E., Barth, H. G., and Lochmüller, C. H., *Anal. Chem.* **54**, 342R (1982).
28. Harrison, K. H., Miller, V. I., and Yates, T. L., "Comprehensive Guide to Reversed Phase Materials for HPLC," Figs. 1 and 8. The Separations Group, Hesperia, California, 1982.
29. Peaden, P. A., Lee, M. L., Hirata, Y., and Novotony, M., *Anal. Chem.* **52**, 2268 (1980).
30. Karger, B. L., Su, S. C., Marchese, S., and Peresson, B. A., *J. Chromatogr. Sci.* **12**, 678 (1974).
31. Peresson, B. A., and Karger, B. L., *J. Chromatogr. Sci.* **12**, 521 (1974).
32. Kessner, L., and Muntroyler, E., *Anal. Chem.* **9**, 1964 (1966).
33. Weast, R. C., and Astle, M. J., eds., "Handbook of Chemistry and Physics," 59th ed. CRC Press, West Palm Beach, Florida, 1978–1979.
34. Dean, J. A., ed., "Lang's Handbook of Chemistry," 11th ed. McGraw-Hill, New York, 1973.
35. Bombaugh, K. J., *in* "Modern Practice of Liquid Chromatography" (J. J. Kirkland, ed.), pp. 237–285. Wiley, New York, 1971.
36. Halpup, H., Krebs, K. F., and Hauk, H. E., *HRC CC, J. High Resolut. Chromatogr. Chromatogr. Commun.* **3**(5), 215 (1980).
37. Sherma, J., "TLC Technical Series 1." Whatman Chemical Separation, Inc., Clifton, New Jersey, 1982.
38. Snyder, L. R., "Principles of Adsorbtion Chromatography," pp. 194–195. Dekker, New York, 1968.
39. Snyder, L. R., *in* "Modern TLC Practices of Liquid Chromatography" (J. J. Kirkland, ed.), p. 143. Wiley (Interscience), New York, 1971.
40. Bidlingmeyer, B. A., *J. Chromatogr. Sci.* **18**, 525–550 (1980).
41. Hansen, C., and Beerbower, A., *Kirk-Othmer Encycl. Chem. Technol. 2nd Ed.* Supplement, p. 889 (1971).
42. Majors, R. E., *Anal. Chem.* **45**, 755 (1973).
43. Schoenmaker, P. J., Billiet, H. A. H., and de Galen, L., *J. Chromatogr.* **185**, 179 (1979).
44. Tanaka, N., Goodell, N., and Karger, B. L., *J. Chromatogr.* **158**, 233 (1978).
45. Bakalyar, S. R., McIlwrick, S. R., and Roggenderf, R., *J. Chromatogr.* **142**, 353 (1977).
46. Glaph, J. L., Kirkland, J. L., Squire, J. J., and Minor, K. M., *J. Chromatogr.* **199**, 57 (1980).
47. Antle, P. E., *Chromatographia* **15**(5), 277 (1982).
48. Snyder, L. R., Glajch, J. L., and Kirkland, J. J., *J. Chromatogr.* **218**, 299 (1982).
49. Glajch, J. L., Kirkland, J. J., and Snyder, L. R., *J. Chromatogr.* **238**, 269 (1982).
50. Snyder, L. R., *J. Chromatogr. Sci.* **16**, 223 (1978).
51. Glajch, J. L., and Kirkland, J. J., *Anal. Chem.* **55**, 319A (1983).
52. Goldberg, A. P., *Anal. Chem.* **54**, 342 (1982).
53. Wallin, H. F., and Eiceman, G. A., *NBS, Spec. Publ. (U.S.)* **519**, 185–190 (1979).
54. Edwards, R. W., Nonemaker, K. A., and Colter, R. L., *NBS Spec. Publ. (U.S.)* **161D**, 87–94 (1979).
55. Bristol, D. W., *J. Chromatogr.* **188**, 193–204 (1980).
56. Saner, W. A., Jadamec, J. R., Sager, R. W., and Killeen, T. J., *Anal. Chem.* **51**, 217–223 (1979).
57. Van Vliet, H. P. M., Bootsman, T. H. C., Frei, R. W., and Brinkman, V. A., *J. Chromatogr.* **185**, 483–495 (1979).
58. Werkhoven, C. E., Brinkman, V. A. T., and Frei, R. W., *Anal. Chem.* **53**, 2072–2080 (1981).

59. Slelzenbach, K. J., Jensen, S. L., and Thompson, G. M., *Environ. Sci. Technol.* **16**(5), 250–254 (1982).
60. Rice, J. R., and Kissinger, P. T., *Environ. Sci. Technol.* **16**, 263–268 (1982).
61. Cassidy, R. M., and Elchuk, S. J., *J. Chromatogr. Sci.* **18**, 217–223 (1980).

RECOMMENDED BIBLIOGRAPHY OF HPLC REFERENCE MATERIAL

Horváth, C., ed., "High Performance Liquid Chromatography: Advances and Perspectives," Vol. 2. Academic Press, New York, 1980.
Karger, B. L., Snyder, L. R., and Horváth, C., "An Introduction to Separation Science." Wiley, New York, 1973.
Snyder, L. R., "Principles of Adsorption Chromatography." Dekker, New York, 1968.
Snyder, L. R., and Kirkland, J. J., "Introduction to Modern Liquid Chromatography," 2nd ed. Wiley, New York, 1979.

7 INFRARED SPECTROPHOTOMETRY OF POLLUTANTS IN WATER SYSTEMS

Fred Katsumi Kawahara

Environmental Protection Agency
A. W. Breidenbach Environmental Research Center
Cincinnati, Ohio

WATER ANALYSIS, VOL. III

I. INTRODUCTION

Infrared spectroscopy is the study of the interaction of infrared light of known frequency and intensity with molecules. Measurement is usually made in the region of fundamental rotation–vibration in the wavelength range of 2.5 to 25 μm or frequency range of 4000 to 400 cm^{-1}. The relative masses of the atoms in the molecule under irradiation, the force constants of the bonds, and the geometric arrangements of the atoms in the molecule have a strong influence on the frequency or wavelength of absorption in the spectrum.

Most of the discussions in this chapter involve the region of 2.5 to 25 μm because it is the region of greatest utility. While this region is of great interest, there has been increasing attention given to the near-infrared region 14,290–400 cm^{-1} (0.7–2.5 μm) and the far-infrared region, 700–200 cm^{-1} (14.3–50 μm). As indicated above, the more convenient frequency term $\bar{\nu}$, expressed as units of reciprocal centimeters (cm^{-1}), may be used. This term is obtained by dividing ν, the frequency in cycles per second, by c, the velocity of light.

$$\bar{\nu} = \frac{\nu}{c} \, \text{cm}^{-1} = \frac{\text{cycles/s}}{\text{cm/s}}$$

By using Planck's relationship $E = h\nu$, the energy equivalent to a quantum of light is related to the frequency, and the very low frequencies characteristic of the infrared region could cause only very small excitations that are not capable of chemical-bond breaking or alteration of molecular configurations. The wave number also may be taken as a unit of energy; 1 cm^{-1} is equivalent to 11.959×10^7 erg/mol or 2.858 cal/mol. Therefore, at 4000 cm^{-1} the energy is 11.433 kcal/mol, which is insufficient for molecular alteration in the bonds or configuration [1]. The wavelength λ is the reciprocal of $\bar{\nu}$.

When it absorbs infrared light, the molecule becomes excited, resulting in the initiation of vibrational movements of the atoms about their equilibrium positions. Only certain discrete energies are permitted for the molecule and absorption of light is related to a transition between two of these energy levels. On the other hand, rotational energy results from the absorption of photons by the molecules with the complete conversion of photon energy into energy of molecular rotation. The energy differences in rotational spectra are ~1% of those found in vibrational spectra; the small rotational energies only widen the vibration band or are observed in the far-infrared region [1].

II. RECORDING OF INFRARED SPECTRA

A spectrophotometer consists of a light source (S), a dispersing prism (P) or grating to separate the frequencies, focusing optics, and a detector. Figure 1 illustrates schematically the primary elements of a simple, single-beam infrared spectrometer. Light, providing radiant energy from source (S), which may be a heated Nichrome wire or heated refractory substance, is focused by a mirror M_1 through a narrow slit S_1. The mirror M_2 collimates the slit beam prior to its passage through the prism P, which has excellent dispersive and transparency properties in the spectral region of 2.5 to 25 μm. The infrared light passing from the prism P is refracted so that mirror M_3 focuses only a narrow range of frequencies through the second narrow slit S_2. The M_4 mirror focuses the monochromatic light of selected frequency onto a detector D, such as a bolometer, thermocouple, or Golay detector. The mirror M_3 may be rotated slowly for scanning the spectral region and, more specifically, for the desired frequency passing through the slit S_2. A spectrum is obtained when the detector response is a function of the rotation angle of the mirror and hence of the frequency of the incident radiation reaching the detector. The spectrum taken from a simple, single-beam instrument is uncorrected for the light-intensity distribution of the source; variations in reflectance or transmission of radiation by mirrors occur also with wavelengths. Therefore, the absorbed spectrum is somewhat distorted. As a result, the response of the instru-

Fig. 1. A schematic optical path of an infrared spectrometer.

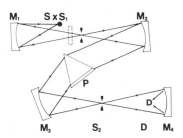

ment with no sample is not flat. Signals are distorted accordingly when sample is present.

In the more elaborate optical-null, double-beam infrared spectrophotometer, a higher degree of accuracy is ensured by eliminating errors through the use of superior source, detector, and optical components. The previously mentioned nonflatness is eliminated by ratioing sample and blank responses. Further information may be found in Conley's text, which is an excellent introduction to infrared spectrophotometry in theory, measurement, and interpretation [1].

III. THEORETICAL CONSIDERATIONS

Spectra arise from the absorption of infrared radiation, which promotes transitions in the molecules between rotational and vibrational energy levels in the ground electronic energy state. In the classical model of the atom, an electron radiates continuously because of its acceleration in the Coulomb field of the nucleus; the electron then returns to its ground state when it loses energy.

Niels Bohr postulated the existence of stationary electron orbits that have discrete energy states. Of all the elliptical orbits transversed by a single electron moving in the Coulomb field, only those orbital paths in which the major axis is proportional to the square of the number n are allowed. The letter n is the principal quantum number and may be 1, 2, 3,.... The minor axis is proportional to the product of n and $l + 1$, where l is the azimuthal quantum number. For a given value of n, l can be 0, 1, 2,..., $n - 1$. The magnetic quantum number m_l is related to the orbital magnetic momentum, and the rotational momentum of the electron is characterized by the spin quantum number m_s. These four quantum numbers are needed in order to characterize or describe the electron; however, according to Pauli's exclusion principle, no two electrons can have the same set of quantum numbers. Thus, for a given n, l, and m_l in an orbit, there are at most two electrons with opposite spins.

According to the rules of quantum mechanics, no electromagnetic radiation is emitted by an atom during the time the electron is in its permitted orbit. However, when an electron loses energy in going from one energy orbit E_i to a final energy orbit E_f, where E_f is less than E_i, energy is emitted as a wavelength of light and equals the difference of the two energy states; $E_i - E_f = h\nu$ (frequency rule).

For absorption of radiation, the transition occurs when energy is absorbed in the system and when the electronic transition occurs from a lower to a higher energy level. Here, E_f is greater than E_i, and the transitions occur in accordance to the selection rules while all other transitions are forbidden. The wave number of the absorbed and emitted radiation can be expressed as a function of the final and initial states (quantum numbers).

Various energy levels (e_i) attained by the particle under absorption or emission of infrared radiation are determined by solutions of the Schrödinger equation describing the motion of the electron:

$$\frac{\partial^2 \psi}{\partial x^2} + \frac{\partial^2 \psi}{\partial y^2} + \frac{\partial^2 \psi}{\partial z^2} + \frac{8\pi^2 n}{h^2} (E - V)\psi = 0$$

where x, y, and z are the coordinates necessary to describe its position, V is the potential energy, m is mass of the particle, h is Planck's constant, and E is the kinetic energy of the electron. The motion of the electron, determined by the function $\psi(x, y, z)$ is represented by a solution that is single valued, continuous, and finite. Specific eigenvalues E_i and E_f, which are solutions from the Schrödinger equation, are differentiated by possessing different quantum-number levels or energies of the electrons at their respective stationary states. At each quantum level, the electron energy is said to be quantized or to exist in a discrete energy state.

When infrared electromagnetic radiation interacts with a molecule, the interaction that occurs is represented by $\mathbf{M} \cdot \mathbf{E}$, where $\mathbf{M}$ is the electric dipole moment of the molecule system, and $\mathbf{E}$ is the electric vector of the infrared wave. The effect of infrared radiation on the motion of the molecule is a change of the dipole moment manifested by a change in the energy of the system. The probability for the transition or change to occur between the two states as a result of the interaction energy is said to be proportional to the square of $\mathbf{R}^{nm}$, the transition moments. The transition moments $\mathbf{R}^{nm}$ can be calculated from the eigen functions ψ_n and ψ_m for the electron states n and m. The intensity of the resulting spectral line can be calculated from the above. These topics are developed in more detail by Avram and Mateescu [2].

A. Molecular Spectra

A molecule is a complex system consisting of two or more atoms. The mass of a molecule is located in each of the atoms, each of which can be considered pointlike, thus giving a molecule as many degrees of freedom as the total degrees of freedom of its individual atoms. Since each atom needs three degrees of freedom to describe its position relative to other atoms, a molecule of n atoms has $3n$ degrees of freedom. For nonlinear molecules, three of the degrees of freedom are for rotation and three describe translation; the remaining $3n - 6$ degrees of freedom are vibrational and fundamental in nature. For linear molecules, there are $3n - 5$ vibrational degrees of freedom because linearity requires only two degrees of freedom for rotation.

Molecules have electronic, translational, rotational, and vibrational energy. Thus, the total energy of a molecule E can be approximated as $E = E_e + E_v + E_r + E_t$, where E_e is the energy of the electron, E_v is the vibrational energy, E_r is the rotational energy, and E_t is the translational energy. Translational energy cannot be quantized because translation is continuous in nature, but the molecule can possess quantized electronic, vibrational, and rotational states. Thus, types of energy involved in emission or absorption of radiation are only E_e, E_v, and E_r.

When a molecule absorbs radiation, it can rise from a lower energy level to a higher level; the rise in energy terms is equal to the difference in energy between the two states. Because E_t cannot be quantized, it is not involved in emission or absorption of light quantum. Rotational energies of a molecule E_r are small and E_r is equal to the difference between the two rotational energies necessary to go from level E_{r_i} to level E_{r_f}, where E_{r_f} equals rotational energy for the final state and E_{r_i} equals rotational energy for the initial state. The small value of E_r reflects a small frequency value, $\bar{\nu} = E_{r_f}/hc - E_{r_i}/hc$, and is seen in widening of narrow fundamental absorption bands or may be designated in the far-infrared region of ~ 30 μm or greater.

As mentioned, vibrational energies of a molecule E_v are approximately 100 times greater than the rotational energies; E_v is equal to the differences between the two vibrational energies needed for the vibrating molecule to go from the initial to the final energy level, or from E_{v_f} to E_{v_i}. The resulting difference in energy is also much larger than in the case for rotational transitions, or $\bar{\nu} = E_{v_f}/hc - E_{v_i}/hc$. The radiation absorbed in the vibrational change belongs to the infrared region of 2.5 to 25 μm or sometimes slightly into the near-infrared region.

For the electronic transitions E_e to occur, a much higher energy of radiation is required since, for $\bar{\nu}'' = E_{e_f}/hc - E_{e_i}/hc$, the $\bar{\nu}''$ values are much higher as E_{e_f} and E_{e_i} values are quite large. The differences are also much greater than in the cases for rotation and vibration. The radiation quanta involved in the electronic transition belong in the visible region of 0.4 to 0.8 μm or in the ultraviolet region (<0.4 μm). Thus, $\bar{\nu}''$ is far greater than $\bar{\nu}'$, which is greater than $\bar{\nu}$, and the frequencies encountered in the ultraviolet region are far greater than those found for the infrared region, which in turn are a little higher than those observed for the far-infrared region. For a more elaborate and theoretical discussion of energy levels and spectra of rigid and nonrigid rotators, of harmonic and nonharmonic oscillators, and other advanced topics in rotation–vibration motions, the reader may refer to Chapter 2 of Avram and Mateescu [2]. For basic spectroscopy, the reader is referred to Conley [1] and H. A. Szymanski [3].

Band intensities may be expressed as transmittance T or as absorbance A. Transmittance is the ratio of the radiant power transmitted by a sample to the radiant power incident on the sample. Absorbance is the logarithm, to the base 10, of the reciprocal of the transmittance [4].

B. Band Positions

Because polyatomic molecules are much more complicated than an atom or diatomic molecules, the number of degrees of freedom increases with an increase in the number of atoms in the molecule. Moreover, the mathematical treatment of the infrared spectra becomes more complicated; vibrational frequencies and a complete analysis for these larger molecules are obtained essentially on an empirical basis. Band location, or the wavelength or frequency of absorption in the infrared region, depends on the relative masses of the atoms, the force constants of the bonds, and the geometry of the atoms. The wavelength λ in micron units is used in infrared spectrometry. It has also been convenient to use the wave-number unit in terms of reciprocal centimeters (cm^{-1}) since it is directly related to energy terms. The conversion from wavelength unit to wave numbers is readily accomplished by dividing by 10,000.

C. Types of Vibrations

The two types of molecular vibrations are stretching and bending. A stretching vibration is a rhythmical movement along the bond axis so that the interatomic distance between the two end atoms is increasing or decreasing. For example, as shown in Fig. 2, in the linear CO_2 molecule with three atoms, a symmetrical stretching vibration (1) is inactive in the infrared because there is no change in the dipole moment. However, the asymmetrical stretching vibration (2) is active in the infrared since a change is produced in the dipole moment [4].

The other type of molecular vibration is the bending vibration, which consists of a change in the bond angles formed by bonds with a common atom, or it may involve twisting, rocking, and torsional vibrations (all of which involve movement of a group of atoms with respect to the remainder of the molecule without movement of atoms in groups). Examples (3) and (4) in Fig. 2 are those of the bending vibrations; they have the same frequency.

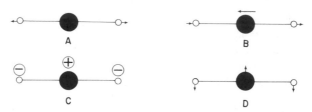

Fig. 2. Stretching and bending vibration of CO_2. Plus and minus indicate movement perpendicular to the plane of the page. (A) Symmetric stretching (ν_s CO_2), 1340 cm^{-1}; (B) asymmetric stretching (ν_{as} CO_2), 2350 cm^{-1}; (C) scissoring (bending) (δ_s CO_2), 666 cm^{-1}; (D) scissoring (bending) (δ_s CO_2), 666 cm^{-1}.

In symmetric stretching, the vibrations of the oxygen atoms in CO_2 are symmetrical with respect to the central carbon atom and move along the line of the valence bonds. In the asymmetric stretching (2), one oxygen atom departs and one oxygen atom approaches the carbon (there is a change in the dipole and the mode is active in the infrared). In examples (3) and (4), the deformation vibration involves the change of the angle formed by oxygen–carbon–oxygen atoms. With the resultant changing dipole, this vibration mode is active in the infrared. Examples (3) and (4) have the same frequency and are called "doubly degenerate." In CO_2 there are four fundamental vibrations calculated from the $3n - 5$ rule, which gives $(3 \times 3) - 5 = 4$. Thus, CO_2 has four possible vibrational modes, two of which are doubly degenerate.

For the case where the molecule is totally asymmetric, all modes of vibration are active. Molecules with strong dipoles provide strong absorption bands in the infrared. In contrast, highly symmetric molecules have fewer vibrational modes active in the infrared and also have degenerate vibrational modes. Thus, the correlation can be noted that highly symmetrical molecules have simple spectra. Hence, the development of mathematical correlation between symmetry of a molecule and infrared activity of its vibrational mode is important. With the use of group theory, one may determine the number of active vibrations and selection rules in the infrared for a molecule. The reader is referred to page 66 of the text by Avram and Mateescu [2] for discussion on symmetry elements, symmetry operations, and point-group theory.

D. Interactions

The theoretical number of fundamental vibrations or absorption frequencies is not usually observed, because some interactions increase the number of absorption bands found in the spectrum. Such increases are caused by multiples of a given vibrational frequency; these multiples may be overtones or the sum of two other vibrations or even a combination of tones. Other phenomena may occur that reduce the number of bands, such as weakness of intensity, coalescence of several infrequencies, or the inadequate induction of dipole moments, etc. Readers may find additional information in Silverstein *et al.* [4] and Colthup *et al.* [5].

Assignments of the stretch frequencies can be mathematically approximated via the use of Hook's law:

$$\nu = \frac{1}{2\pi c} \left[\frac{F}{\dfrac{M_x \cdot M_y}{M_x + M_y}} \right]^{1/2} = \frac{1}{2\pi c} \frac{F}{\mu}$$

where ν is the vibrational frequency (cm^{-1}); F is the force constant of bond (dynes/cm); c is the velocity of light (cm/sec); M_x and M_y are the masses of atoms x and y; and $M_x M_y / M_x + M_y = \mu$ = reduced mass.

Calculations of frequency assignments are illustrated in reference [4]. It should be noted that in general strong absorption in the infrared spectrum is caused by strong dipolar characteristics of functional groups, such as ester, carboxyl, hydroxyl, amino, and sulfonic groups.

1.   Coupling

When a strong coupling [4] or interplay occurs, some shifting of the stretching and bending vibrational frequencies is manifested; this interaction is referred to as *vibrational coupling*. Vibrational coupling is the interaction of two vibrating units having coincidental frequencies that can split the band into two components, one above and the other below the common frequency [1]. For example, in the CO_2 molecule, which has two bond oscillators with a common bond (O=C=O), there are two fundamental stretching vibrations—an asymmetric and a symmetric stretching mode (refer to Fig. 2). Because the symmetric mode produces no change in the dipole moment of the molecule, it is inactive in the infrared but observable in the Raman region near 1340 cm^{-1}. In contrast, the asymmetric stretching mode produces a change in the dipole moment, and therefore it is active with an infrared absorption band at 2350 cm^{-1}; the normal carbonyl absorption frequency of the carbonyl group occurs at 1715 cm^{-1}. In the illustration observed for the ethanol molecule, coupling of the asymmetric stretch vibration occurs at 1053 cm^{-1} as a result of the "C—C—O" stretch. However, in methanol, the normal stretching absorption frequency "C—O" occurs at 1034 cm^{-1} with no coupling.

2. Hydrogen Bonding

Another type of shift observed in the infrared is found in the hydrogen-bonding system [4] caused by a weak chemical reaction that occurs between the s orbital of a proton and an overlap of the π orbital of the X acceptor group of a H–X molecule. This weak chemical bonding occurs in organic acids, alcohols, amines, amides, etc., with oxygen, nitrogen and halogen atoms serving as proton acceptors. In an infrared spectrum, the H—X bending vibration shifts to a shorter wavelength, while the stretching vibration shifts to longer wavelengths with *intermolecular* hydrogen bonding (where the proton donor and acceptor are in two separate molecules).

In an intramolecular hydrogen-bonding system, the proton donor and acceptor are found within the same molecule and the subsequent formation of a quasi five- or six-membered ring is possible because of the overlap of the s orbital of hydrogen and the π orbital of the acceptor group. An example is salicylic acid in which the phenolic hydrogen overlaps the carbonyl oxygen of the carboxyl. A six-membered ring is then formed:

In dilute solutions one may observe orthohydroxyaryl acids and esters displaying this shift phenomenon.

3. Fermi Resonance

Fermi resonance arises when there is an interaction between fundamental vibrations and overtones or combination tones. As in coupling, two new modes of vibration are formed and these occur at higher and lower absorption frequencies than the absorption frequency that is observed in normal vibration when this interaction is absent. In a polyatomic molecule, two different vibrational levels (or combination of levels) may have nearly the same energy. When the two vibrational states have the same energy levels, they are said to be degenerate. One of the energy levels may be shifted toward lower energies and the other energy level may be shifted toward higher energies. This phenomenon may show a substantial increase in the intensity of some of the bands. For example, the symmetrical stretching band of CO_2 appears in the Raman region near 1340 cm^{-1} (7.46 μm). Two bands result from the splitting; one appears at

1286 cm^{-1} (7.78 μm) and the other is at 1388 cm^{-1} (7.20 μm). The mean value is very close to that expected for the first overtone of the fundamental 2 $\times$ 666 cm^{-1} = 1332 cm^{-1}, which is comparable to 1340 cm^{-1}, the normal fundamental that is split. Because the energy level of the first level of the fundamental is close to that of the overtone, a Fermi resonance occurs with a shift of the first level toward a higher level and intensity and a shift of the second level toward a lower energy level [4].

IV. FUNDAMENTALS OF SPECTRAL INTERPRETATIONS

The infrared spectrum of an organic compound shows a number of absorption bands characteristic of the structural features of that molecule. The interpretation of an infrared spectrum requires certain conditions as follows [4]:

1. The spectrophotometer must be initially calibrated with a reliable standard such as polystyrene film. Once calibrated for the day, bands are observed at their correct frequencies or wavelengths.
2. Sample handling with respect to solvent, concentration, and cell thickness must be stated.
3. The spectrum obtained should show proper and adequate resolution of peaks and adequate intensity of the peaks.
4. The spectrum should be of a pure compound.

On introducing a sample such as acetone into the radiation pathway, the spectrum is drawn as a plot of emitted radiation transmission (as percentage) versus wavelength. Absorption characteristics are drawn and are unique for the compound under investigation. Many compounds have been studied, and the resulting curves can be distinguished from one another by the presence or absence of certain bands. There are certain basic postulates that are established for the chemist using infrared spectroscopy for the determination of structural information, as summarized in Conley's text (p. 7) [1]:

1. Organic substances exhibit characteristic group frequencies in the infrared region.
2. The absorption spectrum of a given substance is generally specific for that and only that substance.
3. The intensity of an absorption band is related to the concentration of the substance that absorbs the incident radiation in accordance to Beer's law $A = ELc$, where A is the absorption, L is the cell path length, c is the concentration of the substance, and E the specific absorption coefficient.

4. The absorption spectrum of a mixture is generally additive; that is, it is the sum of the individual spectra representative of each of the components in the mixture.

Absorption bands in an infrared spectrum are at frequencies corresponding to the vibrational frequencies of the molecule involved. Because these frequencies depend on the masses of the atoms, the bonds holding the atoms in the molecule and the spatial geometry of the atoms in the molecule, slight differences in the structure and/or in the position of functional groups in one molecule lead to vibrational–rotational differences in the spectral pattern of that molecule when compared to those of a similar molecule. Differences in a spectrum are large for cases where a greater dissimilarity exists in structure of one organic molecule compared to another.

A cursory examination of two general areas is useful in the preliminary interpretation of an infrared spectrum [4]. These regions are 4000–1300 cm^{-1} (2.5–7.7 μm) and 909–650 cm^{-1} (11.0–15.4 μm). In the former, shorter-wavelength region of the infrared, the characteristic stretching frequencies occur for important functional groups such as OH, NH, C=O, C—OH, etc. The absence of absorption in the 1850–1540 cm^{-1} region, for example, usually indicates that a carbonyl moiety is absent. That is, ketones, aldehydes, carboxylic acids, and acid chlorides are absent. Similarly, alcohols show their presence with broadband stretching absorption at 3620–3640 cm^{-1}, whereas phenols absorb strongly at 3610 cm^{-1}.

In the region 909–650 cm^{-1}, strong sharp absorption bands may indicate aromatic and heteroaromatic compounds caused by out-of-plane C—H bending and ring-bend absorption. Conversely, lack of absorption in this region indicates a nonaromatic structure for the compound. An examination of Appendix B of Silverstein [4] or Colthup [5] shows that napthalenes, benzenes, and mono-, di-, and trisubstituted benzenes display very strong characteristic absorptions in the region 909–650 cm^{-1} (this is often referred to as the "fingerprint" region). However, Conley refers to the region 1430–830 cm^{-1} (7–12 μm) as the fingerprint region [1]. More detailed discussions of characteristic absorptions are presented in following sections.

V. SPECTRAL INTERPRETATIONS

A simple organic molecule can yield a fairly complex spectrum, which is excellent evidence for identity when a peak-by-peak correlation can be made between an authentic sample of known structure and an unknown compound. This reliability in frequency assignments for band groups is well established, since various functional groups or atoms repeatedly give rise to absorption at or near the same frequency regardless of the structure of the rest of the molecule. This repeatability feature allows one to gain structural information on an unknown molecule by the simple process of inspection and then referring to charts containing characteristic group frequencies.

Generally, stretching vibrations are found at higher frequencies, while bending vibrations are found at lower frequencies. From this information, we know that there is more energy required to produce a stretching vibration than is necessary for the bending vibrations.

A. Group Frequencies

This section covers the characteristic group frequencies that are pertinent in the analysis of water pollutants. In order to supplement this section, the reader should refer to books by Bellamy [6], Colthup et al. [5], Conley [1], and Silverstein et al. [4]. These books are the sources for the group frequencies given in the following sections.

In the interpretation of infrared spectra, the approach is similar to characterization of organic compounds or determining their mechanistic changes. Thus, one may consider classifying organic compounds by their functional groups or by the substituting reactions these compounds undergo. As stated by Conley, the infrared spectrum of an organic molecule should provide both positive and negative information regarding molecular structure [1]. Positive information is derived from characteristic bands appearing in a spectrum relating functional groups present in the molecule. Negative information is provided when functional groups are considered absent in the molecule as a result of the absence of characteristic group frequencies in an infrared spectrum.

In the qualitative analysis of organic molecules, it is necessary to correlate the mass of spectral information available by considering portions of the molecule as a two-atom vibrating unit or combination of units. In addition, one should acquire a good knowledge of the structural elements of the saturated CH (alkanes) and the unsaturated CH (alkenes and aromatics) groups, which are present in most of the organic molecules. For further reading, one may refer to Conley [1].

B. Alkanes

1. Carbon–Hydrogen Stretching Vibrations: Methyl Groups

Two bands are visible in the infrared spectra of saturated hydrocarbons with methyl groups; one is located at 2962 cm^{-1} (3.38 μm) and caused by

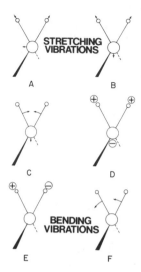

Fig. 3. Vibrational modes for a methylene group. Plus and minus indicate movement perpendicular to the plane of the page. (A) Asymmetric stretching (ν_{as} CH$_2$); (B) symmetric stretching (ν_s CH$_2$); (C) in-plane bending or scissors (δ_s CH$_2$); (D) out-of-plane bending or wagging (ω CH$_2$); (E) τ out-of-plane bending or twisting (CH$_2$); (F) in-plane bending or rocking (ρ CH$_2$).

an asymmetric stretch; the second arises at 2872 cm^{-1} (3.48 μm) and results from symmetric stretching (Fig. 3).

2. Carbon–Hydrogen Stretching Vibrations: Methylene Groups

Two bands are produced from methylene groups of the normal paraffinic type. One is the result of asymmetric stretching vibrations occurring at 2926 cm^{-1} (3.43 μm); the other arises from methylene symmetrical stretching vibrations and occurs at 2853 cm^{-1} (3.15 μm) (see Fig. 3). According to Silverstein *et al.* [4], the variation in the positions of these stretching bands is ±10 cm^{-1} for aliphatic and cyclic hydrocarbons.

3. Carbon–Hydrogen Bending Vibrations: Methyl Groups

Two bending vibrations can occur for a methyl group: an asymmetric vibration near 1450 cm^{-1} (6.90 μm) and a symmetrical vibration near 1375 cm^{-1} (7.28 μm). The latter symmetrical bending of carbon–hydrogen bands appears very stable in position at 1375 cm^{-1} when a methyl group is attached to another carbon atom (see Fig. 3) [4].

4. Carbon–Hydrogen Bending Vibrations: Methylene Groups

Carbon–hydrogen bending vibrations for a methylene group (CH$_2$) are shown in Fig. 3. These vibrations of the bending type are rocking, scissoring, twisting, and wagging. The scissoring–bending vibrations are in conflict with the asymmetrical methyl bending vibration (1450 cm^{-1}) because the former occurs at 1465 cm^{-1}. For long-chain n-paraffinic compounds, the methylene rocking–bending vibration appears at ~720 cm^{-1} (13.9 μm). Thus, absorption band at ~720 cm^{-1} strongly indicates the presence of a chain of at least four or more methylenic units in a molecule [4].

C. Branched-Chain Hydrocarbons

We stated that the symmetric methyl bending vibration occurred at 1375 cm^{-1} (7.28 μm). This 1375-cm^{-1} methyl bending vibration is split by resonance into two frequencies when branching occurs or when there are two or three methyl groups on a single carbon atom. Therefore, an isopropyl group gives a strong doublet at 1385–1380 cm^{-1} (7.22–7.25 μm) and at 1370–1365 cm^{-1} (7.30–7.33 μm). The tertiary butyl split occurs at 1395–1385 cm^{-1} (7.17–7.22 μm) and 1365 cm^{-1} (7.33 μm). Confirmation for the branching is obtained by other secondary skeletal vibrations. Isopropyl branching shows a band at 1170 cm^{-1} (8.55 μm) with a shoulder at 1145 cm^{-1} (8.73 μm). Tertiary butyl branching is indicated by presence of bands at 1255 cm^{-1} (7.97 μm) and 1210 cm^{-1} (8.27 μm). However, absorption characteristics change when there is one methyl group next to an oxygen or carboxyl group. When a gem dimethyl group occurs at an internal position, band positions appear at 1385–1375 cm^{-1} (7.22–7.28 μm) and at 1372–1362 cm^{-1} (7.29–7.35 μm).

D. Olefins

1. Carbon–Carbon Double Bond Stretching Vibration

In the isolated, nonterminal double bond, an absorption peak occurs at 1680–1620 cm^{-1} (5.95–6.17 μm). Terminal double bonds with alkyl groups absorb more strongly than internal double bonds and absorb at 1658–1648 cm^{-1} (6.03–6.07 μm).

For cycloolefins the C=C stretching frequency is in the same general range as that for the linear olefins. Thus, for example, cyclohexene shows a C=C absorption at 1648 cm^{-1} (6.08 μm). However, in strained ring systems, as in cyclobutene, the frequency of absorption is reduced to 1566 cm^{-1} (6.39 μm).

In the case of an aliphatic conjugated C=C system, the olefinic absorption band is split. One of the bands occurs near 1600 cm^{-1} (6.25 μm), and the other occurs near 1650 cm^{-1} (6.06 μm) and is more intense than the 1600-cm^{-1} absorption band. A smaller shift of the double bond absorption band is also observed when the double bond is conjugated with a carbonyl group or with an aromatic ring.

If the alkyl group is replaced by a halogen atom and attached directly to the carbon atom of the olefinic linkage, the frequency of the C=C stretching is lowered except when the halogen is fluorine, in which case this effect is opposite [1].

2. Olefinic Carbon–Hydrogen Stretching Vibrations

The olefinic methylene group $\left[\begin{array}{c} {}^H \\ =C{\Large\diagup} \\ {}_{\diagdown H} \end{array}\right]$ has an asymmetric methylene stretching vibration at 3090–3070 cm^{-1} (3.24–3.26 μm), whereas the symmetric methylene stretching vibration appears at 2985–2965 cm^{-1} (3.35–3.37 μm). The important point is that the aromatic or aromatic–olefinic C—H stretching vibration occurs above 3000 cm^{-1} (or below 3.33 μm), but any C—H stretching band for an alkyl or nonaromatic group occurs below 3000 cm^{-1} (above 3.33 μm).

3. Olefinic Carbon–Hydrogen Bending Vibrations

The olefinic C—H bond may bend in the same plane as the carbon–carbon double bond as in scissoring (out-of-phase) or as in rocking (in-phase). The more significant vibrational band of the olefinic C—H bending is the out-of-plane bending vibration perpendicular to the plane as in twisting, which is an out-of-phase, out-of-plane bending, or as in wagging, which is an in-phase, out-of-plane bending. These vibrations yield strong characteristic peaks in the 1000–800 cm^{-1} (10–12.5 μm) region; the absorption frequency of the out-of-plane carbon–hydrogen bending band of the alkenes is dependent on position and number of substitutions about the C=C double bond. From Table I, example (1), the monosubstituted vinyl shows sharp-bending symmetrical absorption at 995–985 cm^{-1} (10.05–10.15 μm) and also at 910–905 cm^{-1} (10.99–11.05 μm). For the disubstituted vinyl group the symmetric bending vibration is at 895–885 cm^{-1} (11.17–11.3 μm), as shown in example (3) [1].

Halogen substitution on an olefinic carbon shifts the out-of-plane bending vibrations of cis olefins to a higher frequency region when compared to unsubstituted cis olefins. However, for trans halogenated olefins, the shift is toward a lower frequency. The alkene correlations are given in Table II [1]. Functional groups for which characteristic absorptions are given include nonconjugated, conjugated, vinyls, substituted vinyls, trans olefins, trans–trans conjugated dienes, and multiple-conjugated olefins.

TABLE I

Out-of-Plane Carbon–Hydrogen Bending Bands of Alkenes[a]

Structure	Bands	Structure	Bands
R^1, H / C=C / H, H	995–985 cm^{-1} (10.05–10.15 μm) 910–905 cm^{-1} (10.99–11.05 μm)	R^1, R^3 / C=C / R^2, H	840–790 cm^{-1} (11.90–12.66 μm)
R^1, H / C=C / H, R^2	980–965 cm^{-1} (10.20–10.36 μm)	R^1, R^2 / C=C / H, H	~690 cm^{-1} (14.49 μm), ambiguous in certain cases
R^1, H / C=C / R^2, H	895–885 cm^{-1} (11.17–11.3 μm)	R^1, R^3 / C=C / R^2, R^4	No C—H deformation

[a] Reproduced with permission from Conley (1972).

E. Aromatics

The modes of vibration of benzene give rise to a spectrum quite different from the alkanes, but it is somewhat similar to the spectra of the alkenes.

1. Aromatic Carbon–Hydrogen Stretching Vibrations

The characteristic features of the aromatic type of structures are the stretching vibrations observed in the 3030-cm^{-1} region (3.3 μm) and the ring skeletal vibrations found at 1650–1450 cm^{-1} (6.06–6.8 μm). With high resolution instrumentation, the 3.3 μm aromatic absorption vibration for the aromatic C—H stretching shows three bands, but it usually appears as a single band under normal conditions. As can be noted in the characteristic infrared group frequencies, aromatics show absorption at 3030, 1600, and ~1500 cm^{-1}; alkenes as the conjugated dienes present only the first two absorption bands [1].

2. Aromatic Carbon–Hydrogen Bending Vibrations

Aromatic ring systems show out-of-plane bending below 900 cm^{-1} (>11.11 μm) and in-plane bending vibrations between 1275 and 960 cm^{-1} (7.85 and 10.42 μm). The out-of-plane type is the more important for determining the number of hydrogens in an aromatic nucleus. Vibration is quite intense and when the substituent is an electron-attracting group,

TABLE II

Summary of Typical Vibrational Frequencies of a Number of Common Alkenes[a]

Molecule	νCH	$\nu C{=}C$	δCH (in-plane)	$\delta' CH$ (out of plane)	Overtone of δ'
A. Monosubstituted alkenes					
Propylene	3082	1646	1417	996, 919	1831
	3013	—	—	—	—
1-Butene	3087	1645	1420	992, 911	1832
1-Pentene	3075	1647	1420	992, 915	1835
1-Hexene	3083	1642	1416	994, 909	1820
1-Heptene	3082	1645	1400	995, 910	1825
3,3-Dimethyl-	3094	—	—	—	—
butene	3000	1646	1416	1000, 911	1827
B. Cis disubstituted alkenes					
2-Butene	3029	—	—	—	—
	2987	1662	1406	675	—
2-Pentene	3018	—	—	—	—
	2972	1657	1407	692	—
2-Hexene	3012	1654	1407	693	—
3-Hexene	3016	1653	1408	715	—
C. Trans disubstituted alkenes					
2-Butene	3021	—	1302	962	—
2-Pentene	3029	—	1296	965	—
2-Hexene	3027	1668	1300	965	—
3-Hexene	3030	—	1289	965	—
D. Asymmetrically substituted alkenes					
Isobutene	3086	—	—	—	—
	2987	1662	1420	887	1790
2-Methyl-1-butene	3092	1652	1416	890	1788
2-Methyl-1-pentene	3079	—	—	—	—
	2969	1652	1414	890	1787
2-Methyl-1-heptene	3076	1654	1415	888	1790

[a] Reproduced with permission from Conley (1972).

such as a chlorine or nitro group, there is an increase in the frequency of absorption ($\sim$30 cm^{-1} or $\sim$0.6 μm decrease in the wavelength). Table III gives a summary of C—H out-of-plane bending bands in the spectra of mono-, di-, tri-, tetra-, and pentasubstituted benzenes. This table can also be used to determine the number of adjacent hydrogen atoms that are on the aromatic ring [1].

TABLE III

Summary of C—H Out-of-Plane Bending Bands in the Spectrum of
Substituted Benzenes[a]

Phenyl substitution	Frequency (cm^{-1})	Wavelength (μm)
Benzene	671	14.90
Monosubstitution	770–730	12.99–13.70
	710–690	14.08–14.49
Disubstitution		
1,2	770–735	12.99–13.61
1,3	810–750	12.35–13.33
	710–690	14.08–14.49
1,4	833–810	12.00–12.35
Trisubstitution		
1,2,3	780–760	12.82–13.16
	745–705	13.42–14.18
1,2,4	825–805	12.12–12.42
	885–870	11.30–11.49
1,3,5	865–810	11.56–12.35
	730–675	13.70–14.82
Tetrasubstitution		
1,2,3,4	810–800	12.35–12.50
1,2,3,5	850–840	11.76–11.90
1,2,4,5	870–855	11.49–11.70
Pentasubstituted	870	11.49

[a] Reproduced with permission from Conley (1972).

3.   Aromatic Ring Vibrations: Skeletal, C=C

The aromatic, carbon–carbon, double-bond (skeletal) absorption vibration, the aromatic carbon–hydrogen stretching, and aromatic carbon–hydrogen bending vibrations are very useful assignments for determining the presence of an aromatic nucleus. An aromatic conjugated ring system shows the ring vibrations at 1600 (6.25 μm) and 1500 cm^{-1} (6.67 μm). When the aromatic nucleus is conjugated with an unsaturated group, such as aldehyde or vinyl, as in benzaldehyde or styrene respectively, a third band appears in this region at 1580 cm^{-1} (6.33 μm). The intensity of these bands is increased further by this additional conjugation.

Characteristic group frequencies of the alkanes, alkenes, and the aromatic ring vibrations are summarized in reference 1. For examples of aromatic, alkene, and alkane spectra, the reader is referred to Chapter five of Conley [1] or the text by Colthup et al. [5].

F. Alcohols and Phenols

1. Oxygen–Hydrogen Stretching Vibration

The O—H stretching frequency for the hydroxyl group of alcohols and phenols as dilute solutions in carbon tetrachloride is assigned to the 3640–3610 cm^{-1} (2.75–2.77 μm) region. Primary hydroxyl groups absorb at ~3640, secondary at 3630, tertiary at 3620, and phenolic groups at 3610 cm^{-1}. In concentrated solutions, the free hydroxyls give rise to absorption bands located at lower wave number (longer wave lengths) as a result of intermolecular and intramolecular hydrogen bonding (see Section III,D,2).

2. Carbon–Oxygen Stretching Vibrations

The carbon–oxygen stretching vibration of alcohols and phenols is assigned to the 1200–1000 cm^{-1} (8.33–10 μm) region of the spectrum. However, because other absorptions from various functional groups also appear in this region, absorption caused by carbon–oxygen stretching vibrations are of limited value for interpretation. When interferences from various other functional groups are absent, then these absorption bands are of value [1].

3. Hydroxyl Bending Vibrations

Hydroxyl bending vibrations yield an in-plane bending band and an out-of-plane bending band. The latter band occurs near 650 cm^{-1} (15.38 μm) in aliphatic alcohols that are hydrogen bonded; in dilute solutions, no out-of-plane absorption is observed. The in-plane bending vibration occurs at 1500–1300 cm^{-1} (6.67–7.69 μm) in concentrated solutions. In dilute solutions, no association of the hydroxyl occurs, and the broad band is replaced by a narrow sharp band at 1250 cm^{-1} (8.0 μm) [1].

G. Ethers, Epoxides, and Peroxides

1. Ethers

Vibrational characteristics of the C—O—C system produces a strong dipole change with strong intensity from the oxygen atom. However, the presence or absence of an ether linkage may not be firmly established using infrared spectroscopy because the C—O linkage is also common to alcohols, esters, and acids.

Saturated aliphatic ethers exhibit a strong characteristic absorption band at 1150–1085 cm^{-1} (8.70–9.23 μm) from asymmetric C—O—C stretching. The symmetric C—O—C stretching is generally weak, but is observable in the Raman spectrum.

The ether linkage, when conjugated with an olefin or aromatic group, as in vinyl ethyl ether or in phenyl alkyl ether, has the C—O—C asymmetric stretching absorption shifted to 1275–1200 cm^{-1} (7.85–8.33 μm). Weak symmetric stretching vibrations of these unsaturated ethers appear at 1075–1020 cm^{-1} (9.3–9.8 μm) [1].

Of special interest is the OCH$_3$ group attached to aromatics; these vibrational bands occur near the 2850-cm^{-1} region. For aliphatic methyl ethers the stretching vibrations occur at 2830–2815 cm^{-1}.

2. Peroxides

The intensity of the characteristic frequency of the O—O peroxide linkage is not very strong and cannot be assigned with much certainty. Alkyl and acyl peroxides display asymmetric stretching of C—O—O in the 1198–1176 cm^{-1} (8.35–8.50 μm) region. Acyl and aroyl peroxides show two carbonyl absorption bands in the 1818–1754 cm^{-1} (5.5–5.7 μm) region. (Peroxides, in general, are of limited stability and are explosive when concentrated [7,8].)

H. Carbonyl Compounds

Carbonyl compounds [1] show a strong C=O stretching absorption band at 1905–1550 cm^{-1} (5.25–6.45 μm). Compounds that possess this functional group are ketones, aldehydes, carboxylic acids, esters, amides, lactones, lactams, anhydrides, etc. Their carbonyl stretching vibrations are relatively free from interferences and fairly constant with respect to band position, but subject to small shifts. Generally, higher frequencies are observed when these compounds are in the vapor state than when they are in the liquid state. In nonpolar solvents, such as carbon tetrachloride or carbon disulfide, carbonyl stretching frequencies are higher than those recorded in polar solvents. Internal factors that cause band shifts are (1) nature of the substituents, (2) hydrogen bonding, and (3) vibrational coupling. Let us consider (a) the ketone, (b) the ester, (c) the acid chloride, and (d) the amide.

a   b   c   d

For comparisons, diethyl ketone is arbitrarily chosen as a representative ketone; it exhibits a carbonyl stretching band at 1715 cm^{-1} (5.83 μm) similar to other dialkyl ketones. The electronic distribution of the carbonyl group for a ketone is the total contribution representative of each of the following resonance structures:

$$\begin{array}{cccc}
\underset{R'}{\overset{R}{>}}C{=}O & \underset{R'}{\overset{R}{>}}\overset{+}{C}{-}O^{-} & \underset{R'}{\overset{\overset{+}{R}}{>}}C{-}O^{-} & \underset{R'}{\overset{R}{>}}C{-}O^{-} \\
I & II & III & IV
\end{array}$$

(R = R' = CH$_3$CH$_2$)

The contribution of each structure depends on the ability of R and R' to attract or repel electrons [1]. If structure II is significantly contributing to the entire molecule, then the frequency could be higher than 1715 cm^{-1}; if the other two structures are significant contributors, then the frequency is decreased from 1715 cm^{-1}. Structure II probably contributes the greatest change in dipole moment and charge separation.

Using this approach, it is possible to approximate the various carbonyl types. For an ester, acid chloride, and amide, any substituent that increases the positive charge on the carbon of the carbonyl group tends to increase the carbonyl stretching frequency in a manner similar to structure II of the dialkyl ketones. Thus, any electromeric polarization that does the same will also cause this shift to an increased frequency. This effect of the ester alkoxyl group increases the carbonyl stretching frequency to the 1736-cm^{-1} (5.76 μm) region. The chlorine atom of the acid chloride causes the absorption to occur at 1810 cm^{-1} (5.53 μm). However, because of its basic nitrogen, the amide group has an opposite effect and decreases the carbonyl stretching frequency to the 1680-cm^{-1} (5.95-μm) region.

Conjugation, in addition to electronic factors, also affects the position of the carbonyl absorption. That is, a vinyl or aromatic group lowers the frequency ~35 cm^{-1}; a smaller decrease is noted for the aromatic grouping.

Alpha chlorinated ketones, such as α-chloroacetone, absorb at 1724 cm^{-1} (5.80 μm), whereas acetone absorbs at 1715 cm^{-1} (5.83 μm). This shift caused by a halogen atom increases with the increasing dipole of the carbon–halogen bond (F > Cl > Br > I).

The carbonyl stretching absorption frequency is decreased when a carbonyl group is hydrogen bonded with a hydroxylic compound; the association of a hydrogen with the carbonyl group tends to decrease the double bond character of the carbonyl. The free carboxylic acid carbonyl stretching frequency (in dilute solutions) is near 1760 cm (5.68 μm) for most organic aliphatic acids, while in liquid state (where hydrogen bonding occurs) the carbonyl stretching frequency is 1700 cm^{-1} (5.88 μm).

Vibrational coupling is the interaction of two vibrating units that, when suitably located, can split a band into two frequencies (one above and one below the common initial frequency). This topic was briefly discussed for the CO_2 example in Section III,D.

1. Aldehydes and Ketones

Aldehydes and ketones show almost the same carbonyl absorption frequencies. However, the aldehydic C—H stretching region can be used as a good characteristic group frequency for distinguishing aldehydes from ketones. The aldehydic C—H doublet shows near 2820 and 2720 cm^{-1} (3.55 and 3.67 μm); ketones, of course, contain no O=C—H stretching. We have already discussed the effect of conjugation with a double bond or with an aromatic ring that shifts the carbonyl vibration to a lower frequency [1].

2. Carboxylic Acids and Salts

The O—H stretching vibration of an organic acid monomer is at 3595 cm^{-1} (2.78 μm). The dimer gives a broad O—H stretching band at 3000–2500 (3.33–4.00 μm). The monomeric carbonyl frequency occurs in the 1760-cm^{-1} region (5.68-μm); that of the dimer is in the 17.0-cm^{-1} (5.85-μm) as a consequence of hydrogen bonding and the resulting decrease in carbonyl character from association of hydrogen and the carbonyl group. A third absorption band based on O—H out-of-plane bending of the dimer provides confirmation for an acid structure [1].

The carboxylate anion gives rise to a strong asymmetric stretch band at 1610–1550 cm^{-1} (6.22–6.45 μm) and a symmetric stretch band from 1400 to 1300 cm^{-1} (7.15–7.0 μm).

3. Esters

Ester carbonyl stretching absorption frequencies are higher than those of ketones. Ester groups exhibit their carbonyl frequencies at 1735 cm^{-1} (5.76 μm) while ketones exhibit theirs at ~1700 cm^{-1}. Characteristic asymmetric and symmetric vibrations of the C—O—C group of an ester are distinctive at 1300–1050 cm^{-1} (7.69–9.52 μm). No vibration for ketones is found in this region.

With the incorporation of conjugated unsaturation, such as in ethyl benzoate, the carbonyl absorption band is shifted to a lower frequency

(1720 cm^{-1}). If the unsaturation is on the C—O—C portion, as in vinyl acetate, the carbonyl band is shifted to higher frequencies of ~1760 cm^{-1} (5.68 μm). Other correlations for ketones, etc., are shown in references 4, 5, and 6.

It should be noted that water-soluble esters and methyl and ethyl esters of low molecular weight acids (about four to six carbon atoms) readily hydrolyze to free acids and alcohols. Esters, such as phenyl formate, also hydrolyze readily; phenolic esters usually hydrolyze faster than the alcoholic types. Thus, certain esters contained in a water environment have limited lifetimes.

4. Acyl and Aroyl Halides

Acid chlorides exhibit a carbonyl stretching frequency near 1800 cm^{-1} (5.56 μm). Unsaturated acyl and aroyl halides show a carbonyl absorption in the region 1780–1750 cm^{-1} (5.62–5.71 μm) [1].

However, little or no acyl or aroyl halide, after discharge in an effluent, is found in rivers and streams, as these halides are rapidly decomposed to the free acid in water (which serves as a suitable nucleophile). In a basic aqueous medium, acid chlorides hydrolyze quickly to the salt of the corresponding organic acid.

5. Amides

a. Carbonyl Stretching

Of the following classes of organic compounds, anhydrides, acyl peroxides, acid halides, acids, esters, aldehydes, ketones, and amides, the amide group generally has the shortest stretching frequencies (or longer wavelengths) since resonance effects increase the carbonyl bond and decrease the frequency of vibration. Primary amides exhibit a strong band at ~1694 cm^{-1} (5.9 μm) in the solid phase, while in dilute solution the absorption occurs near 1690 cm^{-1}. Linear secondary amides exhibit a band near 1640 cm^{-1} (6.1 μm) in the solid state, but this absorption occurs at 1680 cm^{-1} (~5.95 μm) in dilute solution. Tertiary amides show an absorption band at 1680–1630 cm^{-1} (5.95–6.13 μm) in solid or dilute solution [1].

b. Nitrogen–Hydrogen Stretching

In solid samples of primary amides, the N—H stretching frequencies are observed near 3350 (2.99 μm) and 3108 cm^{-1} (3.15 μm). In dilute, nonpolar solvents, the two N—H stretching frequencies of primary amides are found near 3520 and 3400 cm^{-1} (2.84–2.95 μm).

In solid samples of secondary amides, the asymmetric and symmetric N—H stretching vibrations occur as multiple bands at 3330–3060 cm^{-1} (3.00–3.27 μm). Under concentrated conditions, dimeric cis and trans configurations are produced, resulting in multiple bands. In dilute solutions of secondary amides, the two N—H stretching frequencies are observed at ~3500–3400 cm^{-1} (~2.86–2.94 μm). No free N—H stretching frequencies are observed for tertiary amides because no free H is available.

c. Nitrogen–Hydrogen Bending

In solid samples, the N—H bending vibration for primary amides occurs at ~1655–1520 cm^{-1} (~6.04–6.17 μm). In dilute solution, this band shows at 1620–1590 cm^{-1} (6.17–6.29 μm).

For secondary amides under solid sample examination, the N—H bending vibration occurs at 1570–1515 cm^{-1} (6.37–6.60 μm). This band occurs in dilute solution at 1550–1510 cm^{-1} (6.45–6.62 μm). Conley summarizes the various amide vibrations [1].

Primary or unsubstituted amides may be hydrolyzed in presence of acidic or basic catalyst; the products are the free acid and ammonium ion or the salt of the acid and ammonia in the presence of a basic catalyst. N-substituted and N,N disubstituted amides also may be hydrolyzed in a similar manner. However, in water alone, most amides do not hydrolyze readily. This stability is in contrast to the ease with which some esters hydrolyze in water.

6. Anhydrides

As a result of vibrational coupling manifested by two carbonyls, two carbonyl bands are exhibited by acid anhydrides. These bands, separated by ~60 cm^{-1}, absorb at 1860–1800 cm^{-1} (5.38–5.56 μm) and at 1800–1750 cm^{-1} (5.56–5.71 μm). For linear anhydrides (e.g., acetic anhydride), the higher frequency band is more intense than the lower one. For cyclic anhydrides (e.g., succinic anhydride), the reverse is true. Conjugation shifts the two bands to lower frequencies [1].

In open-chain anhydrides, the C—O—C stretching vibration is exhibited at 1175–1045 cm^{-1} (8.51–9.57 μm); in the strained cyclic anhydrides, the C—O—C stretching vibration is located at 1310–1210 cm^{-1} (7.64–8.27 μm).

Although anhydrides are usually more difficult to hydrolyze than acyl halides, water is sufficiently strong as a nucleophile to effect their hydrolysis. The decomposition of anhydrides can be accomplished by either acid or base. Thus, anhydrides are not isolatable from water systems after prolonged contact and are not common water pollutants.

7. Lactams

Solid lactams absorb strongly near 3200 cm^{-1} (3.12 μm) for the N—H stretching vibration. Carbonyl absorption occurs near 1650 cm^{-1} (6.06 μm) for six-membered lactams; five-membered lactams absorb at 1750–1700 cm^{-1} (5.71–5.88 μm) region, and four-membered lactams absorb from 1760 to 1730 cm^{-1} (5.68–5.78 μm). Lactams undergo hydrolysis in a manner similar to amides. Since amides are more resistant to hydrolysis than esters, acid or base catalysis is needed for decomposition to take place. Amino acids are formed from hydrolysis of lactams [1].

8. Lactones

Carbonyl absorption of six-membered, saturated, delta lactones (cyclic esters) occurs in the normal carbonyl region 1750–1735 cm^{-1} (5.71–5.76 μm). A double bond α to the C=O reduces the carbonyl absorption frequency to 1760 cm^{-1} (5.68 μm).

Saturated γ lactones (five-membered ring) absorb at wavelengths shorter than those observed for the open-chain esters 1795–1760 cm^{-1} (5.57–5.68 μm). A double bond located α to the C=O reduces the frequency of the carbonyl absorption to 1750 (5.71 μm), whereas unsaturation alpha to the —O— increases the carbonyl absorption frequency to 1800 cm^{-1} (5.56 μm) for the respective unsaturated γ lactones. Cyclic lactones undergo hydrolytic reactions in a manner similar to normal esters; hydroxy acids are formed.

I. Amines

1. Nitrogen–Hydrogen Stretching Vibration

Primary amines, both aliphatic and aromatic and in dilute solutions, display an asymmetric N—H stretching band near 3490 cm^{-1} (2.87 μm). The symmetric N—H stretching band occurs near 3400 cm^{-1} (2.94 μm). Amino groups can be differentiated from hydroxyl groups by the broadband intensity and position of the latter group.

Secondary amines show only one band at 3450–3310 cm^{-1} (2.90–3.02 μm). For aromatic and heterocyclic aromatic amines, the N—H stretching absorption, which is intense, is observed at 3450 cm^{-1}; aliphatic amines and cyclic amines absorb weakly at 3350–3310 cm^{-1}. Imines (C=N—H) exhibit their N—H stretching band from 3400 to 3300 cm^{-1} (2.94 to 3.03 μm). This latter group also exhibits a weak C=N stretching vibration in the intense carbonyl C=O stretching absorption region 1690–

1640 cm^{-1} (5.92–6.10 μm). Tertiary amines do not absorb in this region since they contain no N—H [1].

2. Nitrogen–Hydrogen Bending Vibrations

N—H bending vibrations of medium to strong intensity for primary amines absorb at 1650–1580 cm^{-1} (6.06–6.33 μm). A second broad diffuse absorption can be seen at 900–650 cm^{-1} (11.11–15.38 μm). Secondary amines absorb weakly.

3. Carbon–Nitrogen Stretching Vibrations

In the case of aliphatic primary, secondary, and tertiary amines, moderately weak absorption bands are observed at 1250–1020 cm^{-1} (8.0–9.80 μm). In the case of aromatic amines, strong C—N stretching vibrations appear for the primary amines at 1340–1250 cm^{-1} (7.46–8.00 μm), the secondary amines at 1350–1280 cm^{-1} (7.41–8.81 μm), and the tertiary amines at 1360–1310 cm^{-1} (7.35–7.63 μm).

J. Amine Salts

Amino groups may be converted with hydrogen chloride to salts of the type, RNH_3^+, $R_2NH_2^+$, R_3NH^+, R_4N^+. Spectral absorptions of the corresponding amine salts can then aid in amine identifications [1]. Primary amine salts exhibit a strong, broad absorption between 3000 and 2800 cm^{-1} (3.33 and 3.57 μm), from asymmetric and symmetric stretching of the NH_3^+ group. A prominent band among the multiple-combination bands occurs near 2000 cm^{-1} (5.0 μm).

Salts of secondary amines also absorb at 3000–2700 cm^{-1} (3.3–3.7 μm) region with multiple combination bands occurring at $\sim$2733 cm^{-1} ($\sim$4.0 μm). At $\sim$2000 cm^{-1} ($\sim$5.0 μm), a medium band appears as in the primary amine salts. Tertiary amine salts exhibit absorption at 2700–2250 cm^{-1} (3.7–4.4 μm) [4].

K. Amino Acids and Salts

Free amino acids (zwitterions) show a strong, broad NH_3^+ stretching band at 3100–2600 cm^{-1} (3.23–3.85 μm). A strong band also appears at $\sim$2222–2000 cm^{-1} ($\sim$4.50–5.00 μm). Two bands from asymmetric and symmetric bending also appear at $\sim$1660–1610 (6.03–6.21 μm) and 1550–1485 cm^{-1} (6.45–6.73 μm). The asymmetric and symmetric C—O stretching occurs at $\sim$1600–1590 cm^{-1} ($\sim$6.25–6.29 μm) and at $\sim$1400 cm^{-1} ($\sim$7.14 μm) [4].

L.   Other Nitrogen Compounds

1.   Nitriles and Isonitriles

Aliphatic nitriles (RCN) exhibit a characteristic of C—N stretching vibration at ~2260–2240 cm^{-1} (~4.42–4.46 μm). Aromatic nitriles, such as benzonitrile, or nitriles with conjugation show a more intense band at lower frequencies (2230–2210 cm^{-1}, or 4.48–4.53 μm). With chlorine or oxygen attached to the alpha carbon, the intensity of the band is reduced.

Nitriles, upon contact with strong acids or bases, are gradually converted to the corresponding acid and ammonia. Alkylisonitriles (R—N≡C:) display strong absorption at 2175–2150 cm^{-1} (4.6–4.65 μm) region. Aromatic nitriles are shifted to lower frequencies. Isonitriles, upon contact with aqueous acids, are readily decomposed to amines and formic acid. They are not affected by alkalis, but on heating, isonitriles undergo rearrangement to form nitriles [1].

2.   Nitro Compounds

Aliphatic nitroparaffins (RNO$_2$) exhibit their strong asymmetric and symmetric NO$_2$ stretching bands at 1615–1540 cm^{-1} (6.19–6.49 μm). A strong characteristic symmetric stretching band is also noted between 1390 and 1320 cm^{-1} (7.20 and 7.58 μm). For conjugated nitroalkenes, the NO$_2$ asymmetric stretching frequencies are decreased ~40–100 cm^{-1}, while the NO$_2$ symmetric stretching frequencies are decreased ~15–30 cm^{-1}.

Aromatic nitro compounds exhibit their NO$_2$ asymmetric stretching frequency from 1548 to 1508 cm^{-1} (6.45 to 6.63 μm), and the NO$_2$ symmetric stretching frequency is observed in the region from 1356 to 1340 cm^{-1} (7.37 to 7.46 μm) [1].

3.   Nitrates

Organic nitrates (RONO$_2$) exhibit strong NO$_2$ asymmetric and symmetric stretching vibrations at ~1640 (6.10 μm) and 1285 cm^{-1} (7.78 μm). The O—N stretching vibration occurs at ~855 cm^{-1} (11.30 μm). The variation for these frequencies is ~40 cm^{-1}. Many nitrates are highly explosive when dry so care must be taken when working with them.

4.   Nitrites

The ease with which the alkyl nitrites decompose in water to nitrous acid and alcohol is noteworthy, and indicates that few if any of these compounds will be identified as water pollutants.

M. Organohalogens

1. Carbon–Fluorine Compounds

Monofluoro alkanes exhibit an intense absorption in the 1100–1020 cm^{-1} (9.09–9.80 μm) region. With further fluorine substitution, the frequency rises and a complex absorption pattern appears at 1400–1050 cm^{-1} (7.15–9.52 μm). The CF$_3$ and CF$_2$ groups absorb at 1350–1120 cm^{-1} (7.41–8.93 μm) [1].

2. Carbon–Chlorine Compounds

The aliphatic C—Cl absorption frequency is observed between 850 and 550 cm^{-1} (11.76 and 18.18 μm). Upon further chlorine substitution, this frequency shifts to 850 cm^{-1} because of interaction with other groups. Characteristic frequencies of the halogenated compounds are summarized by Conley [1].

3. Other Halogenated Compounds

Bromine compounds absorb at 690–515 cm^{-1} (14.49–19.42 μm) and organo iodine compounds absorb at 600–500 cm^{-1} (16.67–20.00 μm). Conley [1] summarizes the characteristic frequencies of the halogenated compounds.

N. Organophosphorus Compounds

The stretching vibration of the P=O group occurs at 1315–1180 cm^{-1} region (7.60–8.49 μm). Its sulfur analog P=S has its weak stretching vibration at 800–650 cm^{-1} (12.50–15.38 μm). When the aromatic substituent type is P—O phenyl, the P—O—C absorption occurs in the region 950–875 cm^{-1} (10.53–11.43 μm). However, the nonaromatic substituent type, P—O methyl, appears at 1050 cm^{-1} (9.52 μm). Phosphine, being readily oxidized from PH to P—OH, will not be of concern here. The characteristic absorptions of the organophosphorous compounds are given in Table IV [1]. These are important because of the many organophosphorous pesticides that can be found as water pollutants.

O. Organosulfur Compounds

Organic substances, such as aliphatic mercaptans and thiophenols, possess the S—H group and absorb weakly at 2600–2550 cm^{-1} (3.85–

TABLE IV

Summary of Characteristic Absorptions Attributed
to Organophosphorus Compounds[a]

Functional group	Frequency (cm^{-1})	Wavelength (μm)	Remarks
P—H	2425–2325	4.12– 4.30	P—H stretching vibration (sharp, medium intensity)
	1250–950	8.00–10.53	P—H bending (very weak)
P=O	1315–1180	7.60– 8.49	P=O stretching vibration (strong; position affected by the number of electronegative substituents)
P=S	800–650	12.50–15.38	P=S stretching vibration (weak absorption)
P—O—C	1100–950	9.00–10.53	Where C = CH, strong sharp band at 1050 cm^{-1} (9.52 μm); a sharp weak band near 1190 cm^{-1} (8.40 μm) also present from P=O stretching (higher alkyls absorb similarity)
			Where C = phenyl, a strong band is present at 950–875 cm^{-1} (10.53–11.42 μm)
P—OH	~2600	3.85	Hydrogen-bonded OH stretching (strong, very broad absorption)
	~1050	9.52	O—H bending

[a] Reproduced with permission from Conley (1972).

3.92 μm). The S—H stretching absorption does not shift as much as the hydroxyl or amino stretching band since S—H forms only a very weak hydrogen bond.

1. Carbon–Sulfur Vibrations

The carbon–sulfur stretching vibration is extremely weak and is exhibited from 700 to 590 cm^{-1} (14.28 to 16.95 μm). The thioether stretching vibration occurs at 695–655 cm^{-1} (14.39–15.27 μm). Compounds with the thiocarbonyl (C=S), which is less polar than the carbonyl (C=O), have a less intense band than the carbonyl [1].

The weak thiocarbonyl stretching band occurs at 1200–1050 cm^{-1} (8.33–8.69 μm). In thioamides, where the C=S is attached next to a nitrogen, the stretching vibration occurs at 1400–1290 cm^{-1} (7.12–7.75 μm). Because the stretching vibration is very weak, identification of this group is not certain by infrared methods.

2. Sulfides and Disulfides

The sulfide and disulfide stretching vibrations are very weak and both bands are of little value in structure determination.

3. Sulfur–Oxygen Vibrations

The intense sulfur–oxygen stretching vibration (S=O) is exhibited at 1080–1000 cm^{-1} (9.26–10.00 μm). Dimethyl sulfoxide, for example, shows a strong absorption at 1050 cm^{-1} (9.52 μm). Aromatic methyl sulfoxides absorb in nearly the same frequency region [1].

Sulfones (—SO$_2$—) show asymmetric and symmetric stretching vibrations at 1340–1300 (7.46–7.69 μm) and 1160–1135 cm^{-1} (8.62–8.81 μm). Unsaturation has little effect on the position of this absorption.

Sulfonyl chlorides (—SO$_2$Cl), sulfonyl esters (—SO$_2$OR), and amides (—SO$_2$NH$_2$) may be hydrolyzed to the sulfonic acids. However, the sulfonamides are somewhat more resistant to hydrolysis than the first two sulfonyl compounds. Sulfonamides exhibit absorption from 1370 to 1335 cm^{-1} (7.30 to 7.49 μm) and from 1170 to 1155 cm^{-1} (8.55 to 8.66 μm). Three absorption bands are observed for the sulfonic acids: 1250–1160 cm^{-1} (8.00–8.62 μm, intense), 1080–1000 cm^{-1} (9.26–10.00 μm), and 700–610 cm^{-1} (14.29–16.39 μm).

VI. INFRARED TECHNIQUES

A suitable discussion relating to laboratory techniques, sample preparation, and application of techniques for spectral examination of organic compounds and mixtures may be found in Chapter 4 of Conley's *Infrared Spectroscopy* [1].

A. Quantitative Analysis

For quantitative determination of a specific component using an infrared spectrophotometric method, a relationship between the radiation detected after passage through sample and the concentration of the component is needed [1]. This relationship is known as the Beer–Lambert law. An explanation of this law and its application follows.

Let I represent the amount of radiation passing through a square centimeter of cell area per second. The decrease in the amount of radiation ($-dI$) passing across a small differential segment (db) of the cell is proportional to the available amount of radiation (I) per second per square centimeter and also to the number of molecules (c) present for absorption of radiation in a square centimeter. The number of molecules absorbed in that area (or the decrease in intensity) is proportional to the concentration (c) of the absorbing molecules, the amount of radiation, and the differential length db as described by

$$-\mathrm{d}I = a'cdbI$$

where a' is a proportionality factor dependent on the molecule at a certain frequency.

Thus a' can vary with each given wavelength [1], and

$$\int_{I_0}^{I} \frac{\mathrm{d}I}{I} = -a'c \int_{0}^{b} \mathrm{d}b$$

or

$$\log e \frac{I}{I_0} = a'cb$$

or

$$\log 10 \frac{I}{I_0} = -abc = \log T$$

where T is the percentage transmission.

For analytical determination, the term A is used for absorbance

$$A = -\log \frac{I}{I_0} = \log \frac{1}{T} = abc$$

Here absorbance A is a measured quantity taken directly from the spectrophotometer using the absorbance scale rather than the transmission scale [1]. When the plot of the absorbance A versus the concentration c is a straight line, then the Beer–Lambert law is obeyed. As many compounds and mixtures do not yield a linear relationship throughout all ranges, an experimentally derived calibration curve is applied to provide corrections for the nonlinear portions of a plot.

B. Measurement of Absorbance

Two general methods employing peak heights for the quantitative determination are employed in practice. These are the "baseline" method and the "cell-in, cell-out" method commonly employed in ultraviolet measurements. The reader is referred to Chapter 6 on quantitative analysis in Conley's text [1]; possible baseline constructions are provided.

Advantages for the use of the baseline method are rapid and simple computations, minimized cell and other absorptions, and fairly rapid determinations. Routine procedures for expediting an analysis are cited in Conley for a simple mixture consisting of para, meta, and ortho xylenes [1].

An example of the baseline method used for a heavy residual fuel is given in Fig. 1 of reference 8a, illustrating measurements from seven peaks of the infrared spectrum. Peak and baseline distances for each peak were measured and the net absorbances calculated.

C. Accuracy of Infrared Analysis

Quantitative measurements with best precision may be obtained in the 25–50% transmission range (0.3–0.6 absorbance units); here the variation of absorbance is ~1%. Ultimate accuracy in analytical infrared spectroscopic determinations may be obtained by use of the differential analytical method, which compares an unknown sample directly against a proper reference, using matched cells with the same path lengths.

VII. PESTICIDE ANALYSIS

Pesticides have long been used for the control of disease and pests that destroy agricultural crops and forests; however, their toxicity and persistence in the environment render necessary their restrictive use and monitoring in order to insure public health. Thus, it has become essential for regulatory agencies to identify and quantify pesticides present in waters, in air, and on the land. In general, the determination of organic pesticides and similar compounds in the water matrix involves collection of water, sediments, plant and animal tissues, extraction, and separation plus cleanup followed by identification and quantitation.

A. Sampling and Storage

In order to obtain an average pesticide residue profile, the sample taken is generally a composite of several subsamples selected from various locations and depths of the water, soil, and sediment matrices. Details on collection and storage of the sample are discussed in Chapters 1 and 2 of the C.R.C. text on analysis of pesticides [9]. Sample stability and potential interferences that may be encountered in glassware, solvents, and reagents are also discussed in that text.

B. Extraction

For water samples, especially discharge samples of 1-L size, extraction of pesticides can be accomplished using a separatory funnel with several portions of an organic solvent such as hexane, benzene, or chloroform. Successive extractions are combined, dried using an anhydrous salt (e.g., Na_2SO_4), and carefully concentrated to a small volume using a "keeper" or retainer.

Another technique involves extracting the sample using a mechanical stirrer and a nonpolar solvent such as benzene [10]. The sample is then transferred to a separatory funnel and the extract is separated, dried, and concentrated. This stirring method is recommended when an extraction is necessary for samples containing suspended particles [9].

Many factors affect the efficiency of extraction of pesticides from soil samples. Pertinent factors include the method of extraction, solvent system, and moisture content. The foregoing factors are critically considered elsewhere [9]. Sediment samples with high water content can also be extracted with good recovery [11]. For the removal of organochlorides present in biota samples, the reader is referred to Table 8 of the C.R.C. text [9].

C. Cleanup and Fractionation

Usually it is necessary to remove coextractives, such as fats, phospholipids, and waxes, which commonly elute with the pesticides. These coextractives usually interfere with the end analysis involving gas chromatography, infrared spectroscopy, and mass spectrometry. The cleanup step may involve liquid–liquid partitioning [12–17], using acetonitrile/hexane or petroleum ether as the partitioning solvent system. Though these methods are generally excellent for most nonfatty and fatty samples, an alternative approach is suggested for fatty samples containing aldrin, hexachlorobenzene, mirex, heptachlor, and PCBs in order to minimize fatty material in the final pesticide residue. For such samples, the use of a partitioning–adsorption column [18,19] with modifications is suggested. A Florisil column also can be employed for the cleanup of extract residues obtained from water, soil, sediment, fish, tissue, and grain samples. Recoveries of organochlorides and polychlorobiphenyls from fish oil are quantitative when the Reynolds and Cooper [20] procedure is employed.

High pressure liquid chromatography also can be applied to the cleanup of pesticide residues from water and fish [21], and it may be employed as well for the extraction and separation of organochlorides in fats and fatty food residues [22,23].

The column-enrichment technique is also of considerable interest in that it may be used not only for the extraction and cleanup of pesticides from water, but also for the concentration, collection, and fractionation of these compounds and oils [24,24a,b]. For example, pesticides in untreated water samples can be concentrated when bypassing a sample through a reversed-phase micro Bondapak-C_{18} column and then eluting them from the column using a polar solvent. Fractionation of pesticides is attained through solvent programming and the appropriate fractions can be collected for subsequent analysis, such as gas chromatography, infrared spectrometry, etc.

The presence of sulfur and organosulfur compounds in the extracts obtained from bottom sediments of lakes and rivers interferes with gas chromatographic and spectrometric analysis. Numerous methods for removing sulfur interference have been developed; these involve use of metallic mercury [25], activated copper dust [26,27], Raney Nickel [28], copper–aluminum alloy [29], etc. Of these, copper powder appears to be the most attractive when efficiency of conversion, ease of operation, time, etc. are considered.

D. Analysis

For pesticide analysis, gas chromatography provides a means of separation and quantitative analysis with a degree of qualitative inference; for identification or confirmation, infrared spectroscopy, as one of several techniques, can be used for an isolated pure compound. Microtechniques involving 0.1–1 μg of sample are described by Blinn [30]. For the infrared determination of organochlorine insecticides in soil and water samples, the reader will find pertinent references in a review article by Pionke and Chesters [31].

The sensitivity of infrared methods is usually inadequate for the quantitative measurement of pesticides at trace levels in water unless adequate concentration and purification procedures are employed. For trace analysis, the ultimate in detection limit of instrumental methods is necessary. Detection limits are 10^{-8}–10^{-12} g (gas chromatography), 10^{-12}–10^{-15} g (chemical ionization mass spectrometry) and ion-probe MS, 10^{-9} g (ultraviolet–visible detector liquid chromatography), and 10^{-6} g of pure sample (standard infrared spectroscopy), or 10^{-9} g (Fourier transform infrared spectroscopy). Therefore, the applicability of IR spectrophotometer measurements for trace, routine analysis in water has received limited attention.

An early illustration (1965) of the combined use of gas chromatography and infrared spectrophotometry is the determination of methyl and ethyl parathion in the presence of common chlorinated hydrocarbon pesticides, dieldrin, endrin, aldrin, heptachlor, DDT, and DDE [32]. Spectra of a sample taken from the spill drainage ditch near a chemical warehouse and of the reference, ethyl parathion, compared favorably; both had the common characteristic gas chromatographic and infrared absorption peaks, thus lending support to the identification of the parathion sample, which was an obvious mixture of substances spilled in the Missouri River during a chemical fire. A water sample collected 55 miles south of the Omaha Water Treatment Plant intake 11 h after the fire and the chemical spill contained 37 ppb of parathion, 1–7 ppb of dieldrin, endrin, and aldrin, and lesser quantities of DDT, DDE, and heptachlor. Techniques such as electron-capture detector and microcoulometric gas chromatography, infrared spectrophotometry, and thin-layer chromatography for initial cleanup provided the means for the separation, quantitation, and confirmatory analysis of spill components found in the Missouri River. A laboratory equipped with a gas chromatograph interfaced to a computer-controlled mass spectrometer could analyze the above samples with much less time and expense [33].

A new separation and analysis method was developed by Gomez-Taylor, Kuehl, and Griffiths [34]. They used dual-beam Fourier infrared spectroscopy to identify pesticides and other water pollutants that were separated by gas chromatography or high performance liquid chromatography. Using this procedure, solutions containing as little as 50 ppb of organochlorine pesticides could be analyzed after eluting them from XAD-2 resin.

In agriculture, the loss of farm animals from chemical or pesticide poisoning represents considerable economic loss to the owners [35]. Prevention of poisonings of this type requires proper precaution in pesticide use and application. However, when poisoning has occurred, a rapid method of analysis is needed in order to minimize the potential damage that can be inflicted on the animals. A case in point is the following: a farmer had applied the contents of an unlabeled jug, assumed to be DDT, as a dilute wash spray for his 29 yearling calves and 1600-pound prize bull. After 2 h had lapsed, the farmer observed that some of his calves were dying. The veterinarian recognized organophosphate poisoning. Plasma and red blood cell cholinesterase activity in the animals were found to be markedly inhibited from the anticholinesterase activity caused by the organophosphate pesticide, which has a rapid dermal-absorption characteristic. Subsequent infrared spectrophotometric analysis confirmed the concentrate to be demetron.

As another example, in 1971, routine sampling of fish, water, and soil taken from Arroyo, Colorado, and analysis by gas chromatography using an electron-capture detector showed 1 ppb of dimethyltetrachlorotere-phthalate (DCPA) in the water samples taken over a 2-year period. However, this chemical in fish varied from 1 ppb to 8 ppm as determined by electron-capture detector gas chromatography after separation by thin-layer chromatography. An OV-210 GLC column proved to be of great importance in the separation of DCPA from heptachlor epoxide; these two compounds elute with almost the same relative retention time on other columns. The presence of DCPA and not that of heptachlor epoxide was demonstrated by infrared spectrophotometry. The nanogram amount of DCPA needed for analysis required the extraction of 20 gal of water, followed by concentration. GC–MS analyses confirmed the presence of DCPA in water. The authors state that the most critical differentiation was between heptachlor epoxide and DCPA [36].

VIII. PHENOL ANALYSIS

A classic determination of phenolic materials by infrared spectropho-tometry is described by Beynon *et al.* at CONCAWE [37]. Following sampling, the determination required three preparative steps. First, phe-nols in water were brominated in the ortho and para positions; second, the brominated phenols were extracted from water; and third, they were measured at 2.84 μm. Ortho substituted phenols, such as *o*-toluol, do not respond to the desired bromination method, which is similar to the method of Simard *et al.* [38]. Sensitivity of this method is in the low parts per billion.

Infrared spectrophotometry also can provide a powerful technique for preliminary identification of coal coking products spilled in waterways as illustrated in the following case [39]. Coking conversion products contain finely divided carbon that interferes in the analytical procedures.

In identification, possible source samples were taken to determine the party responsible for the spill occurring at the confluence. The concen-trates of source and spill samples were separately dissolved in chloro-form; the colloidal mixtures were centrifuged, decanted, and evaporated under vacuum at 40°C. Figures 4 and 5 are the infrared spectra of the two residues. Figure 4 shows the infrared spectrum of the liquid-coal conver-sion products (minus the colloidal carbon) taken at the confluence of Peter's Creek and the Monongahela River in Pennsylvania on June 29, 1971.

From these carbon-free samples, the Fuson–Shriner separation proce-dure [40] yielded the respective phenolic weak acid fractions in Figs. 6 and 7. The broad absorption bands at 3250 and 1600 cm^{-1} indicate phe-

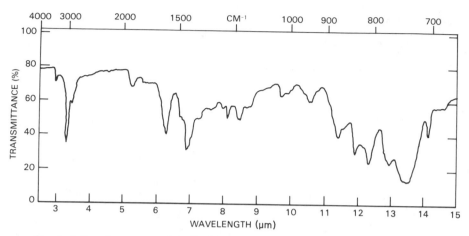

Fig. 4. Infrared spectrum of the liquid-coal conversion products (spill) (minus colloidal carbon).

nols. The two spectra are similar, indicating that the phenolics of the spill may be similar to those from the suspected source. The Fuson–Shriner separation procedure as applied to the original carbon-free residues of the spill and suspected source also yielded the basic fractions soluble in 5% HCl (Figs. 8 and 9) which again have similar spectra. These are aromatic amines as shown by the strong broad absorption occurring at 1600 and 3050 cm^{-1}.

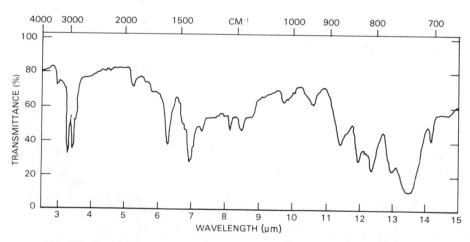

Fig. 5. Infrared spectrum of a possible source sample (minus colloidal carbon).

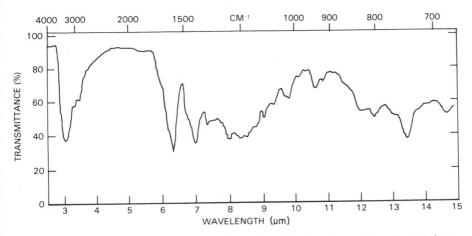

Fig. 6. Infrared spectrum of the phenolic weak-acid fraction from spill products (minus colloidal carbon).

IX. PETROLEUM ANALYSIS

Aquatic biota may be affected by the presence of oil in water at the part per billion level. Also, at these trace levels, bad tastes and odors are experienced in drinking water and fish. Infrared spectrophotometry is a technique that can be employed to quantitatively monitor oil levels in waterways, bays, and industrial effluents.

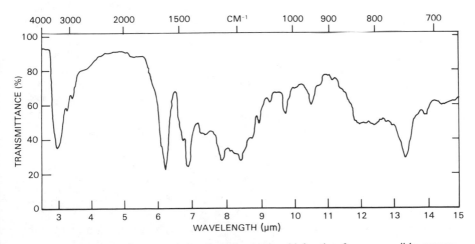

Fig. 7. Infrared spectrum of the phenolic weak-acid fraction from a possible source (minus colloidal carbon).

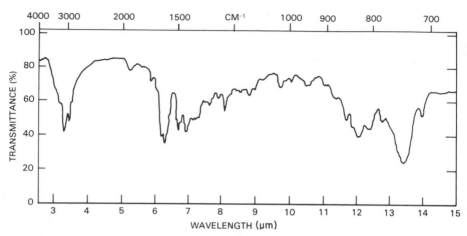

Fig. 8. Infrared spectrum of the basic fraction from spill products (minus colloidal carbon).

Beynon *et al.* [37] described one of the earliest uses of infrared spectroscopy for trace contamination of oil in water on a quantitative basis. They recommended that the optical density be determined at 2925 cm^{-1} whenever possible, especially for the petroleum distillates. However, they stated that in certain cases the determination at 2925 cm^{-1} leads to erroneous results because of the composition of the hydrocarbons and resolution of the spectrometer used. For certain cases, it was recommended that the optical density be determined at 2962 or 2860 cm^{-1}. If the hydrocarbon composition of crude oils remains unknown or uncalibrated, a more accurate method for quantification than that provided by the single-band method (with absorption measured at 2930 cm^{-1}) was devised by using all three wave numbers: 2925, 2962, and 2860 cm^{-1}. Thus, the saturated paraffinic hydrocarbons and cycloparaffins were quantitatively determined; essentially, the aromatic components of petroleum were left unquantified in this improved method. This is one of the earliest papers documenting erroneous findings from compositional differences in petroleum hydrocarbons.

Hughes *et al.* [41] reported on the evaluation of three extraction procedures using spiked, salted, distilled water with 3.8–188 ppm of Bass Strait crude oil. Recoveries with mean values of 102% were obtained only when the carbon tetrachloride-spiked oil–water mixture was subjected to stirring plus ultrasonic dispersion; NaCl was also necessary to optimize the extraction of oil from the water phase. Any combination of two of the three techniques was inadequate in obtaining high extraction efficiencies because the extraction recovery was ~80–90%. These data suggest that,

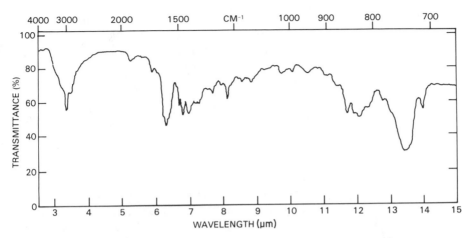

Fig. 9. Infrared spectrum of the basic fraction from a possible source (minus colloidal carbon).

at several ppm of crude oil in water, the oil exists as solubilized oil in water, and high efficiency in recovery cannot be realized without a high surface-contact area for the extracting solvent. Because of the solubilized nature of the oil in water, the salting-out procedure must be considered.

Gruenfeld [42] extracted defined, calibrated oils with 1,1,2-trichloro-1,2,2-trifluoroethane (Freon 113), a solvent less toxic than carbon tetrachloride. Four different oils, No. 2 fuel oil, No. 6 fuel oil, and high- and low-viscosity crude oils, were used with infrared measurements taken at 2930 cm^{-1}. Solutions of the four representative oils in the two solvents, carbon tetrachloride and Freon 113, were prepared. These solutions simulated solvent extracts of dispersed oil-in-water samples. Accurate quantitative determination of water-dispersed oils by single-point analysis with either solvent can be accomplished, according to Gruenfeld, in a concentration range of 2 to 40 ppm of oil in water using 10 mm path-length cells. In the procedure, 1-L oil-in-water samples can be extracted and the extracts analyzed using 100 mm pathlength cells without ordinate-scale expansion. However, accurate quantitation by the single-point analysis is not possible without ordinate-scale expansion. Some differences were noted in the absorptivities among the South Louisiana and Bachaquero crude oils. These two crude oils are solvent dependent for their respective absorptivities; these values appear to be reasonably stable despite loss of volatile oil components via weathering.

Large differences in absorptivities (such as those characteristic of "aromatic"—type crude oil), for which there is no direct quantitative method, versus those (such as "paraffinic- or naphthenic"-type crude oil), for which there is a quantitative method, may present difficulties in attaining true values for uncalibrated oils. When aromatics are absorbing at 3050 cm^{-1}, measurements taken at 2930 cm^{-1} do not account for the presence of the aromatics. Thus, there is a strong need for the development of a direct method that determines unknown oils with varying percentages of aromatics.

In order to circumvent part of this problem, Suzuki *et al.* [43] measured carbon tetrachloride extracts between 1900 and 1400 cm^{-1}; the concentration of mineral oil, fatty oil, and fatty acids were calculated from absorbances at 1750, 1710, and 1400 cm^{-1} respectively. This method has provided for the analysis of fatty oils (triglycerides) and fatty acids, in addition to one type of petroleum oil (mineral oil). Approximately 30 min is required for the simple method, which includes addition of 20 g of NaCl, shaking for 5 min, and allowing 5 min for the oil–water separation. The detection limit is 0.3 ppm when 700 mL of an oil–water sample is used.

Thus to insure high recoveries, oil solubilized in the water phase must be decreased through the use of NaCl or any other suitable salt. Extraction of desolubilized oil by the carbon tetrachloride phase must be facilitated by prolonged contact and shaking.

Manning [44] extracted 100 parts of a water sample containing traces of oil with 2 parts of carbon tetrachloride; the extract was then placed in the spectrophotometer and scanned through the hydrocarbon-absorption region (CH$_2$) at ~2930 cm^{-1}. Excellent absorption bands were obtained when the extracts from water containing 1 and 10 ppm of petroleum oil were analyzed using fused-silica cells of 10-mm path length. For greater sensitivity, Manning used a 100 mm pathlength cell and demonstrated that 0.1 ppm of oil gave a response of ~20% of full scale, on a recorder operated with a double-beam spectrophotometer.

The use of a long 100-mm pathlength cell was a significant improvement in petroleum oil quantification of known oils that have been previously calibrated. However, unknown, uncalibrated crude oils and oil products containing large amounts of aromatic and naphthenic–aromatic components cannot be measured with much accuracy by using a single absorption band at 2930 cm^{-1}. This band (CH$_2$) is useful only for the measurement of saturated aliphatic hydrocarbons.

A new approach and development for providing continuous monitoring of oil in wastewaters employs an optical fiber that has been precoated by chemical reaction with an organophilic grouping, such as octadecylatrichlorosilane [24a,b]. This coating enables it to absorb oil from the water matrix. The amount and type of oil adsorbed on the organophilic fiber affect the refractive index relationship at the fiber surface. A suitable optical sensor then provides the means for measuring oil dispersed in water. Seventeen pure hydrocarbons, diesel fuel, and crude oil were used as contaminants and tested with the new coiled capillary tube sensor cell. Crude oils can be measured at a few parts per million in water, and solvent extraction is not necessary.

X. PETROLEUM IDENTIFICATION

The application of infrared spectroscopy in the petroleum industry can be found in earlier reviews that deal with the composition and analysis of petroleum and its fractions [45–47]. Later, in the field of oil spill pollution problems, infrared spectroscopy has found application by providing for source identification of spills.

Some of the more pertinent absorption bands are covered briefly; these are useful in the study of the medium and heavy fuel products. However, light-petroleum products, such as jet fuels and gasoline, are so volatile that identification by infrared spectroscopy can be effected successfully only when the spill is fresh. But a weathered spill of gasoline can be traced to the source, at times, when flame-ionization detector gas chromatography (GC/FID) is employed. For example, the gasoline vapors sealed in a quart jar and vaporized from twigs, porous rocks, and water can be identified with the proper technique [48]. In this instance, only the heavier components of the gasoline can be matched to the known source gasoline, which also must be subjected to simulated weathering conditions.

With volatile products, the details of the aromatic fingerprint portion of the infrared from 900 to 670 cm^{-1} are likely to be missing or weak. In the case of the heavy residual fuels, weathering may make the source identity of a sample somewhat difficult, but the identity still can be established when infrared data are used in conjunction with metal analysis, fluorescence, gas chromatography, etc.

A simple infrared spectrum can determine whether an oil or grease sample is of mineral or vegetable origin or a mixture of both. For example, the presence of a strong carbonyl absorption band at 1740 cm^{-1} plus a very broad absorption at ~1150 cm^{-1} usually suggest whether the oil is of animal or vegetable origin (the presence of both indicating vegetable origin). A GC/FID determination also indicates whether it is of mineral, vegetable, or animal origins as their retention times are distinctive. Occasionally, although a crude oil may possess absorption in the carbonyl region, the absorption at 1150 cm^{-1} is small in contrast to a vegetable or fat triglyceride. This latter characteristic differentiates petroleum from triglycerides.

Infrared spectrometry has also been used to determine or estimate mixtures of hydrocarbons, such as petroleum and petroleum products, whose composition contains many components [49,50]. Extensions of these methods for the analysis of branching in paraffins were made and applied to determine the number of $CH_3 : CH_2$ groups [51].

In the analysis of hydrocarbon mixtures, the method of differential infrared spectroscopy has also been useful. When a hydrocarbon mixture with solute is placed in the sample beam and the hydrocarbon mixture is placed in the reference beam, analysis with a double-beam spectrometer yields the solute composition [52]. The differential method allows for the determination of the minor component in a mixture of major components, as in the following examples: deicers in gasoline, additives in lube oil, and insecticides in kerosene [53]. Such trace components, dissolved in petroleum oils, are different in chemical structure and polarity from the major petroleum components. Therefore, the identity and amount of such trace materials are rather easy to determine. However, when one attempts to identify and determine the amount of cyclic hydrocarbon components, which essentially are of one petroleum type, the analysis is very difficult.

For aromatic moieties, characteristic vibrations occur at 3000–3100 cm^{-1} for the aromatic C—H stretching, at 1600 cm^{-1} for the skeletal-ring stretching, and at ~800 cm^{-1} for the out-of-plane wagging. Thus, it is possible to determine the percentage of aromatic relative to other hydrocarbons in petroleum and its products. The ratio of the intensity of the 1600-cm^{-1} band for aromatics to the intensity of the 720-cm^{-1} band from paraffins can give the percentage of aromatics relative to n-paraffins [54–57]. The ratio of the intensity of the aromatic 810-cm^{-1} to the intensity of the 720-cm^{-1} band from paraffinic-type compounds can give a relative percentage of trisubstituted benzenes (symmetric and unsymmetric) and naphthalenes, relative to n-paraffinic-type molecules [57]. This ratioing technique has been applied to the characterization of spilled petroleum oils and their products with a high degree of precision when the infrared spectra are used in conjunction with linear discriminant function analysis (LDFA) [58,59] and computer techniques. Furthermore, after characterization, the source of the spilled oil can be identified using an extension of Kawahara's and Yang's classification scheme [60] with a high degree of certainty.

The nature of lubricating oil was classified by proposing the ratio of intensities at frequencies 1600 and 720 cm^{-1} as a measure of the aromatic or paraffinic character of crude oils [61]. For petroleum oils, the stretching vibrations of the aromatic C—H at 3050 cm^{-1} appears as a shoulder adjacent to aliphatic CH_2 and CH_3 bands, at 2930 and 2960 cm^{-1}, respectively. This aromatic C—H vibration and the trisubstituted aromatic vibration at 810 cm^{-1} can be useful when applied as a ratio with 720 cm^{-1} in order to provide information concerning petroleum-type aromaticity relative to paraffinicity in oils. For the study of oxidation of crude oils, the absorptive carbonyl band occurring at ~1720 cm^{-1} [62] can be used because it reflects the presence of aldehydes and ketones. Aldehydes and ketones reflect the autoxidative deterioration of the oil because they are formed as secondary decomposition products, which result from the induced decomposition of hydroperoxides generated by autoxidation of primary labile compounds in crude oils.

To determine the presence of sulfur–oxygen moieties resulting from oxidation of thioethers present in petroleum products, bands at 1300 and 1110 cm^{-1} can be considered [63,64]. Increases were noted for absorption bands at 1740, 1030, and 1175 cm^{-1} in an oxidation study [57]. These increases were presumably caused by the formation of carbonyls, sulfoxides, and sulfones in a No. 6-fuel oil–asphalt mixture that was exposed to ambient temperature for 8 days in a Lake Michigan asphalt spill.

XI.   CHARACTERIZATION AND IDENTIFICATION OF SPILLED PETROLEUM OILS

By controlling pollution, human health and the ecological integrity of man's life-support systems can be protected. This control is being achieved through legislation aimed at the reduction of automotive exhaust, control of stack emissions, regulated use of pesticides, management of solid waste disposals, and other legislative devices. A very real and significant problem, and one which is exceedingly difficult to regulate, is the discharge of oil, accidental or intentional, into surface waters. Such discharges affect the biota and can render domestic water unfit for use. Moreover, oil in water is known to concentrate fat-soluble poisons, such as insecticides and other organic toxicants, to a level many times higher than would normally occur in water alone. Thus, oil spilled into water can permit inordinate amounts of pollutants to be ingested by organisms in the food chain. Therefore, procedures developed for the identification of spilled petroleum products and the successful application of these identification techniques are significant contributions to environmental improvement; responsibility for violations of water quality standards can then be established.

The procedure of characterization followed by identification may be clarified by the following definitions: Characterization of a product relates the features of composition and structure of a material. Identification of a product is the demonstration of sameness in the composition and structure found in an organic material spill when compared to a source sample under consideration.

Tracing (or identifying) an oil spill to its source becomes difficult when the spill product is weathered, or a mixture of two or more products is involved, or there is no source specimen. In this last instance, the spill specimen must first be characterized before the possible origin of the spill can be investigated and the source identified. This procedure [65] is illustrated by the 1970 and subsequent oil spills into the Ohio River at Aliquippa, Pennsylvania. Characterization of organic weak acids, metallic particles, waste oils, etc. found in the spill specimens made possible the identification of industrial waste products discharged by a steel mill, which was the source of this pollution. Compounds were characterized by infrared analysis, gas chromatography, and other methods.

Table V illustrates the matching comparison of the phenolic fractions of the pump oil and that sample taken at the confluence of the outfall waste stream and the river. Phenols were characterized by comparing the retention times of the unknowns as their pentafluorobenzyl ethers (PFB) with synthesized, known, PFB ethers of phenol, p-cresol, and di-t-butylcresol [66]. The octanethiol was identified by comparing its PFB derivative with the synthesized PFB thioether of the known compound. The two gas chromatograms of the source and spill appeared similar, except that the river sample contained less of the phenols as a result of solubility losses.

The infrared spectra of the source oil and spill oil matched fairly well. Further evidence of a good match between the source and spill oils is illustrated by the types and amounts of the metals found in the two waste lubricating oils (Table VI). Except for the nickel and cobalt content of the

TABLE V

Electron Capture Gas Chromatograms of Lube
Oils Retention Distances (cm)

Sample 2 source oil (before entering river, Log 169)	Sample 5 (spill, Log 172)
1.35—	1.36—
2.25 Phenol	2.28 Phenol
3.02 p-Cresol	3.07 p-Cresol
5.07—	5.12—
7.73 Octanethiol	7.67 Octanethiol
11.37 di-t-Butylcresol	11.45 di-t-Butylcresol

river sample, it and the pump oil match fairly well. Thus, metal analysis, infrared spectra, and the electron-capture detector gas chromatograms of the two samples strongly indicated that the pump oil was the source of the oil spilled near the confluence of the outfall waste stream and the river. The infrared findings were confirmed by two other techniques that required considerably more time and effort. Infrared spectrometric analysis requires only a few minutes to prepare, analyze, and assess.

The technique of infrared spectrophotometry has provided useful data for characterizing and identifying unknown petroleum product specimens with known groups and sources. The infrared data were produced; then relevant infrared bands and peaks of spectra were measured in order to characterize the petroleum products. For example, bands at 720, 810, 1375, 1600, 2925, and 3050 cm^{-1} were measured for the 1967 Lake Michigan oil spill near Chicago [57]. Ratios from these absorbances were used to classify the spilled oil as an asphalt with some heavy residual oil present. The analysis and assessment of seven unknowns, two source asphalts, and two source No. 6 fuel oils demonstrated that the unknown spill was mostly asphaltic in nature. This conclusion was confirmed by a study of spectra from eight commercial asphalts and eight No. 6 fuel oils taken from nearby sources. A linear relationship was obtained by plotting the 810/1375 versus 810/720 ratios (Fig. 10). Ratio values for asphalts were located near the lower portion of the plot, whereas those values representing No. 6 fuel oils were located much farther from the asphalts represented in the plot. This finding also indicated that the No. 6 fuel oils

TABLE VI

Metals Found in Waste Lube Oils

Metals	Sample 2 source oil (before entering river, Log 169) (μg/sample)	Sample 5 (spill, Log 172) (μg/sample)
Iron	1200	1200
Manganese	200	150
Zinc	50	40
Lead	35	25
Barium	12	12
Aluminum	10	12
Chromium	6	6
Vanadium	10	10
Nickel	6	110
Cobalt	6	180

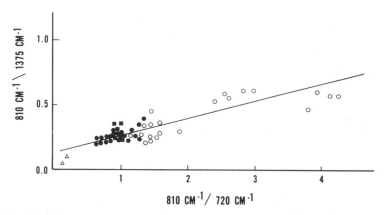

Fig. 10. Linear relationship (ratio of infrared absorbances, 810/1375 versus 810/720) for lubricant oils (△), No. 6 fuel oils (heavy residual fuel oil) (○), and two asphalt spill samples (●, industrial samples; ■, samples from North Star Transport Case).

were more aromatic in composition relative to straight- and branched-chain characteristics than were the asphalts. The positive slope denotes proportional decrease of aromatics responsible for absorption at 810 cm^{-1} for points located near the origin. In contrast to residual oils, asphalts are characterized by relatively larger intensities caused by carbon-methyl and methylene chains. Other oil pollution samples retrieved during the spills were subjected to the new method of identification.

This ratio method was tested in a later study [67] on 20 known asphalts and 21 heavy residual fuel oils. The 810/1375 versus 810/720 ratios were again plotted as a straight line. Nineteen of the heavy residual fuel oils and 15 of the asphalts were classified into their respective groups or categories. Only 5 of the asphalts and two of the heavy fuel oils overlapped on this plot using the above two ratios.

A third class of oils was added and found to be located in a distinctive region after the six absorbance ratios were determined. Unknown oil samples from spills were classified as lubricating oils, No. 6 fuel oils, No. 5 fuel oils, and asphalts, using the key ratios of 810/720 and 810/1375. Confirmation required the use of other ratios.

The usefulness of the ratio method was further demonstrated when both gas chromatography (GC/FID, GC/ECD, *n*-paraffin analysis, and capillary-column GC) and infrared spectroscopy were used to match two different oils derived from two different sources that were spilled near the same area at about the same time [68]. In this study, No. 5 and cutter stocks were classified or characterized. Both spilled oils were matched or identified to their respective known sources after determining the six

ratios of infrared absorbances (RIA). A grating instrument of higher resolution was used to confirm the lower resolution results of the infrared
prism instrument. In this study, the infrared cell was of path length 0.025
mm for all No. 6 fuel oils; for the asphalts, a sandwich of suitable thickness (~0.025–0.035 mm) was used. The infrared results were again confirmed by each of the four GC techniques: flame ionization detection,
electron capture detection, normal paraffinic analysis, and use of a tubular column of 150 ft × 0.02 in. stainless steel coated to 0.05-μm thickness
with DC-200. Each gas chromatogram from the four GC techniques applied to the spill specimens matched the suspect source gas chromatogram.

The RIA method was modified and used in conjunction with an established mathematical treatment [8a] in order to classify an oil as a heavy
residual fuel oil or an asphalt. Absorbances of the following bands were
used in the mathematical analysis: 720, 810, 870, 1027, 1375, 1460, and
1600 cm^{-1}, and all possible 42 ratios of absorbances of these bands were
calculated. A combination of data treatment and LDFA with computer
assistance has resulted in effecting a more precise and accurate method of
distinguishing between these heavy-petroleum products. Two separate
discriminant functions are obtained as follows:

$$Za = b_0 + b_1 X_1 + b_2 X_2 + \ldots + b_{36} X_{36}$$
$$Zo = b_0' + b_1' X_1 + b_2' X_2 + \ldots + b_{36}' X_{36}$$

where Za and Zo designate the asphalt and heavy residual-oil functions,
respectively, and X_is are the absorbance ratios (variables), and b_i and b_i'
are the coefficients determined from asphalts (or heavy residual fuel oils).
These coefficients are determined by maximizing the ratio of the "between" group variance (heavy residual oils and asphalts) to the "within"
group variance (heavy residual fuel oils or asphalts). Using b_i (or b_i')
values determined from known samples, an unknown is classified as an
asphalt or heavy residual fuel oil by measuring the variable values of an
unknown sample, calculating the appropriate absorbance ratios, and substituting these values for X_i in the equations. The larger value obtained
after substitution in the two equations provides the correct classification,
with a very high degree of certainty, without the possible overlap found in
the earlier, simpler use of the ratio method.

This ratio method has been applied successfully for not only the characterization but also the identification (fingerprinting an unknown spill to
its source) of petroleum pollutant spills. Applying LDFA to relevant infrared spectrophotometric data, 99% of known and unknown oil samples

have been correctly classified and identified [8a,60]. Unknown samples included weathered crude oil, heavy residual fuel oils, and asphalts, which were correctly identified by extension of the discriminant function analytical treatment. Noteworthy is the fact that results of this nonsubjective method were confirmed by chemical analysis. In this study, the two-fold objectives were to characterize correctly an unknown oil spill sample from among several classes of petroleum products studied, and subsequently to identify (fingerprint) correctly that unknown to a specific source. In the first step, the unknown oil spill was characterized by one of the three LDFAs. Each linear discriminant function (LDF) was calibrated with known No. 6 fuel oil, asphalts, or crude oils. In the second step, the characterized unknown was finally identified to a source by using the appropriate set of LDFs for one of the three groups selected in step 1. For example, assume that the unknown spill oil was initially characterized in step one as an asphalt. In step 2, the 19 LDFs for asphalt alone were used to identify which 1 of the 19 different asphalts matched the characterized (step 1) unknown. The one-step LDFA procedure for identification was tried and failed as there were a large number of misidentifications. The powerful nature of this mathematical analysis is illustrated in Fig. 11 in which two-dimensional plots of $U_{ij} = \lambda_1(X_{ij} - \bar{X} \ . \ . \ .)$ and $V_{ij} = \lambda_2(X_{ij} - \bar{X} \ . \ . \ .)$, λ_1 and λ_2 are the first 2 eigenvectors of the 42 eigenvectors from the oil groups of 19 asphalts, 6 crude oil residues, and 21 No. 6 fuel oils; these were well separated into three distinct groups. The unknown petroleum products (W's, X's, Y's, and Z's) were assigned correctly as asphalt, crude oil residue, or No. 6 fuel oil group, among the proper discrimination clusters, of A's (asphalt), B's (crude oil residue), and C's (No. 6 fuel oil). There were only two misclassifications; the overall characterization was 99% correct as shown in Table VII.

Procedures for identification were the same as those for characterization; identification was performed for each oil group. The four unknown oil spill samples W, X, Y, and Z were correctly identified or fingerprinted to their respective sources with a probability of 1.000. Of the 72 + 24 + 80 replicates, only one known replicate, a No. 6 fuel oil, was misidentified as shown in Table VIII. The 18 variables involved in identifying the crude oil residues are obviously fewer than the 42 variables necessary for identifying No. 6 fuel oils and asphalts. Similarly, only one eigenvector was necessary, in this situation, to account for more than 99% of the total variability of the crude oil residues, whereas 5 and 6 eigenvectors were needed for 95 and 96% of the total variability for the asphalt and No. 6 fuel oil dispersion, respectively. These differences in the ease of mathematical resolution probably existed because the crude oil residues, limited in number for this study, are more dispersed and, therefore, easier to iden-

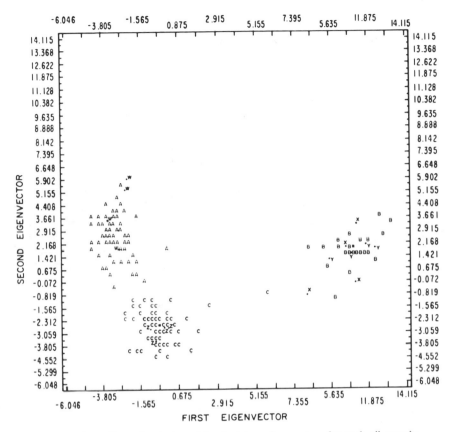

Fig. 11. Two-dimensional plot of the first two eigenvectors for each oil sample.

tify. Unknown samples W, X, Y, and Z were identified to their respective sources by chemical means as well as by discriminant analysis fingerprinting, as shown in Table IX.

For the spectrophotometric analysis of the crude oil residues and for all No. 6 fuel oil samples, a sample cell path thickness of 0.025 mm in the demountable cell was satisfactory since the absorbance magnitudes were in the linear part of the curve for absorbance values versus sample thickness measurements taken for peaks at wavelengths 720, 810, 870, 1020, 1375, 1420, and 1600 cm^{-1}. The absorbances from thin samples (i.e., 0.02 mm) were small; the dispersion of absorbance values would be too large to show any meaningful relationship between samples.

In a similar but more extensive study of a data set of 194 oils, Mattson *et al.* [59] subjected six classes of petroleum products of 62 crude oils, 60 No. 2 and diesel fuel, 12 No. 4 fuels, 22 waste crankcase lubes, 10 No. 5

TABLE VII

Results of Characterization and Frequency Distribution by Probability of Correct Classification

Oil groups	Number of samples correctly classified	Number of samples misclassified	Percentage correct classification	Correct classification probability distribution			Number of unknown samples correctly classified into groups
				$p = 1.00$	$.99 < p < 1.00$	$.89 < p < .99$	
Asphalt	72	1	98.6	65	4	3	4
Crude residue	24	0	100.0	24	0	0	8
No. 6	80	1	98.8	77	3	0	4

TABLE VIII

Results of Identification

		Known									Unknown	
Oil group	Number of subgroup	Number of samples correctly classified	Number of samples misclassified	Percentage correct classification	Number of variables used	Correct identification probability distribution					Correctly classified	Mis-classified
						1.00	.999	.996	.976	.662		
Asphalt	19	72	0	100.0	41	72	0	0	0	0	4	0
Crude residue	6	24	0	100.0	18	24	0	0	0	0	8	0
No. 6	21	80	1	98.8	41	76	1	1	1	1	4	0

TABLE IX

Chemical and Instrumental Identification

Unknown	Sample Number	Mode of Analysis
W	Log 126	Electron capture gas chromatography, analysis for asphaltene content, metal analyses, ultimate analysis (C, H, and N), solubility, density, etc. [69]
X	Log 201	Metal analyses, flame ionization detector gas chromatography
Y	Log 204	Metal analyses, flame ionization detector gas chromatography
Z	Log 164	Electron capture gas chromatography

fuel oils, and 28 No. 6 fuel oils to pattern classification using LDFA. Several decision tree schemes were tested in order to develop a high predictive capability resulting in a recognition power of 97.5% with very high probability of 84 to 90%. A high-resolution, sophisticated, computer–spectrometer system was employed to select variables among 21 wavelengths with experimental precision, and to choose between parametric and nonparametric pattern-recognition techniques for the discrimination of six classes of oils. The linearity of absorbances of the instrument, KBr window cleanliness, freedom from water, and proper baseline method were emphasized. His classification scheme was not extended and adapted for identifying or fingerprinting the "characterized" unknown to its suspected source or party. In the earlier fingerprinting (identification) method developed by Mattson [70], 40 different crude and residual oils yielded digitized infrared-transmission spectra and each of the digital spectra was read directly into a computer, which converted the spectra into units proportional to absorbance and smoothed spectra. From the spectra visually displayed, 11 bands were chosen for the fingerprint evaluation. The operator selected endpoints for integrating each band, and the integration was computerized. The computer calculated the baseline correction and normalized each at 1456 cm^{-1} in order to eliminate sample thickness problems. Study of observed band areas, ratios of such areas, and their standard deviations led to the elimination of several bands. The fingerprint was then obtained for each sample by first dividing the range of band areas of each of the eight selected bands observed for all 40 samples into eight equal increments. Each increment was assigned an integer value from one to eight. Each spectrum was then fingerprinted by assigning the appropriate integer to each of the eight bands, giving an eight-character string for the spectrum. Thus, the intensity of each band was coded for the fingerprint of each oil.

The third fingerprinting method is that of Brown *et al.* [71–73]. These investigators used 21 selected bands between the 1200- and 650-cm^{-1} regions. A cell of 0.1 mm thickness is not acceptable for analysis of heavy residual fuel and crude oils since measurements would yield absorbances in the nonlinear range [74]. Measurements should be conducted at one thickness rather than at random thicknesses since absorbances vary widely at different path lengths as shown in Figure 12.

Kawahara and Julian [74] observed that the cell thickness for the infrared analysis of crude oil residues, obtained after removal of volatile components via heating to 260°C (to simulate weathering), should be 0.025–0.035 mm. In this thickness range, measurements were made in the linear part of the curves drawn by plotting absorption values recorded at various cell thickness for each wavelength used (Fig. 12). At some cell thicknesses, the absorption values changed rapidly with thickness as with 1375, 3050, and 720 cm^{-1}; however, for some wavelengths as with 870 and 1600 cm^{-1}, the absorption values were less variant at different cell thickness. Thus, it appears that exact sample thickness should be employed for all types of oil products.

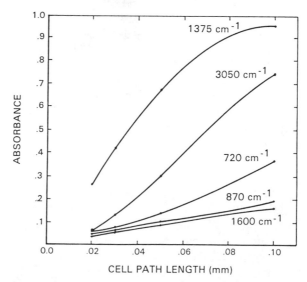

Fig. 12. Relationship of optical absorbance versus cell path length at a certain wave number; Devonian crude residue after distillation to 270°C.

To obtain a fingerprint of a sample, the percentages transmittances of the background for the spectrum of the selected bands are read into a computer. Absorbances are calculated from transmittance values and converted into pseudoabsorptivities by dividing by either the exact or estimated sample thickness. The writer feels that an exact sample thickness is necessary for all types of oil products; however, because asphalts are hard and tacky, an exception might be necessary for this one product only.

To match the data from an unknown sample to data from one of the known samples stored in the file, the percentage transmittance for each of the bands of the unknown is read into the computer, which calculates pseudoabsorptivities for the unknown in the same manner as described for the knowns. A comparison of known absorptivities is made, and the ratio of absorptivities of each known to that of the unknown is calculated; the average ratio for each known to unknown comparison is then calculated. Next each ratio is divided by the average ratio. The computer then lists the number of bands for each known with a calculated difference (ratio minus average) of 0.05, 0.10, 0.15, and 0.20. In the comparison, an ideal match has the ratio of absorptivities of known to unknown with a value of 1.0.

Brown *et al.* studied the effects of 30 days of weathering in the fingerprinting region of 950 to 650 cm^{-1} [72]. In that study it was stated that the infrared band and shapes change to some extent, but the infrared region remains unchanged from 950 to 650 cm^{-1}, with band differences occurring at 10% or less. The writer feels that this finding is true only for neat No. 6 fuel oil or asphalt, which start to boil at ~600° F; therefore, because of the high boiling point, there is very little loss of volatility under weathering conditions. However, a No. 2 fuel oil exposed to ambient temperature evaporates readily to ~20% of its original volume in 30 days, leaving a residue totally different in infrared spectrum from its original state. The residue resulting from evaporating No. 2 fuel oil was observed to be similar to lube oil [74].

The fourth fingerprinting method [75] is that being developed by the U.S. Coast Guard at Groton, Connecticut. For matching an oil spill to a suspect source, the Coast Guard Research and Development Center finds that a simple overlay of the region 700–1200 cm^{-1} is sufficient when the oil slick is discharged from the bilge of a ship. Since the bilge discharge usually contains lube oil, diesel oil, and water, the oil slick, having been previously water washed, is easy to match to the ship's discharged oil mixture. The Coast Guard Research and Development Center is also seeking a rough digital code classification system that others [67,68] have found necessary for the rapid characterization of the oils in surface waters. Classification of multicomponent petroleum oils (crude oils, lubricant oils, distillates, and residual fuel oils) is a necessary task because it can reduce the amount of sampling required for oil spill characterization and identification, because the analytical technique provides information for aliphatic, aromatic, carbonyl, and organosulfur composition as well as for autoxidation (weathering) of an oil. Among the methods tried, the LDFA method is probably more objective than the other three since all values used for the LDFA computerization were measured experimentally. Approximate cell thicknesses were not employed in the LDFA calculation except for asphalts, and the cell thickness employed reflects the linear range of the curve for cell path lengths versus absorptivities.

Attenuated Total Reflectance Spectroscopy

The technique of attenuated total reflectance (ATR) and its technical advantages in many areas are given in the original papers by P. J. Wilks [76–78]. It was applied advantageously to the analysis of water pollution for measuring spectra of sediments and algae [79]. ATR was used to illustrate the difference between natural seeps and well-blowout samples in the Santa Barbara well-blowout of 1969; absorption bands at 1700 cm^{-1} suggested carbonyl compounds that arise from the autoxidation of natural seep samples. Spectral differences were also noted below the 1200-cm^{-1} region, which Mattson and Mark [79] favored for qualitative purposes.

More extensive use of ATR is found in the identification of surface films. Baier [80–82], using the prism-dip technique, assessed the surface quality of water in Lake Chautauqua, New York. He calculated the film thickness as a function of infrared transmittance obtained from microsampling of the surface waters of the lake during the boating season. There was an increase of oily surface pollutants during weekends followed by decrease of such pollutants during the midweek collections. Degradation and decrease of surface films were attributed to ultraviolet irradiation and to a lesser extent to bacterial action. Autoxidation with sunlight appears to play the prominent role for the elimination of surface-film pollutants via oxidative degradation.

Analysis of surface emulsions showed infrared bands at 3400 and 1050 cm^{-1} from alcoholic groups; bands at 1640 and 1530 cm^{-1} are possibly caused by amides. Thus, it appears that these bubbles are of biological origin. These biological bubbles enhance autoxidation of the petroleum film and its products by introducing oxygen from air into the oil-pollutant system of thin films. Ultraviolet irradiation initiates free-radical autoxidation of the hydrocarbons via hydrogen abstraction followed by addition of O_2 (air) and subsequent formation of the hydroperoxide.

XII. LINEAR DISCRIMINANT FUNCTION ANALYSIS

For a growing number of problems in analytical chemistry, classification or identification techniques involving statistical analysis have been applied. These are sometimes necessary because of the increasing difficulties encountered in interpreting the results of analytical methods involving a large number of parameters [83]. Methods involving multicomponent analysis, including gas chromatography, infrared spectrometry, activation analysis, etc., can generate large amounts of data. In order to obtain optimal combination of all parameters, better identification of an impending disease, oil spill, or individual compounds, all or most of the information should be used for these multiple-variate problems. Use of LDFA provides answers to some multivariate problems.

An LDF is a function that is linear in the variables and discriminates or classifies data into one of the groups under consideration [84]. A separate LDF is obtained for each product or compound type. For example, forms for two functions are

$$z_a = b_0 + b_1X_1 + b_2X_2 + b_3X_3 + b_4X_4 + b_5X_5 + \ldots + b_nX_n$$

and

$$z_0 = b_0' + b_1'X_1 + b_2'X_2 + b_3'X_3 + b_4'X_4 + b_5'X_5 + \ldots + b_n'X_n$$

where z_a and z_0 designate the asphalt and heavy residual fuel oil functions, respectively. The X_ns are the variables used in the problem. The coefficients (primed and unprimed bs) are determined by maximizing the ratio of the variance between groups (heavy residual fuel oils and asphalts) to the variance within groups (heavy residual fuel oils or asphalts). To evaluate both discriminant functions, the variable values of an unknown sample are used after the coefficients are established using known samples.

One area in which stepwise LDFA (SLDFA) has been applied is in petroleum analysis. Kawahara and Yang [60] used SLDFA for the characterization and identification of petroleum pollutants on the basis of infrared-spectrophotometric data with a 99% correct classification of numerous known and unknown oil samples. The known samples constitute three learning groups, namely asphalts, No. 6 fuel oils, and crude-oil residues. Identity and characterization of unknown samples including weathered oils, asphalts, and simulated weathered crudes were established by testing these unknowns into the calibrated SLDF equations. Identities were also confirmed by other instrumental analysis methods, such as gas chromatography and metals analysis. The 42 discriminating parameters (formed by ratios of absorption values at various wavelength positions) resulted in only two misclassifications for 178 known subsamples and none for the 16 unknown subsamples.

Earlier studies of Kawahara *et al.* [8a] supported the fact that SLDFA permits more laboratory information to be used. This statistical technique results in a superior discrimination of petroleum products. Mattson *et al.* [59] applied a modified SLDFA method and also used a higher degree of sophistication in data acquisition that was employed previously. A larger sampling of 194 oil samples over six classes was attempted with 97.5% of them correctly classified. Included were lube oils, diesel fuels, No.-4 fuels, No.-5 fuels, No.-6 fuels, and crude oils.

Thus, the use of infrared spectrophotometry coupled with computerized LDFA has been shown to be an effective means of correctly classifying and identifying water-borne petroleum products to their source. Because the multiple-point "crooked" baseline method requires manual measurement, it is cumbersome and time consuming. If the single-point method can be tested and found to be equivalent to the multiple-point method, the former method would be of great interest because it is more adaptable to instrumental automation.

The author has used both the single- and multiple-point baseline measurement methods to generate parallel matched data sets [85]. SLDFA was performed for parallel sets of infrared absorption peaks taken at selected wavelengths; the performance of the two measurement methods was then evaluated and compared by tests of significance. Results of the statistical analysis show the two methods to be approximately equivalent. That is, the probability of correct discrimination for the single-point method of measurement approximately equals the probability of correct discrimination for the multiple-point method of measurement when employing SLDFA.

XIII. FOURIER TRANSFORM INFRARED SPECTROSCOPY

The recent application of Fourier transform to infrared spectroscopy (FTIR) has permitted spectacular advances to be made in analysis of chemical compounds. A much shorter measurement time with a sensitivity 10 times greater than that of the grating spectrometer are among the benefits of using FTIR. With FTIR and computer techniques, traces of compounds, which may provide very weak signals, can be easily identified.

In the FTIR technique, all of the nuclei are excited at the same time and the emitted frequencies are scanned as these nuclei fall back to their equilibrium energy-distribution positions [86]. An interference pattern containing all the information of a normal spectrum can be obtained in

~1 s. However, the measurement of an interferogram can be accomplished rapidly and accurately only by the use of the Fourier transform; without that mathematical treatment, the calculation would be almost impossible to complete [87]. Another advantage in the use of the Michelson interferometer for spectroscopy is that a greater throughput of radiation is possible when compared against the monochromator of the optical-null grating spectrometer.

In mid-range FTIR, a major breakthrough was realized with the application of the Cooley-Tukey [88] fast Fourier transform (FFT) algorithm to interferometry. Use of this mathematical procedure enabled computations that formerly took hours to be completed in a few minutes.

In general, FTIR is favored for measurements where data acquisition time is limited (as in experiments requiring the use of on-line GC–IR) spectrometry of transient species (where data-acquisition time using a grating spectrometer is very long) or with samples that provide very low transmittance [89]. FTIR also is advantageous in situations where a large number of samples are analyzed. However, the conventional grating spectrometer also can be interfaced with minicomputers that permit spectral manipulations previously possible only by FTIR [89].

For further reading in the theory, instrumentation, and applications of FTIR, the reader is referred to the text by Griffiths [90]. An excellent paper illustrating the application of FTIR to the identification of trace organics in water also has been written by Griffiths et al. [91].

Identification of trace quantities of organic compounds in water at the parts per billion level has been accomplished using a Digitlab FTS-14 spectrometer with dual parallel light pipes and an MCT detector interfaced to a gas chromatograph. Effluent from the gas chromatograph is passed through one light pipe, and when a compound enters the interferograms are averaged until the compound leaves the light pipe. Measurement times are of the order of 10 s. The FFT of each interferogram is computed after subtracting a dual-beam interferogram of the empty cell from the dual-beam interferogram corresponding to each gas chromatographic peak.

Although the absorption of chlorinated pesticides is very weak when compared to those of anisole, butyl ether, diethyl malonate, etc., these pesticides can be analyzed by the gas chromatographic–infrared technique at the 50-ppb level after concentration and recovery from 20–50 mesh XAD-2 resin [91].

Cournoyer et al. [92] reported a technique that permitted the determination of an infrared spectrum on <1 ng of material; the spectrum was signal averaged for an hour in a beam condenser of an FTIR spectrometer. With the Fourier spectrometer, good measurements also can be obtained with very thin samples, but the presence of interference fringes in spectra of such films can obscure weak absorption bands. Several methods have been developed for eliminating these fringes. Hirschfeld and Mantz [93] used a patch device in order to remove the spike from sinusoidal fringes. Cournoyer *et al.* [92] employed a sufficiently thin sample that the fringes lie only at higher than the 4000-cm^{-1} absorption region.

Two new devices are available commercially for measuring the infrared spectra of gas-chromatographic fractions without isolating the analyte from the carrier gas. In the optimized gas chromatograph equipped with a heated infrared gas cell, the flow is stopped while the spectrum is measured [94,95]. The second device employs a capillary light pipe containing the sample so that its spectrum can be measured with a Fourier spectrometer fitted with an MCT detector. The light pipe is a long glass tube that has its interior surface coated with a gold film, a technique developed by Azarraga [96]. It is adaptable to most chromatographs and can detect peaks that are measurable with a thermal conductivity detector [96–98].

High performance liquid chromatography (HPLC) is rapidly becoming a popular technique for separating components in complex mixtures characterized by compounds of low volatility or of thermal instability; such compounds are not amenable to separation by gas chromatography. For species separated by HPLC, the on-line identification by FTIR spectroscopy is more difficult to achieve than for the GC case. The problem in the measurement of LCIR spectra occurs because the mobile-phase solvent itself interferes in the infrared absorption spectrum of the analyte [91]. A review by McDonald provides up-to-date information regarding FTIR spectrometry, GCIR, LCIR, etc. [99].

Fourier transform spectroscopy has provided the spectroscopist with a powerful technique for measuring infrared spectra at a higher sensitivity and/or at greater speed than that previously possible. With GC–FTIR and LC–FTIR techniques, the chemical analyst possesses two powerful tools for the identification of trace water pollutants. A combination of the above methods provides complementary information of greater usefulness than either alone [91]. It is quite likely that a combination of FTIR and mass spectrometry interfaced with GC or LC will make possible identification of a very high proportion of all chromatographically separated compounds. These combinations may be of great value for the rapid identification of trace water pollutants in the future.

REFERENCES

1. Conley, R. T., "Infrared Spectroscopy." Allyn & Bacon, Boston, Massachusetts, 1972.
2. Avram, M., and Mateescu, G. H., "Infrared Spectroscopy." Wiley (Interscience), New York, 1972.

3. Szymanski, H. A., "Theory and Practice of Infrared Spectroscopy." Plenum, New York, 1964.
4. Silverstein, R. M., Bassler, G. C., and Morrill, T. C., "Spectrometric Identification of Organic Compounds." Wiley, New York, 1974.
5. Colthup, N. B., Daly, L. H., and Wiberley, S. E., "Introduction to Infrared and Raman Spectroscopy." Academic Press, New York, 1964.
6. Bellamy, L. J., "The Infrared Spectra of Complex Molecules." Methuen, London, 1958.
7. Tobolsky, A. V., and Mesrobian, R. B., "Organic Peroxides." Wiley (Interscience), New York, 1954.
8. E. G. E. Hawkins, "Organic Peroxides." Van Nostrand-Reinhold, Princeton, New Jersey, 1961.
8a. Kawahara, F. K., Santner, J. F., and Julian, E. C., *Anal. Chem.* **46**, 266 (1974).
9. Chau, A. S. Y., and Afghan, B. K., eds., "Analysis of Pesticides in Water," Volume II, Chapters 1, 2, and 3. CRC Press, Boca Raton, Florida, 1982.
10. Kawahara, F. K., Eichelberger, J. W., Reid, B. H., and Stierli, H., *J. Water Pollut. Control Fed.* **39**, 592 (1967).
11. Chau, A. S. Y., unpublished results.
12. Biros, F., McMahon, B. M., Sawyer, L. D., and Corneliussen, P. E. (eds). "Pesticide Analytical Manual," Vols. I, II, and III. Food and Drug Administration, Office of the Associate Commissioner for Compliance, Rockfield, Maryland, 1967–1974.
13. "Official Methods of Analysis." Association of Official Analytical Chemists, The Association, Washington, D. C., 1975.
14. "Analytical Methods For Pesticides Residues in Foods," rev. ed. Health Protection Branch, Department of Health and Welfare, Ottawa, Canada, 1973.
15. Johnson, L. D., Waltz, R. H., Ussary, J. P., and Kaiser, F. E., *J. Assoc. Off. Anal. Chem.* **59**(1), 174 (1976).
16. Griffith, K. R., and Craun, J. C., *J. Assoc. Off. Anal. Chem.* **57**, 168 (1974).
17. Bong, R. L., *J. Assoc. Off. Anal. Chem.* **58**, 557 (1975).
18. Langlois, B. E., Stemp, A. R., and Liska, B. J., *J. Agric. Food Chem.* **12**, 243 (1964).
19. Braun, H. E., "A Comparison of the Mills and Langlois cleanup procedures for the analysis of chlorinated hydrocarbon residues in fatty and non-fatty food samples." *Proc. Semin. Pestic. Residue Anal., 1st, 1968* p. 20 (1968).
20. Reynolds, L. M., and Cooper, J., *ASTM Spec. Tech. Publ.* **STP 573**, 196 (1975).
21. Larose, R. H., *J. Assoc. Off. Anal. Chem.* **57**, 1046 (1974).
22. Zimmerli, B., and Marek, B., *Mitt. Geb. Lebensmittelunters. Hyg.* **66**, 362 (1975).
23. Rohleder, H., Staudacher, H., and Soemmermann, W., *Z. Anal. Chem.* **279**, 152 (1976).
24. "Waters Associates Technical Bulletin, H63." Water Associates subsidiary of Millipore Corp., Milford, Massachusetts, 1976.
24a. Kawahara, F. K., Fiutem, R. A., Silvus, H. S., Newman, F. M., and Frazar, J. H., *Anal. Chim. Acta* **151**, 315 (1983).
24b. Kawahara, F. K., Silvus, H. S., and Newman, F. M., American Society of Civil Engineers, 1984 National Conference on Environmental Engineering, Paper No. 167, June 25–27, 1984, Los Angeles, California.
25. Goerlitz, D. F., and Law, L. M., *Bull. Environ. Contam. Toxicol.* **6**, 9 (1971).
26. Thompson, J. F., ed., "Analysis of Pesticide Residues in Human and Environmental Samples." U.S. Environ. Prot. Agency, Research Triangle Park, North Carolina, 1977 (revision).
27. Nash, R. G., and Harris, W. G., *J. Assoc. Off. Anal. Chem.* **55**(3) (1972).
28. Ahnoff, M., and Josefsson, B., *Bull. Environ. Contam. Toxicol.* **13**, 159 (1975).
29. Schutzmann, R. L., Woodham, D. W., and Collier, C. W., *J. Assoc. Off. Anal. Chem.* **54**, 1117 (1971).

30. Blinn, R. C., and Gunther, F. A., *Residue Rev.* **2**, 99–152 (1963).
31. Pionke, H. B., and Chesters, C., *Soil Sci. Soc. Am.* **32**(6), 749–759 (1968).
32. Kawahara, F. K., Lichtenberg, J. J., and Eichelberger, J. W., *J. Water Pollut. Control Fed.* **39**(6), 446 (1967).
33. Budde, W. L., Eichelberger, J. W., and Harris, L. E., *J. Chromatogr.* **134**, 147 (1977); *Anal. Chem.* **51**, A567 (1979).
34. Gomez-Taylor, M. M., Kuehl, D., and Griffiths, P. R., *Int. J. Environ. Anal. Chem.* **5**, 103–117 (1978); *Pestic. Abstr.* **78-1473** (1978).
35. Watson, M., Benson, W. W., and Gabica, *J. Arch. Environ. Health* **22**(5), 582 (1971).
36. Miller, F. M., and Gomes, E. D., *Pestic. Monit. J.* **8**(1), 53 (1974).
37. Beynon, L. R., Kashnitz, R., and Rijnders, G. W. A., "Methods for the Analysis of Oil in Water and Soil," Stichting, CONCAWE, *in* "Hydrocarbons in Water or Soil by Infrared Spectrophotometry," CONCAWE, The Hague, 1968.
38. Simard, R. G., Hasegawa, I., Bandaruk, W., and Headington, C. E., *Anal. Chem.* **23**, 1384 (1951).
39. Kawahara, F. K., Memorandum to Comm. R. E. Anderson, U.S. Coast Guard, Captain of the Port, Pittsburgh, Pennsylvania, March 15, 1972.
40. Shriner, R. L., Fuson, R. C., and Curtin, D. Y., "The Systematic Identification of Organic Compounds," 4th ed., p. 347, Wiley, New York, 1956.
41. Hughes, D. R., Belcher, R. S., and O'Brien, E. J., *Bull. Environ. Contam. Toxicol.* **10**(3), 170 (1973).
42. Gruenfeld, M., *Environ. Sci. Technol.* **7**, 636 (1973).
43. Suzuki, R., Yamaguchi, N., and Matsumoto, R., *Bunseki Kagaku* **23**, 1293–1303 (1974).
44. Manning, R. J., *Environ. Pollut. Manage.* Beckman Instrument Co., p. 265 (1962).
45. Gill, F., and Gordon, R. R., *Petrol (London)* **8**, 11 (1945).
46. Speight, J. G., *Appl. Spectrosc. Rev.* **5**, 211 (1971).
47. Coggeshall, N. D., "Symposium on Spectroscopy in the Petroleum Industry." *Am. Chem. Soc.,* Div. Petrol. Chem., Washington, D. C., 1955.
48. Kawahara, F. K., Memorandum to Comm. R. T. Sommer, U.S. Coast Guard, Captain of the Port, Cincinnati, Ohio, March 16, 1976.
49. Rose, F. W., *J. Res. Natl. Bur. Stand. (U.S.)* **20**, 129 (1938).
50. Braittain, R. R., Rasmussen, R. S., and Cravath, A. M., *J. Appl. Phys.* **14**, 418 (1943).
51. Pushkina, R. A., and Kuklinskii, A. Y., *Khim. Tekhnol. Topl. Masel* **T6**, 55 (1971).
52. Hammer, C. F., and Roe, H. E., *Anal. Chem.* **25**, 668 (1953).
53. Powell, H., *J. Appl. Chem.* **6**, 488 (1956).
54. Lafaix, A., *Bull. Soc. Chim. Fr.* **8-9**, 1639 (1962).
55. Le Pera, M. E., U.S. Clearing House Federal Science Technical Information, 1967, AD 663816. Available as CF STI from *U.S. Govt. Res. Develop. Rep.* **5**, 73 (1968).
56. Luther, H., and Oelert, H. H., *Erdoel Kohle, Erdgas, Petrochem.* **24**, 216 (1971).
57. Kawahara, F. K., *Environ. Sci. Technol.* **3**, 150 (1969).
58. Kawahara, F. K., and Santner, J. F., *Anal. Chem.* **48**, 2023 (1976).
59. Mattson, J. S., Mattson, C. S., Spencer, J. J., and Spencer, F. W., *Anal. Chem.* **49**(3), 500 (1977).
60. Kawahara, F. K., and Yang, Y. Y., *Anal. Chem.* **48**, 651 (1976).
61. Bhattacharya, S. N., *Rev. Inst. Fr., Pet. Ann. Combust. Liq.* **14**, 321 (1959).
62. Botneva, T. A., and Shulova, N. S., *Geol. Nefti Gaza* **12**, 24 (1968).
63. Brown, A. B., and Knobloch, J. O., *ASTM Spec. Tech. Publ.* **STP 224**, 213 (1967).
64. Stewart, J. E., *J. Res. Natl. Bur. Stand. (U.S.)* **58**, 265 (1957).
65. Cincinnati Post Times, Memorandum from F. K. Kawahara to I. L. Dickstein, Director of Regulatory Programs, Ohio Basin Region, p. 47, Dec. 4, 1970.

66. Kawahara, F. K., *Environ. Sci. Technol.* **5**, 235 (1971).
67. Kawahara, F. K., and Ballinger, D. G., *Ind. Eng. Chem. Prod. Res. Dev.* **9**, 553 (1970).
68. Kawahara, F. K., *J. Chromatogr. Sci.* **10**, 629 (1972).
69. Kawahara, F. K., *Environ. Sci. Technol.* **10**, 761 (1976).
70. Mattson, J. S., *Anal. Chem.* **43**, 1872 (1971).
71. Lynch, P. F., and Brown, C. W., *Environ. Sci. Technol.* **7**, 1123 (1973).
72. Brown, C. W., Lynch, P. F., and Ahmadjian, M., *Environ. Sci. Technol.* **8**, 669 (1974).
73. Brown, C. W., Lynch, P. F., and Ahmadjian, M., *Anal. Chem.* **46**, 183 (1974).
74. Kawahara, F. K., and Julian, E. C., unpublished results (1974).
75. Brown, C. W., Lynch, P. F., and Ahmadjian, M., *Appl. Spectrosc. Rev.* **9**(2), 223 (1975).
76. Wilks, P. A., Jr., *Appl. Spectrosc.* **22**, 872 (1968).
77. Wilks, P. A., Jr., *Appl. Spectrosc.* **23**, 63 (1969).
78. Gilby, A. C., Cassels, J., and Wilks, P. A., Jr., *Appl. Spectrosc.* **24**, 539 (1970).
79. Mattson, J. S., and Mark, H. B., Jr., *Environ. Sci. Technol.* **3**, 161 (1969).
80. Baier, R. E., *Proc.—Conf. Great Lakes Res.* **13**, 114 (1970).
81. Baier, R. E., *J. Geophys. Res.* **77**, 50 (1972).
82. Baier, R. E., Goupil, D. W., Perlmutter, S., and King, R. W., *J. Rech. Atmos.* **8**(3-4), 571–600 (1974).
83. Coomans, D., Massart, D. L., and Kaufman, L., *Anal. Chim. Acta* **112**, 97–122 (1979).
84. Kendall, M. G., and Stuart, A., "The Advanced Theory of Statistics," Vol. III, Chapter 44. Hafner, New York, 1946.
85. Kawahara, F. K., Santner, J. F., and Dunn, J. R., *Abstr. Pap.—Chem. Congr. North Am. Cont., 2nd, 1980* Div. Anal. Chem. Pap. No. 138 (1980).
86. Michelson, A. A., *Philos. Mag.* [5] **34**, 280 (1802).
87. Fellgett, P., *Aspen Int. Conf. Fourier Spectrosc. [Proc.], 1970* AFCRL-71-0019, p. 139 (1971).
88. Cooley, J. W., and Tukey, J. W., *Math. Comput.* **19**, 297 (1965).
89. Griffiths, P. R., Sloane, J. J., and Hannah, R. W., *Appl. Spectrosc.* **31**(6), 485 (1977).
90. Griffiths, P. R., "Chemical Infrared Fourier Transform Spectroscopy." p. 205. Wiley (Interscience), New York, 1975.
91. Gomez-Taylor, M. M., Kuehl, D., and Griffiths, P. R., *Int. J. Environ. Anal. Chem.* **5**, 103 (1978).
92. Cournoyer, R., Shearer, J. C., and Anderson, D. H., *Anal. Chem.* **49**, 2275 (1977).
93. Hirshfeld, T., and Mantz, A. W., *Appl. Spectrosc.* **30**, 552 (1976).
94. Shaps, R. H., and Varano, A., *Ind. Res.* **19**, 86 (1977).
95. Shaps, R. H., Simons, W., and Varano, A., *Am. Lab.* **9**, 95, 98 (1977).
96. Azarraga, L. V., and McCall, A. C., *U.S. Environ. Prot. Agency, Off. Res. Dev. [Rep.] EPA* **EPA-660/2-73-034** (1974); *Chem. Abstr.* **84**, 173356H (1976).
97. Griffiths, P. R., *U.S. Environ. Prot. Agency, Off. Res. Dev. [Rep.] EPA* **EPA-600/4-76-061** (1976); *Chem. Abstr.* **87**, 62134B (1977).
98. Wall, D. L., and Mantz, A. W., *Appl. Spectrosc.* **31**, 552 (1977).
99. McDonald, R. S., *Analytical Chemistry* **56**[5], 349R (1984).

INDEX